Fruit and Plantation Crops

NIPA® GENX ELECTRONIC RESOURCES & SOLUTIONS P. LTD.
New Delhi-110 034

About the Authors

Dr Om Prakash is working as an Assistant Professor in Department of Fruit Science, College of Horticulture at Banda University of Agriculture and Technology, Banda, 210001 (U.P.). He did his B.Sc. (Agriculture) from Govind Ballabh Pant University of Agriculture and Technology, Pantnagar (U.K.); M.Sc. in Horticulture from Narendra Dev University of Agriculture and Technology, Ayodhya (U.P.); Ph.D. in Fruit Science from Chandra Shekhar Azad University of Agriculture and Technology, Kanpur (U.P.); Qualified ICAR- JRF and ASRB-NET. Author has more than nine years of experience in the field of teaching, research & extension. Published 23 research papers, 25 abstracts and popular articles and 05 practical manuals. Guided 05 M.Sc. students as major advisor and 07 students as co-advisor. Received application award from Vice Chancellor, BUAT Banda in the year 2019 and 2023. Acting as co-coordinator for ICAR-JRF Preperation Cell in the University.

Dr Akhilesh Kumar Srivastava is working as an Professor and Head in the Department of Fruit Science College of Horticulture, Banda University of Agriculture and Technology, Banda. He did B.Sc. Ag and M.Sc. Ag (Horticulture) from the ANDUAT, Ayodhya and completed Ph.D from IAS, BHU, Varanasi. He got an outstanding academic career and worked under AICR on Arid Zone Fruits and AICRP on fruits for 24 years. He qualified for ICAR-JRF, NET and has been recipient of University gold medal at graduation level. He published 40 research papers in NAAS rated journals, 5 Books/Bulletins, 4 book chapters and 100 popular articles and papers in seminar and symposia. He is Executive Councillor of Indian Society of Arid Horticulture since 2017; Executive Councillor for Indian Society of Citriculture, Nagpur Since 2014 and Editor (Crop Production) for Indian Journal of Arid Horticulture since 2020. Contributed for promotion of Arid Horticulture in Hot Arid Zone B of Rajasthan and currently for dry land areas of Bundelkhand in Uttar Pradesh. He is recipient of Fellowship of International Society for Noni Sciences, Chennai Fellow award of ISAH, Bikaner, 2019-20. He received appreciation award from Vice Chancellor, BUAT Banda in the year 2019, 2020.

Mr. Siddharth Kumar is pursuing his Ph.D. in Fruit Science at Banda University of Agriculture and Technology, Banda (U.P.). He did his B.Sc. in Agriculture and M.Sc. in Horticulture from Mahatma Jyotiba Phule Rohilkhand University, Bareilly (U.P.). He has qualified ASRB-NET in horticulture (Fruit Science). He has awarded with fellowship Alliance of Bioversity International & CIAT. He has published about, 3 research papers, 12 articles and 5 book chapters.

Fruit and Plantation Crops

Om Prakash
Assistant Professor
Department of Fruit Science, College of Horticulture
Banda University of Agriculture & Technology
Banda-210001 (U.P.)

A.K. Srivastava
Professor and Head
Department of Fruit Science, College of Horticulture
Banda University of Agriculture and Technology
Banda-210001 (U.P.)

Siddharth Kumar
Ph.D. Research Scholar
Department of Fruit Science, College of Horticulture
Banda University of Agriculture and Technology
Banda-210001 (U.P.)

NIPA® GENX ELECTRONIC RESOURCES & SOLUTIONS P. LTD.
New Delhi-110 034

NIPA® GENX ELECTRONIC
RESOURCES & SOLUTIONS P. LTD.

101,103, Vikas Surya Plaza, CU Block
L.S.C.Market, Pitam Pura, New Delhi-110 034
Ph : +91 11 27341616, 27341717, 27341718
E-mail:newindiapublishingagency@gmail.com
www: www.nipabooks.com

For customer assistance, please contact
Phone: + 91-11-27 34 17 17
Fax: + 91-11-27 34 16 16
E-Mail: feedbacks@nipabooks.com

ISBN: 978-81-19072-26-2

Composed and Designed by NIPA.

Dedicated

to

My Dear Students

who

inspired me

to

Fulfil this commitment

Vice Chancellor Office

Banda University of Agriculture & Technology, Banda-210001, U.P. Tele Fax No:- 05192-232305
Ph-05192-232305; website: buat.edu.in; email-vc.buat@gmail.com, vcbuat.bn-up@gov.in

Ref.:- VC/2023/168 **Date: 28 /08/2023**

Narendra Pratap Singh
Vice Chancellor

Foreword

The fruit and plantation crops play a significant role in the global food industry. These crops provides essential vitamins, minerals, and antioxidants, which are necessary for maintaining good health. The cultivation of fruit and plantation crops offer a source of income and employment to millions of people worldwide, particularly in regions where other crops may not be viable. Moreover, advancements in farming practices continue to expand the scope of fruit industry along with meeting the growing demand for healthy and flavourful food products.

India, the second-largest fruit producer after China, has been continuously expanding its fruit production over the past few years. As per NHB database 2021 at National level, fruits are being cultivated in an area of 6.91 million hectares with production of 102.48. The plantation crops occupies an area of 4.25 million hectares with production of 16.63 MT.

I am pleased to introduce the book entitled **"Fruit and Plantation Crops"** authored by Dr. Om Prakash, Dr. A. K. Srivastava and Mr. Siddhanth Kumar. It covers a broad range of essential aspects related to fruit and plantation crops, starting with an introduction that outlines the importance and uses, composition, nutritional value, origin and distribution, taxonomy and botany, soil, climatic requirements, flowering, varieties, propagation and agro¬techniques. It also covers the latest database on the area and production of fruits and plantation crops at global and national level. In addition, the book provides comprehensive information on plant protection, maturity indices, harvesting and post-harvest management of fruits and plantation crops.

This book will serve as an excellent resource for students, researchers and fruit growers in gaining knowledge and expertise on cultivation, production, and management of fruit and plantation crops. I immensely congratulate authors for valuable compilation.

(Narendra Pratap Singh)
Vice Chancellor

Preface

Fruits are nature's gift to mankind. They are not only delicious and refreshing but are also the chief source of vitamins, minerals and proteins. It may not be exaggeration to say that 80-90 per cent of problems arising from under nutrition or malnutrition can be well managed with indigenous medicines and fruits are an important part of these.

The basic chapter in this book is a collection of recent information and statistical data Global and India related to fruit crops in a single book. The book includes all fruit and plantation crops and their classification. With each crop, with a brief introduction, aspects of botany, climate and soil, varieties propagation, protection and harvesting are covered. Exhaustive information on tropical, sub-tropical, temperate fruits and plantation crops grown commercially in various agro-climatic conditions of India is incorporated.

I must express my appreciation for my wife, Smt. Sushma Gautam, son Omansh Singh and Daughter Mahi Singh, who have been affectionate, understanding and showed their patience during the long hours I spent on the manuscript, which often was very tiring indeed.

However, I have put my best efforts in preparing this book, but if any error or whatsoever has been skipped out, I will by heart welcome your suggestions with the hope that this book will be extremely useful to the students who are pursuing for B. Sc agriculture, B. Sc Horticulture and M. Sc Fruit Science and competitive examinations for JRF, SRF, Ph.D., NET, ARS.

Authors

Contents

Section 3: Temperate Fruit Crops

Section 4: Plantation Crops

Section 1
Basic Chapters

1

Definition and Classification of Fruits

Pomology: Pomology is a branch of horticulture that deals with the cultivation of fruit crops. When given a liberal and flexible translation, the word **"pomology"** means the science of fruit. The term is derived from the Latin word pomum, which means **"Fruit"** and the Greek term **logy**, which means "Science.

Fruit

"A true botanical fruit is defined as a mature or ripened ovary with or without accessory organs."

Classifcation refers to the grouping of individuals with similar identities, related in genetic make up, evolution or climatic requirements. The major units of classification are order, family, genus, sub family, species, variety/cultivars and clone or strain. All fruit crops bear flower and seed (spermatophyte) and belongs to class Angiosperm except in **(*Pinus gerardiana*) of gymnosperm.**

Most of the fruits are covered under the dicotyledonous families of Angiosperm except banana, pineapple, date and coconut which belong to monocotyledonous families e.g. Musaceae, Bromeliaceae and Arecaceae (Palmae) respectively. The families of economically important fruits include Rosaceae (all pome, stone fruits and strawberry), Rutaceae (citrus fruits, bael and woodapple), Musaceae (banana and plantain), Vitaceae (grapes and Anacardiaceae (Mango, cashew and pistachio nut, hog plum) etc.

Nomenclature

In binomial system of nomenclature each scientific name consists of two components i.e., generic name and species name. It is customary to put the name of pioneer author as third component in trinomial system of nomenclature. While **Linneaus** isconsidered the father of plant taxonomy, **De Candolle** is the father of systematic pomology.

1. Based on Botanical Classification

Family	Common Name	Scientific Name	2n	Origin	Fruit type
Monocotyledons					
Bromeliaceae	Pineapple	*Ananas comosus*	50, 75, 100	Brazil	Aggregate berries
Musacae	Banana	*Musa balbisian*	22, 33, 44	Indo-Malayan	Berry
	Plantain	*Musa acuminata*	22, 33, 44	Indo-Burma	Berry
Areaceae/Palmae	Datepalm	*Phoenix dactylifera*	36	West Asia	Drupe
	Coconut	*Cocus nucifera*	32	South-East Asia	Drupe
	Arecanut	*Areca catechu*	32	South-East Asia	Berry
	Oil palm	*Elaeis guineensis*	32	West Africa	
	Palmyra palm	*Borasus flaballifer*	36	Tropical Africa	
Dicotyledonae					
Actinidiaceae	Kiwi	*Actinidia deliciosa*	58	Central America	Berry
Anacardiaceae	Hog plum	*Spondias cytherea*	32	America	Drupe
	Inidan hog plum	*Spondias pinnata*	32	India	Drupe
	Mango	*Mangifera indica*	40	South-East Asia	Drupe
	Pistachio nut	*Pistachia vera*	30	Iran/Iraq	Nut
Annonaceae	Atemoya/ Lakshmanphal	*Annona atemoya*	14	Manmade hybrid	Aggregate berries
	Bullock's heart/ Ramphal	*Annona reticulata*	14	-	Aggregate berries

	Cherimoya/ Hanumanphal	*Annona cherimoya*	14	Bolivia	Aggregate berries
	Custard apple/Sitaphal	*Annona squmosa*	14	West Indies	Aggregate berries
	Sour sop	*Annona muricata*	14	South America	Aggregate berries
Apocynaceae	Karonda	*Carissa carandas*	22	India	Berry
	Natal plum	*Carissa grandiflora*	22	-	Berry
Bombacaceae	Durian	*Durio zibethinus*	56	Malayan region (Borneo)	Berry
Caricaceae	Papaya	*Carica papaya*	18	Tropical America	Berry
Corylaceae/ Betulaceae	Hazelnut	*Corylus colurna*	28	Europe	Nut
	Filbert	*Corylus avellana*	28	Europe	Nut
Dilleeniaceae	Elephant apple	*Dillenia indica*	-	S-E Asia	Fleshy calyx
Ebenaceae	American persimmon	*Diospyros vixgiana*	-	China	Berry
	Persimmon	*Diospyros kaki*	90 (6x)	China	Berry
Ehretiaceae/ Boraginaceae	Lasoda	*Cordiya mixa*	14 or 16	India	Berry
Euphorbiaceae	Aonla	*Emblica officinalis*	28	South-East Asia	Capsule (Drupe)
	Star gooseberry	*Phyllanthus acidus*	-	Madagascar	Berry
	Rubber	*Hevea brasiliensis*	36	Brazil	-
Fabaceae	Tamarind	*Tamarindus indica*	24	Tropical Africa	Drupe
Fagaceae	Chinese chest nut	*Chestua mollissima*	-	-	Nut
	European/ sweet chest nut	*Chestua sativa*	-	-	Nut
Flacourtiacae	Governor'splum	*Flacourtia indica*	-	-	Berry
Guttiferae/ Clusiaceae	Cowphal	*Garcinia cowa*	-	-	Berry

	Malabar tamarind	*Garcinia cambogia*	-	-	Berry
	Mangosteen	*Garcinia mangostana*	28	Malayan Archipelago	Berry
	Kokun	*Garcinia indica*		India	
Lauraceae	Avocado	*Persea Americana*	24	Central America	Berry
Malphighiacae	Barbadaos cherry	*Malphigia punicifolia*	40	Trinidad and Tobago	Drupe
Moraceae	Aini	*Artocarpus hirsute*	56		Sorosis
	Bread fruit	*Artocarpus altilis*	56	Indo-Malayan	Sorosis
	Jack fruit	*Artocarpus heterophylla*	56	India	Sorosis
	Mondey jack	*Artocarpus lakoocha*	56	Western Ghats	Sorosis
	Fig	*Ficus carica*	26	-	Syconus
	Black Mulberry	*Morus nigra*	308	West Asia	Sorosis
	White mulberry	*Morus alba*	28	China	Sorosis
Myrtaceae	Guava	*Psidium guajava*	22	Tropical America	Berry
	Jamun	*Syzygium cuminii*	40	India	Drupe
	Malayan apple	*Syzygium malaccense*	-	-	Berry
	Pineapple guava	*Feijoa sellowiana*	22	South America	Berry
	Rose apple	*Syzygium jambose*	-	-	Drupe
	Watery rose apple	*Syzygium aqueum*	-	-	Berry
	Surinam cherry	*Eugenia uniflora*	22	Tropical America	
Oleaceae	Indian olive	*Olea ferruginea*	-	-	Drupe
	Olive	*Olea europaea*	46	Mediterranean region	Drupe
Oxalidaceae	Bilimbi	*Averrhoa bilimbi*	24	-	Berry
	Carambola	*Averrhoa carambola*	24	Sri Lanka	Berry

Passifloraceae	Passion fruit	*Passiflora edulis*	18	Brazil	Berry
Proteaceae	Macadamia nut	*Macadamia ternifolia*	48	Australia (East)	Nut
Punicaceae	Pomegranate	*Punica granatum*	18	Iran (Persia)	Balausta
Rhammaceae	Chinese jujube	*Ziziphus jujube*	48	China	Drupe
	Indian jujube	*Ziziphus mauritiana*	48	Indo-China (India)	Drupe
Rosaceae	Almond	*Prunus communis*	16	Central Asia (Afghanistan)	Drupe
	Apple	*Malus x domestica*	34	South Western Asia	Pome
	Apricot	*Prunus armeniaca*	16	North Eastern China	Drupe
	Loquat	*Eriobotrya japonica*	34	Central East-China	Pome
	Quince	*Cydonia oblonga*	34	Caucasus region	Pome
	Peach	*Prunus persica*	16	China	Drupe
	Pear	*Pyrus communis*	34	Western China	Pome
	European plum	*Prunus domestica*	48	Europe	Drupe
	Japanese plum	*Prunus salicina*	16	China	Drupe
	Sour cherry	*Prunus cerasus*	16	South East Europe	Drupe
	Sweet cherry	*Prunus avium*	16	Asia Minor	Drupe
	Strawberry	*Fragaria ananassa*	56 (8x)	Manmade hybrid	Etaerio of drupes
Rutaceae	Bael	*Aegle marmelos*	18	India	Amphisarica
	Mandarin	*Citrus reticulate*	18	South Eastern Asia	Hesperidium
	Sweet orange	*Citrus sinensis*	18	Indio-China	Hesperidium
	Acid lime	*Citrus aurantifolia*	18	India	Hesperidium
	Grape fruit	*Citrus paradisi*	18	West Indies	Hesperidium
	Wood apple	*Feronia limonia*	18	India	Amphisarica

	Pummelo	*Citrus grandis/ Citrus maxima*	18	South-East Asia	Hesperidium
	Sweet lime	*Citrus limetta*	18	India	Hesperidium
	Trifoliate orange	*Poncitrus trifoliata*	18	-	Hesperidium
	Tahiti lime/ Persion lime	*Citrus latifolia*	18	-	Hesperidium
	Rangpur lime	*Citrus limonia*	18		Hesperidium
	Lemon	*Citrus limon*	18	South East Asia	Hesperidium
Vitaceae	Grapes	*Vitis vinifera*	38	Western Asia and Central Europe	Berry
Tiliaceae	Phalsa	*Grewia subinequalis*	36	India	Drupe
Sapindaceae	Litchi	*Litchi chinensis*	30	South China	
	Rambutan	*Nephelium lappaceum*	22	Malayan Archipelago	Nut
Sapotaceae	Sapota	*Achras zapota*	26	South Mexico	Berry
Juglandaceae	Walnut	*Juglans regia*	32	Central Asia	Nut
	Peacan nut	*Carya illinoensis*	32	North America	Nut
Sterculiaceae	Cocoa	*Theobroma cocoa*	20	South America	Pod
Salvadoraceae	Pilu	*Salvadora oleoides*	24	India	-
Rubiaceae	Coffee	*Coffea arabica*	44	Ethiopea	Drupe
		Coffea robusta	22	Central Africa	Drupe
Theaceae	Tea	*Camellia sinensis*	30	China	Capsule

2. Based on the Climate Requirements

A-Temperate fruits: Temperate fruit plants have stringent climate requirements. They are only grown in places where winter is bitterly cold, and they must be exposed to a specific chilling temperature for a set period of time in order to flower. These fruit plants are generally deciduous and frost resistant. e.g. **Apple, Almond, Peach, Pear, Plum, Strawberry, Apricot, Persimmon, Pecannut, Walnut, Hazel nut, Cherry, Pistachios, Kiwifruits etc.**

B-Tropical fruits: Tropical fruit plants are typically evergreen and extremely sensitive to cold. They thrive in environments with fewer diurnal temperature fluctuations, as well as light and dark periods., and others require a moist warm climate but can withstand dry weather in some cases. For example, **Mango, Banana, Citrus, Papaya, Sapota, Guava, Grapes, Pineapple, Jackfruit** etc.

C-Sub-tropical fruits: Sub-tropical fruit crops are grown in between temperate and tropical climates. They can be deciduous or evergreen in nature, and can usually withstand cold temperatures but not infrost. They also adapt well to changes in light and darkness during the day and night. Chilling is required for flower bud differentiation in some subtropical fruit plants. For example, **Mangosteen, Avocado, Litchi, Rambutan, Loquat, Durian, Persimmon, Passion fruit,** etc.

Choice of fruit crops for various types of soil and climate condition

Soil/climate condition	Choice of the fruit crop
Climate condition	
Temperate region i. Tree fruits	
ii. Small fruits	Blue berry, goosberry, strwaberry, black berry, raspberry, currant and cranberry
Tropical region Semi arid tropical region	Papaya, datepalm, coconut, banana, sapota, cashew nut, grapefruit, and gooseberry Tamarind and custard apple
Sub tropical region Temperate fruit in sub-tropical region Arid sub-tropical	Mango, citrus, litchi, strwberry, loquat, pomegranate, persimmon, mangosteen and grapes Low chilling peaches, plum, pears, almond and apple Ber, aonla, bael, fig and wood apple
Degraded pastures and grazing land	Jamun, aonla, custard apple, *Annona*, tamarind and fig
Highly shate tolerant	Carambola and bilimbi
Highly shade sensitive	Mango, guava, citurs and coconut

Highly drought tolerant	Ber, aonla, datepalm, Annona, pomegranate and fig
Highly drought sensitive	Banana, litchi, papaya, pomegranate, sapota and coconut
Arid zone	Ber, khair, gonda, khejri and desert date
Semi arid	Custard apple, bael, fig, tamarind, khirni, palmyra palm, phalsa and wood apple
Cold sandy desert	Chilgoza nut
Arid temperate	Oleaster
	Soil conditions
Coastal sand	Ber and coconut
Sandy soils	Coconut, datepalm, cashew nut, phalsa, ber and fig
Loamy soil	Mango, papaya, sapota, grape, banana, apple, citurs and plum
Clayey soil	Jamun and some fig species
Alkaline soil	Guava, datepalm, coconut,, custard apple, olive and phalsa
Acid soil	Strawberry, raspberry, gooseberry, blueberry and fig
Salt affected soil	Datepalm, ber, aonla, bael, guava and coconut
Semi arid temperate	Hazel nut, chestnut and pecan nut
Gullied and Ravinous lands	Ber, custard apple, mulberry, jamun, karonda and palmyra palm
Salt affected lands	Ber (arid), mulberry, jamun (semi arid) and palmyra palm (coastal lands).
Arid region (Rainfed) Arid region (irrigated)	Indian gooseberry, ber, bael, tamarind phalsa, custard apple and karonda. Guava, datepalm, pomegranate, grape, sweet orange, papaya, fig and mango.
Cold desert	Apricot, almond, prunes, apple, sweet cherry, pistachionut, hazel nut and chilgoza.

3. Based on the Woody and Herbaceous Plants

A	**Fruits borne on woody plants**	
i	**Tree fruits**	
	Deciduous	Apple, Pear, Quince, Loquat, Peach, Plum, Apricot
	Evergreen	**Tropical:** Mango, Chironji, Sapota, Guava
		Subtropical: Mangosteen, Litchi, Sweet orange
ii	**Small fruits**	
	Deciduous	Raspberry, Black Berry
	Evergreen	West Indian Cherry
B	**Fruits borne on herbaceous perennial plants**	

	Prostrate growth	Strawberry
	Upright growth	Banana, Pineapple

4. Based on Bearing Habits

Bearing habit refers to the position of fruit bud with respect to vegetative growth. Fruit crops vary in type and position of fruit bud. It also varies with the varieties within the crop. In such crops, the bearing habit of majority of the cultivars is taken as standard for designing various packages and practices in general and pruning and training in particular.

Based on the position of the fruit bud trees are classified as

1. **Shoot bearing:** Bearing on shoot may be terminal, lateral or both terminal and lateral.
2. **Stem or branch bearing (caulifuorous):** Bearing directly from the main stem or branch e.g. jackfruit, cocoa, durian and jaboticaba.
3. **Spur bearing:** Spur is a modified shoot of more than one year which bears fruit bud and remain productive for a few years depending on the cultivar and species.

Fruit Bud

Fruit buds may be simple or mixed. All fruit buds are flower buds but all flower buds are not fruit buds. The staminate flower buds do not bear fruit. Simple bud give rise to flowering or vegetative shoot only depending upon the nature of the bud e.g. cherry, citrus, coconut, datepalm, peach, almond, apricot, plum, mango and loquat. Mixed bud give rise to both flowering and vegetative shoot e.g. apple, pear, ber, cashew nut, grape, guava, pomegranate and olive.

A	**Terminal bearing habits** Fruit buds are borne terminally and give rise to inflorescence without leaves.	
i	**Current season growth** (Bearing on new shoots)	Jackfruit, Loquat, Peacanut
ii	**Old season growth** (Bearing on old shoots)	Mango, Chironji, Banana, Pineapple, Litchi Cherry
B	**Lateral or axillary**	A fruit bud that grows from the axil of a leaf and may develop into a branch or flower.
i	**Current season growth:** (Bearing on new shoots)	Guava, Papaya, Orange, Passion fruit, Coconut, Fig, Mulberry, Phalsa, Karonda, Grape, Ber, Aonla, Sapota
ii	**Old season growth:** (Bearing on old shoots)	Apple, Pear, Peach, Plum, Custard apple
C	Mixed bearing habit	Pomegranate, Citrus, Carambola
D	Cauliflorous bearing habit	Jack fruit, Cocoa, Jaboticaba

5. Based on the Rate of Respiration

Climacteric Fruits (Respiration rate increase after harvesting)	**Non-Climacteric** (Continuous similar or dicrease respiration after harvesting
Mango, Banana, Sapota, Guava, Papaya, Apple, Fig, Peach, Pear, Plum, Annona, Nectarine, Apricot, Avocado, Blueberry, Feijoa guava, Passion fruit, Cherimoya, Kiwi, Cranberry, Raspberry, Durian fruit, Jack fruit.	Citrus spp, Grape, Pomegranate, Pineapple, Litchi, Ber, Jamun, Cashew, Cherry, Rambutan, Strawberry, Phalsa, Loquat, Aonla, Olive,

6. Based on Photoperiodic Responses

Short Day	**Day neutral plant**	**Long Day**
Strawberry, Pineapple and Coffee	Papaya, Guava, Banana and Mango	Passion fruit and Apple

7. Based on Types of Inflorescence

Inflorescence refers to the arrangement of flowers on the stem or axil. It is a flower-bearing shoot, which may be determinate or indeterminate or flowers may be borne in the axil of the leaf. The various types of inflorescence found in fruit and plantation crops are:

	Types	**Fruit crops**
A.	**Raceme**	
i	Solitary	Guava, Peach, Quince, Apricot, Almond, Trifoliate Orange
ii	Raceme	Blackberry, Cape Gooseberry, Raspberry
iii	Catkins	Peacanut, Walnut, Chestnut, Mulberry
iv	Corymbose	Pear
B.	**Cymose**	
i	Panicle	Grapes, Litchi, Mango, Loquat, Pistachio nut, Chironji
ii	Solitary	Papaya, Sapota, Citrus, Phalsa, Persimmon, Strawberry
iii	Fasicle	Sweet Orange, Ber, Plum, Cherry
C	Spadix	Coconut, Arecanut, Date Palm, Banana
D	Hypanthodium	Fig, Pomegranate

8. Based on the Type of Breeding System (which promotes autogamy or allogamy)

Pollination

Pollination refers to the transfer of the pollen from the anthers to the stigma of flower of same or different cultivar or species.

Self Pollination/Autogamy

Pollination in which pollen are transferred from anther to stigma of the same flower is called self pollination or autogamy.

Cleistogamy e.g. Papaya, Grape

Homogamy e.g. Apricot, Citrus, Peach, Phalsa, Dwarf coconut

Cross Pollination/Allogamy

When pollens are transferred to stigma of the other flower or cultivar or species is called cross pollination or allogamy.

2 – Cross - Pollination	
A. Monoecious e.g. Muscadine grape, Banana, Jackfruit, Pecan Nut, Tall Coconut, Hazelnut and Walnut	**B. Dioecious** e. g. Papaya, Kiwifruit, Pistachionut, Date Palm, Kokun, Palmyra Palm and Betal vine
C. Andromonoecious e.g. Mango	D.Gynodioecious e.g. Fig and some varities of papaya

Dichogamy

It refers to the maturity of pollen and receptivity of stigma of the pistil at different periods. The various types of dichogamy observed in fruit crops.

Protoandry	Protogynous	Heterodichogamy	Protogynous Diurnally Synchronous Dichogamy (PDSD)	Duodichogamy
Walnut, Coconut, *A. muricata,* sapota and passion fruit	*Annona spp.*, Except *A. muricata*, fig, banana, plum and pomegranate	Pistachionut, Pecanut	Avocado	Chestnut

Heterostyly

It refers to the condition in which stigma is at different height due to varying length of styles.

I. **Pine type e.g.** Sapota, Litchi, Pomegranate

II. **Thrum type e.g.** Almond, Carambola

Self - incompatibility

It is the inability of the funtional pollens to fertilize the stigma of the same flower. Pollen grains are viable in case of incompatibility but are unviable in sterlity. The various types of self incompatibility observed in fruit and plantation crops.

I. **Sporophytic SI:** Mango, Aonla, Cocoa

II. **Gametophytic SI:** Ber, Pineapple, Apple, Pear, Apricot, Almod, Cherry, Loquat.

9. Based on Mode of Pollination

i. Pollinator

Agency involed in the transfer of pollens (anthers) is called pollinators.

Entomophilous (insect)	Anemophilous (wind)	Ornithophilous (birds)
Citrus, Guava, Litchi, Ber, Annoa, Mango (House fly), Apple, Pear, Peach, Plum,Cherry, Almond, Fig (Wasp), Oil palm (Weevil)	Papaya, Sapota, Jack fruit, Pomegranate, Date palm, Aonla, Chestnut, Cashew nut, Coconut (Tall)	Banana, Pineapple (Humming birds)

Pollinizer: Pollinizers are plants that act as a source of pollen for successful pollination and fertilization.

10. Based on the Hardiness of Skin

Hardiness	Examples
Hard fruits	Bael and Wood apple
Soft fruit	Papaya, Sapota

11. Based on Fruit Morphology

According to thier development fruits are divided into folowing three categories as simple, aggregate and multiple or composite fruits.

	Type of fruit	Examples
A	**Simple fruits:** Fruits developed from single or syncarpous pistil of a single flower with or without accessory organ.	
I	Berry	Banana, Papaya, Grape, Sapota, Guava, Date palm,
a	Modified berry	
i	Balausta	Pomegranate
ii	Amphisarica	Wood apple, Bael
b	Pome	Apple, Pear, Quince, Loquat
ii	Drupe (stone)	Mango, Peaches, Plum, Almond, Ber, Coffee, Apricot, Phalsa, Plum, Jamun, Coconut,
iii	Hesperidium	Citrus
iv	Nut fruit	Litchi, Rambutan, Cashewnut, Walnut, Pecanut, Hazel nut, Chestnut, Pistachionut, Macadamia nut
	Capsule	Aonla, Carambola
B	**Aggregate fruits (Develops from numerous ovaries of the same flower):** Fruits developing from the apocarpous pistil (free carpels) of flower.	
i	Etaerio of berries	Custard apple
ii	Etaerio of drupelets	Blackberry, Longon berry
iii	Etaerio of achenes	Strawberry
C	**Multiple/composite fruits:** Fruits developed as a result of fusion of number of flowers.	
i	Syconus	Fig
ii	Sorosis	Jackfruit, Pineapple, Breadfruit, Mulberry

12. Based on the Edible Part

Fruits are generally put to divere uses. Most of them are used for dessert or table purpuse. Pericarp (developed from wall of ovary) is most common edible part of fruits except nuts and dry fruits where kernel (seed) is consumed fresh or after processing.

Pericarp is furthure divided into epicarp or exocarp (skin of the fruit, outer layer of pericarp), mesocarp (middle layer of pericarp, parenchyma tissues, may be fibrous, juice or pulp) and endocarp (very thin and membranous in some fruits and hard like stone in others.

Edible part	Fruit crops
Mesocarp	Passion fruit, Papaya, Mango, Mulberry, Tamarind, Sapota
Mesocarp Endocarp	Aonla, Apricot, Banana
Mesocarp Epicarp	Olive, Peach, Persimmon, Phalsa, Cherry, Plum, Jamun, Karonda
Pericarp	Custard apple, Avocado, Ber, Date palm,
Pericarp placenta	Grape
Endosperm	Coconut
Fleshy thalamus	Apple, Pear, Loquat, Strawberry, Quince
Cotyledon	Pecan nut, Cashew nut, Almond, Pistachionut,
Succulent placenta	Bael, Wood apple
Juicy placental hair	Citrus
Juicy seed coat	Pomegranate
Thalamus & pericarp	Guava
Bracts & Perianth	Jackfruit, Pineapple
Fleshy aril	Litchi
Fleshy receptacle	Fig

13. Based on Centre of Origin

The concept of centre of origin was given by Vavillov. There are eight broadly recognized centres of origin of plants. Centre of origin refers to the geographical region where a particular species has originated naturally and contain maximum diversity in the form of wild races and relatives.

It is more appropriate to say that all centres of origin are the centres of diversity, but all centres of diversity are not the centres of origin. Centre of origin is of two types namely primary centre and secondary centre of origin. Secondary centre of origin refers to the place where the species has maximum diversity but not originally native to that place.

Some of the important centre of origins of fruit crops includes Mediterranean, Hindustan, Chinese, Asia minor and Tropical America.

Native	Fruit crop
Abyssinia (Ethiopia and Eritrea)	*Coffea arabica*
Asia Minor	Apple, fig, chestnut, pistachionut, almond, walnut and European grape
Near Centre	Peach, plum, hazelnut, fibernut, pomegranate, fig, cherry, *pyrus* and *malus*
Central America(South Mexico)	Papaya, guava, rubber and avocado

Chines(Central and Western China)	Persimmom, quince, litchi, sweet orange, tea and peaches, Apricot, Mandarin, Plum
South America (Peru, Chile and Brazil)	Guava, Pineapple and cashew nut
Central Asia	Pistachio nut, apricot, almond, grape and pear
Hindustan (Burma, Assam, Malayan Archipelago, Java, Borneo, Sumatra and Philippines)	Jackfruit, mango, bael, jamun, ber, karonda, aonla, banana, coconut and wood apple

14. Based on the Type of Placentation

Type of placentation	Fruit crops
Axil	Banana, Citrus, Aonla
Marginal	Litchi
Parietal	Papaya
Basal	Ber

15. Based on the Shade Tolerance

Level	Examples
Highly shade tolerant	Carambola, Bilimbi (Cucumber tree)
Moderate shade tolerant	Banana, Mangosteen
Highly shade tolerant	Citrus, Mango, Guava, Coconut

16. Based on their Respiratory Behavior

Levels	Examples
Highly tolerance	Date palm, Ber, Aonla, Guava, Bael, Coconut, Khirni
Medium tolerance	Pomegranate, Cashew nut, Fig, Jamun, Phalsa
Highly sensitive	Mango, Citrus, Apple, Avocado, Pear, Strawberry

17. Based on the Acid Tolerant

Level	Examples
Slightly tolerant to acid soil	Mango, Citrus, Banana,Guava,Papaya, Apple, Peach, Kiwi fruit
Moderately tolerant to acid soil	Orange, Pineapple, Jack fruit, Avocado, Litchi, Loquat
Highly tolerant to acid soil	Wood apple, Bael, Strawberry
Tolerant to alkaline soil	Guava, Date palm, Aonla, Custard apple, Coconut

18. Based on Growth Pattern

Growth pattern	Examples
Single sigmoid growth curve	Mango, Apple, Date palm, Apple, Pear, Sweet orange, Lemon, Strawberry, Papaya
Double sigmoid growth curve	Stone fruits (Peach, Plum, Nectarine, Cherry, Apricot) Almond, Pineapple, Annona
Triploid sigmoid growth curve	Kiwi fruit

19. Based on the Acid Types

Acids	Fruits
Citric acid	Berries, Citrus, Guava, Pear, Pineapple
Mallic acid	Apple, Banana, Cherry, Plum, Melon
Tartaric acid	Grape
Benzoic acid	Cranberry
Quinic acid	Blueberry
Oxalic acid	Carambola

20. Based on the Bases Storage Behavior

Orthodox seeds	Recalcitrant seeds
Apple, Grapes,Ber, Peach, Plum, Phalsa, Pomegranate, Passion fruit, Custard apple, Date palm, Fig, Guava	Mango, Mangosteen, Citrus, Jack fruit, Loquat, Litchi, Rambutan, Avocado, Barbados cherry, Carambola, Durian, Bread fruit, Rubber, Cocoa, Oil palm

21. Based on the Alternate Bearing Habit

Mango, Apple, Olive, Persimmon, European plum, Pecanut, Date palm

22. Based on the Type of Parthenocarpy

Vegetative	Stimulative	Sternospermocarpy
Banana, Fig and Pine apple	Litchi, Black corianth variety of grape, Bread fruit	Grape and Sindhu variety of mango

23. Apomixis

Parthenogenesis	Recurrent apomixis	Non-rucurrent/Nucellarbudding/ Polyembryony
Mangosteen and Strawberry	Apple Raspberry	Mango-Adventive Citrus-Nucellar Jamun/Javaplum-Nucellar

24. Based on the Aroma of Some Fruits

Fruits	Stage and Compound
Apple	**Ripe:** Ethyl -2-methyl butyrate **Green:** Hexanal
Banana	**Green:** Hexanal **Ripe:** Eugenol **Over ripe:** Isopentanol
Grape fruit	Nootakatone
Lemon	Citral
Orange	Valencene
Grape	Methyl Anthranilate
Pineapple	Anthranilate
Strawberry	Benzyl Acetate, Furaneol

25. Based on the Ethylene Production

Class	μ L C_2H_4 / Kg/hr	Fruit Crops
Very low	<0.1	Grape, Citrus
Low	0.1 – 1.0	Pineapple
Medium	1 – 10	Banana, Mango, Guava, Fig,
High	10 - 100	Apple, Papaya, Avocado, Plum
Very high	>100	Passion fruit, Sapota, Apple

26. Based on the Rate of Respiration

Level of Respiration	Rate of respiration (mg of CO_2/Kg/Hr)	Fruit crops
Very low	<5	Nut, Dried fruits
Low	5 – 1-0	Citrus, Grape, Apple
Medium	10 – 20	Mango, Banana, Pear, Peach, Fig
High	20 - 40	Strawberry, Avocado

27. Based Flower Bud Differentiation

Mango	Oct. – Dec. (but in Dashehari May-June and Sep.-Oct.)
Grape	March-April
Banana	Sep-April
Aonla	First week of March
Pineapple	At 40 leaves stage

28. Based on the Consumer Preference

Very light	50-100gm	Grape, Ber, Banana
Light	100-150gm	Sapota, Pomegranate
Light medium	150-300gm	Mango
Medium	300-350gm	Avocado
Medium to heavy	800-1000gm	Mango
Heavy	1-5kg	Bread fruit, Pineapple
Very heavy	>5kg	Jack Fruit

29. Based on Poly Ploidy Level

Majority of the fruit crops are diploid e.g. some banana (AA, AB genome), citrus fruits, peach, almond, plum, apricot, walnut, pecan nut, cheryy and *Actinidia chinensis*. However, ploidy level of fruit crops varies from diploid to octaploid.

Fruit crops are mostly euploidy except few genotypes of guava which are aneuploids and used as rootstocks. Inter-varietal ploidy differences are also found in some of the fruit species e.g. banana, ber, apple, pineapple etc.

A. Euploidy	
Allo-polyploidy -I	
Allotetraploid/Amphidiploid	Mango
Allo-hexaploid	European plum
Allo-octaploid	Vellaikolamban, Cultivated Strawberry
Auto-polyploidy -II	
Auto-triploid	Cultivated banana, Tahiti lime
Auto-hexaploid	Persimmon, Kiwi
Auto-octaploid	Ber cv. Gola, Illaichi
B. Aneuploidy	
Aneuploid-82	Pusa, Srijan (Guava dwarf rootstock)

30. Based on Nutritive Value

S.No.	Nutrition	Fruits/100g
1	Vitamin-A (Carotene)	Mango (4800 IU) > Papaya (2020 IU) > Persimmom 2710 IU/100g) > Passion fruit (1300-2500 IU/100g)
2	Vitamin-B1 (Thiamine)	Cashew nut (630 mg) > Walnut (450 mg)
3	Vitamin-B2 (Riboflavin)	Bael (1191 mg) > Papaya (250 mg) > Litchi (122.5 mg)

4	Vitamin-C (Ascorbic acid)	Baarbados cherry (1000-4000 mg) >Aonla (600 mg) > Guava (299 mg) > Lime (63mg) > Lemon (39 mg/100g)
5	Carbohydrates	Raisins (77.3%) > Dry Apricot (72.8%) > Dry karonda (67.1 %) > Date (67.8%) > Banana (36.40%) > Bael (30.6%)
6	Protein	Cashew nut (21.2%) >Almond (20.88%) >Walnut (15.60%)
7	Fat	Walnut (64.54%) > Almond (58.9%) >Cashewnut (46.90%) > Avocado (22.80%)
8	Fiber	Fig > Guava (6.9%) > Pomegranate (5.1%) > Aonla (3.40%) > Grape (3.0%)
9	Calcium	Litchi (0.21%) > Dry karonda (0.16%) >
10	Phosphorus	Almond (0.49%) > Cashew nut (0.45%) > Walnut (0.38%)
11	Iron	Dry karonda (39.1%) > Date (10.6%) > Cashew nut (5.0%)
12	Calorific value	Walnut (687 mg) > Almond (655 mg) > Cashew nut (596) > Dates (283 mg)

31. Based on Common Name

S. No.	Fruit crops	Common Name
1	Abiu	Camio, Yellow star apple
2	Almond	Important temperate nut fruit
3	Aonla	Indian gooseberry, Malacca tree, Myrobalan, Amritphal fruit, Scared tree
4	Apple	King of temperate fruits, Symbol of health, Premier fruit of the world, Patient fruit
5	Apricot	Armenian plum, Coppery furit
6	Arecanut	Betel nut, Supari
7	Avocado	Fruit of new world, Butter fruit, Alligator pear, 21st century fruit, Tasteless fruit.
8	Bael	Sirphal, Bengal quince, Indian quince, Golden apple, Symbol of Lord Shiva fruit
9	Banana	Adam's fig, Tree of wisdom, Tree of paradise, Kalpataru, Apple of paradise, Baby fruit, Test tube fruit, Plant of virtue, Antique fruit crop, Moisture loving plant
10	Ber	Poor man's fruit, Chinese fig, Chinese date, King of arid fruits, Summer deciduous fruit, Apple of desert
11	Bilimbi	Cucumber tree, Tree sorel
12	Carambola	Star fruit, Kamrakh, Five corner fruit
13	Cashew	Dollar earning crop, Plough crop, Gold mine of waste land, Wonder nut
14	Citrus	Micro-nutrient loving plant, Fruit of the godly tree, Buddha's hand fruit
15	Cocoa	Food of god

16	Coconut	Kalpavrikhsha, Tree of heaven, Tree of life
17	Coffee	King of beverage crop
18	Custard apple	Sugar apple, Sweet sop, Sitaphal, Sharifa, Fruit of poor people
19	Date palm	Century plant, Oldest plant, Head in fire and foot in water crop
20	Dillenia	Chalta, Elephant apple
21	Dragon fruit	Cacuts plant, Night blooming cactus, Night blooming cereus, Pitaya, Pitahayas, Strawberry pear, Belle of the night, Cinderella plant, Jesus in the cradle
22	Durian	Civat fruit, King of fruit in Indonesia
23	Egg fruit	Canistel, Yellow sapota
24	Fig	Forbidden fruit, Wholesome, Nutritious and delicious fruit, Fruit tree denotes, Peace and Prosperity, Anjir
25	Grape	Sophisticated fruit crop, Symbol of charity
26	Grapefruit	Forbidden fruit, Breakfast fruit
27	Guava	Apple of tropics, Poor man's apple
28	Cherimoya	Hanumanphal, Cherimoyer
29	Jackfruit	Poor man's food, National fruit of Bangladesh, Kathal
30	Jamun	Indian black berry, Black plum, Java plum, Rose apple, Pitangu cherry, Diabetic fighter
31	Jungle Jalebi	Ice cream bean
32	Karonda	Christ thorn
33	Khejri	Life line of Thar desert, Kalptaru of Indian Desert
34	Kiwi fruit	China's miracle, Horticultural wonder of New Zealand, National symbol of New Zealand, Chinese gooseberry, Peach of the Macaque Monkey
35	Kokum	Kokum butter tree, Mangosteen oil tree, Ratamba
36	Lasoda	Gonda, Indian cherry, Lehsua, Cherry of desert
	Litchi	Queen of fruits, Fruit of high commerce
37	Longan fruit	Eye of dragon, Little brother of litchi
38	Loquat	Japanese plum, Japanese medlar
39	Macadamia nut	World's finest dessert nut
40	Mahua	State tree of Gujrat
41	Makhana	Goron nut, Fox nut
42	Mandarin	Fancy fruit, Jacket oranges, Loose skinned orange
43	Mango	King of fruits, Bathroom fruit, National fruit of India, Pride fruit of India, Symbol of love, Pride of garden, Choicest fruit of Hindustan, Ambassador of Indian fruit
44	Mangosteen	Energy tablet, Finest fruit of the world, Fruit of Gods, Queen of tropical fruits, Mystery fruit, Ice fruit, National fruit of Thailand
45	Noni	Cheese fruit, Indian mulberry

46	Oil palm	Small holder irrigated crop, Crop for future
47	Olive	Emblem of United Nations
48	Palmyrah Palm	State tree of Tamil Nadu. Kalpaga thara
49	Papaya	Melon tree, Wholwsome fruit, Wonder fruit of tropics, Subtropics, Backuard fruit, Continuous flowering and fruiting crop
50	Pear	Gift of God
51	Pecanut	Queen of nut
52	Persimmon	Ebony tree, National fruit of Japan
53	Phalsa	Star apple, Dhamani
54	Pilu	Tooth brush tree, Grape of desert
55	Pineapple	Heavenly fruit, Friendship fruit, Symbol of friendship and Hospitality, Dessert fruit, Golden queen
56	Pistachio nut	Smiling nut
57	Pomegranate	Fruit of love, National fruit of Iran, Fruit of paradise
	Rambutan	Hairy litchi, High temperature loving plant
58	Ramphal	Bullock's heart, West Indian custard apple, Anta, Lavni, Krishnabeejam
59	Sapota	Sapodilla plum, Sapodilla bully, Chikoo
60	Seabuckthorn	Gold mine of cold desert
61	Sweet orange	Tight skinned orange
62	Tamarind	Indian date, Imli, Girl fruit
63	Tea	Queen of beverage crops, Golden leaf
64	Walnut	King of nut, Nut of God
65	Waterchestnut	Panifal, Singhara
66	Wood apple	Kavet, Curd apple, Monkey fruit, Curd fruit

2

Plant Propagation of Fruit and Plantation Crops

Sexual plant propagationis multiplication of the plants by fussion of male and female gametes is called sexual plant propagation e. g seed and seedling plant. It is two types:

A. Sexual Propagation

B. Asexual Propagation

Sexual Propagation

Sexual plant propagationis multiplication of the plants by fussion of male and female gametes is called sexual plant propagation e.g seed and seedling plant.

Advantages

Sexual method of propagation has several advantages:

1. Propagation by seeds is simple and easy.
2. Seed propagation is only mean of diversity particularly in the selection of chance seedlings.
3. Seedling plants are long lived, productive and have greater tolerance to adverse soil and climatic conditions and diseases.
4. Seed propagation makes feasible to propagate plants like papaya and coconut in which asexual means of propagation is not common.
5. Hybrids can only be developed by sexual means.
6. Sexual propagation offers opportunities of polyembryony (citrus, mango or Jamun) and apomixis (*Malus sikkimensis*, *Malus hupehensis*, *Malus sargentii*), which produces true- to - type plants.
7. Seed is the source for production of rootstocks for asexual propagation
8. Seeds, if stored properly can be kept for longer duration /period for future use.

Disadvantages

Sexual method of propagation has some disadvantages:

1. Seedling plants are not true to type to the mother plants due to heterozygous nature of fruit plants.
2. Seedling plants have long juvenile phase (6-10 years) and hence flowering and fruiting commences very late in them.
3. Sexually raised plants are generally tall and spreading type and thus are cumbersome for carrying out various management practices like pruning, spraying, harvesting etc.
4. Seeds of many fruits are to be sown immediately after extraction from the fruits as they lose their viability very soon e.g. cashew nut, Jamun, Jackfruit, Citrus, Mango and Papaya.
5. The beneficial influences of rootstocks on scion variety cannot be exploited in sexual propagation.
6. Seedling plants usually produce fruits inferior quality.

B. Asexual Propagation

Advantages

Propagation by asexual means has several advantages over sexual means:

1. Asexually propagated plants are true to type to their mother plants.
2. Asexually propagated plants have short juvenile phase and bear flowers and fruits in the early age (3-4 years) than seedling plants.
3. The vegetatively propagated plants are smaller in stature and hence management operations like spraying, pruning and harvesting etc. become easy.
4. Plants in which seed setting does not take place (e.g. pineapple and banana), asexual propagation serves as a substitute for sexual propagation.
5. Using asexual methods, desirable characters of a mother plant can be perpetuated/ multiplied easily.
6. The benefits of rootstocks and scion are usually exploited through asexual propagation.
7. Repairing of damaged portion of plant is possible through asexual propagation as in case of bridge grafting.
8. It is possible to convert a non-productive local variety into productive improved variety by using asexual methods.
9. It is possible to grow several varieties on one plant or change variety of existing plant by top working

Disadvantages

1. Asexual propagated plants have shorter life-span.
2. Asexual propagation restricts diversity.
3. Sometimes asexual propagation disseminates diseases e.g. *Tristeza* virus in citrus.
4. Technical skill is required.

Micropropagation

Micropropagation is the technique of developing plants from very small portion of plants such as shoot tip root tip, embryo, stem, pollen grain, callus or single cell. Plant tissue culture owes its origin to the famous German plant physiologist, Haberlandt in 1902. This technique has opened a vast scope for improvement of fruits and plantation crops though micropropagation, creating genetic diversity, germplasm conservation, virus elimination, development of somatic hybrids and gene transfer. In India, commercial exploitation of micro-propagation is limited to oilpalm, strwaberry and banana.

List of Commercial Propagation Methods of Fruit and Plantation Crops

S. No.	Common Name	Commercial propagated methods
1	Acid lime	Seed
2	Almond	T-budding
3	Aonla	T-budding/patch
4	Apricot	Tongue grafting
5	Areca nut	Seed
6	Avocado	Air layering
7	Apple	Tongue grafting
8	Bael	Patch budding
9	Ber	Ring & T-budding
10	Banana	Sward sucker
11	Cashew nut	Softwood grafting
12	Cherry	Air layering
13	Cape Gooseberry	Seed
14	Coconut	Seed
15	Chironji	Seed
16	Cocoa	Seed
17	Coffee	Seed
18	Custard apple	T-budding
19	Dragon fruit	Cutting

20	Date palm	Offshoot
21	Fig	Hardwood cutting
22	Grape fruit	T-budding
23	Grape	Hardwood cutting
24	Guava	Stooling
25	Jack fruit	Inarching
26	Jamun	Shield & patch budding
27	Karonda	Seed, hard wood cutting
28	Lasoda	Seed/Budding
29	Litchi	Air layering
30	Loquat	Inarching
31	Lemon	Air, layering, seed
32	Mango	Veneer grafting
33	Mangosteen	Seed, inarching
34	Mandarin	T-budding
35	Mahua	Seed
36	Olive	T-budding
37	Oil palm	Seed
38	Peach	T-budding
39	Papaya	Seed
40	Passion fruit	Seed
41	Phalsa	Seed
42	Pear	Tongue grafting
43	Pecan nut	Patch
44	pineapple	Suckers, slips
45	Persimmon	Crown grafting
46	Pomegranate	Hardwood cutting
47	Pummelo	Seed, T-budding
48	Plum	T-budding
49	Rubber	Forket budding
50	Sapota	Inarching
51	Strawberry	Runners
52	Sweet orange	T-budding
53	Tea	Softwood cutting
54	Wood apple	Seed, in situ budding

3

System of Planting Fruit and Plantation Crops

The selection of a system for planting depends on the conditions and topography of the orchard's location, plant species to be planted and orchard management practices. Generally, six systems of planting are recommended for fruit trees.

1. Square System

This system is considered to be the simplest of all the systems and is widely adopted. In this system, the plot is divided into squares and trees are planted at the four corners of each square, in straight rows running at right angles.

2. Rectangular System

In this system, the plot is divided into rectangles instead of squares, and trees are planted at the four corners of the rectangle in straight rows running at right angles. This system is suitable for high density planting (HDP), viz., mango, aonla.

3. Hexagonal or Equilateral Triangular System

In this system of planting, a tree is planted at each corner of an equilateral triangle. In this way, six trees form a hexagon with a seventh tree in the centre and 15% more trees can be planted in a hexagonal system than in a square system.

4. Quincunx or Diagonal System

This system of planting fruit trees is similar to a square, except that a fifth tree is planted at the centre of each square. The additional tree in the centre is known as "filler". The fillers are usually quick-growing, early-maturing, and erect types of fruit trees like bananas, papayas, pomegranates, etc.

5. Triangular System

The trees are planted in such a way that those in the even rows are midway between those in the odd rows. This system is based on the principles of the isolateral triangle. The distance between any two adjacent trees in a row is equal to the perpendicular distance between any two adjacent rows.

6. Contour System

It is generally followed on the hills with high slopes. It particularly suits land with undulated topography, where there is a greater danger of erosion and irrigation of the orchard is difficult. The main purpose of this system is to minimize land erosion and to conserve soil moisture so as to make the slope fit for growing fruits.

High Density Planting in Fruit Crops

Accommodation of more number of plants per unit known as a high density orchard. In high density orchard uses different techniques like use of dwarfing rootstocks, genetically dwarf scion varieties, by adopting special systems of training and pruning, use of plant growth retardants, and other such other practices. in certain fruit crops like mango, apple, citrus, pear, walnut and guava.

S.No.	Fruit	Variety	Spacing (m)	System of planting	Number of plants
1	Mango	Amrapali	2.5 × 2.5	Triangular	1600
		Dashehari	3.0 × 2.5	Rectangular	1333
2	Banana	Dwarf Cavendish	1.5 × 1.5	Square system	4444
		Robusta	1.5 × 1.5	Square system	4444
3	Pineapple		25 × 60 × 75 cm	Double row system	63000
4	Papaya	Pusa Nanha	1.2 ×1.2	Rectangular	6944
5	Apple	Spur type on rootstock MM111, MM109	4 × 4	Square system	625
6		Non spur type MM106, MM109	5 × 5	Square system	400
		Spur type on MM106, M7	3 × 3	Square system	1111
7	Guava	Aneuploids	3 × 3	Square system	1111
		L-49	6 × 2	Square system	833
8	Acid lime	Kagzi	4.5 × 4.5	Square system	494
9	Sapota	PKM-3			
10	Peach	Red Heaven, Candor			
11	Cherry	Metero			

4

Training and Pruning

In manipulation of plant development, training and pruning are important for which our knowledge about plant development and its phenology has to be complete. These practices are important in fruit crops.

Training

It means developing a desired shape of the tree with particular objectives by controlling habit of growth. Training is start from nursery stage of plant. Some fruit crops like grape vines, ber, fig, guava etc require training.

Objectives of Training

- To direct the growth of the tree so that various cultural operations such as spraying, ploughing, harvesting can be performed easily and at lower cost.
- To admit more light and air to the centre of the tree to expose maximum leaf surface to the sun
- To protect the tree from sun burn and wind damage.
- To secure a balanced distribution of fruit bearing parts of the tree.

Methods of Training

Method of training of a plant is determined by the nature of plant, climate, purpose of growing, planting method, mechanization etc.

1. **Open centre or vase centre leader system:** The main stem is allowed to grow only up to a certain height about 1.5 to 1.8 m and then it cut for development of lateral branches. It allows full sunshine to reach each branch. viz. Peach, Nectarine, Apricot, Ber, Japanese plum,
2. **Central leader system:** In this system the central leader branches are allowed to grow indefinitely so that it will grow more rapidly and vigorously than the side branches and tree became tall. Such a tree bears fruit more near the top. The lower branches are less vigorous and less fruitful. viz. Apple, Pear, Mango, Sapota, Walnut, Pecanut.
3. **Modified leader system:** It is intermediate between the above systems. It is developed by first training the tree to the leader type by allowing the central axil to grow un pruned for the first four or five years. Then central

stem is headed back and lateral branches are allowed to grow as in the open centre system. European plum, Sweet cherry, Pear, Apple, guava etc.

Special Training Systems of Fruit Crops

S.No.	Training systems	Fruit crops
1	Bower system	Grape
2	Telephone system	Grape
3	Espalier system	Apple
4	Cordons	Peach
5	Single stem system	Citrus, Annona, Fig
6	Multiple stem system	Pomegranate
7	Two arm kniffin system	Passion fruit
8	T-Bar system	Kiwi fruit

Pruning

It refers to removal of plant part like bud, shoot, root etc. to strike a balance between vegetative growth and production. This may also be done to adjust fruit load on the tree. "Judicious removal of plant part to obtain better and qualitative yield is termed as pruning". Pruning is started in later part of plant life. When its become capable to produce flower and fruits.

Objectives of Pruning

- To remove the dead, diseased and over aged wood.
- To minimize biannual bearing and consequent riskof die back.
- To regulate the size and quality of the fruits by way of proper distribution of the fruiting area.
- Bringing vigour in old trees.

System of Pruning

- **Heading back (Light pruning):** Only tops of branches are headed back or cut off
- **Thining out:** Complete removed of a branches or a part from the plant
- **Dehorning:** Cutting away the main limbs or thick major branches from the plant. viz. Mango (Vidharbha region of Maharashtra)
- **Notching:** It is a partial ringing of a branch above dormant lateral bud.
- **Smuding:** It is a practice of smoking the tree like mango in Philippines.
- **Girdling:** Removal of 203 mm wide strip of bark around the stem like grapes.
- **Ringing:** Removal of complete ring of bark from a branch or trunk viz. Grapes and Mango
- **Bulk pruning:** Heavy pruning is done of the tree for good fruit production.

5

Leaf Sampling Technique of Major Fruit and Plantation Crops

Fertilizer Requirement of Fruit Trees

Fertilizer requirement of the tree is determined by visual assessment of deficiencies soil nutrient status and plant tissue nutrient status. Among these methods, tissue nutrient analysis is gaining importance primariarily due to fact that it assesses the effective amount of nutrient utilized by the plant. It also detects the hidden deficiency of the nutrient before the symptoms appears on the tree. Standardised tissue nutrient norms for various elements are available for different crops. The precision of such standards depend on the sampiling procedures of the index tissue. The various plant factors affecting the nutrient status are variety, rootstock, position of the leaf, aspect of the tree, height and direction of sampling, sample size, season of sampling etc., the standardised leaf sampling procedure for various fruit and plantation crops.

Crop	Index Tissue	Sampling Technique
Almond	Recently matured leaf	From base of current season growth at bloom stage
Apple and pear	Leaves from middle of terminal shoot growth, recently matured leaf including petiole	8-12 weeks after full bloom 2-4 weeks after formation of terminal buds in bearing trees
Apricot	Fully expanded leaves including petiole from middle of shoot current growth	Early June to mid July from base of the current season growth at bloom stage
Avocado	20 cm^2 section of leaf lamina on both side of midrib in centre of the leaf from non-fruiting terminal	5-7 month leaf from current season growth (spring flush) from centre of the shoot
Banana	20 cm^2 section of leaf lamina from both side of the mid rib in centre of the leaf	3rd fully opened leaf
Ber	6th leaf from apex from secondary or terminary shoot	2 month after pruning
Cashew nut	4th leaf from top of matured branches	At beginning of flowering

Cherry	Fully expanded leaves, mid shoot current growth	July-August
Citrus	3-5-month old leaf from new flush. 1st leaf of the shoot	June
Cocoa	3rd leaf from apex	Blooming initiation
Custard apple	5th leaf from apex	2 months after new growth
Fig	Fully expanded leaves, mid shoot current growth	July-August
Grape	5th petiole from base	Bud differentiation stage for yield forecast, petiole opposite to bloom for quality
Guava	3rd pair of recently matured leaves	Bloom stage (Aug.or Dec.)
Mango	Leaves + petiole	4-7 month old leaves from middle of shoot
Papaya	6th petiole from apex	6 months after planting
Passion fruit	Matured leaf opposite to last open flower	Full bloom
Peach	Mid-shoot leaves, fruiting or non-fruiting spurs	Mid September , rest as for apple
Pecan and walnut	Pair of middle leaflets from the medium leaf on terminal shoots	6-8 weeks after full bloom or 7-21 days after terminal growth has ceased
Pineapple	Middle one third portion of white basal portion of 4th leaf from apex	4-6 months
Plum	Leaves from middle of current seasons extension growth	January-February
Pomegranate	8th leaf from apex	Bud differentiatio. In April for February crop and August for June crop
Sapota	10th leaf from apex	September
Strawberry	Youngest fully expanded matured leaf. Use leaflets and discard main petiole	Peak or harvest period

6

Maturity Indices for Fruits

Maturity (commercial or horticultural): Refers to the stage at which the produce is optimally accepted to the consumer.

Maturity index: The maturity at harvest determines the quality and post-harvest shelf-life of the fresh fruits. The study of maturity indices helps to harvest the crop at right time.

Judging maturity: There are various means of judging maturity but they vary according to the kind of fruits, local soil and climatic conditions, but generally farmers follow visual means (appearance). But this is not a perfect method, many a times it may mislead the farmers.

There are certain limitations for all maturity indices due to variation in nutrition of the crop, fruit shape, size, climate, seasonal factors, moisture, pruning method, use of hormones and other chemicals. We can use various indices to judge maturity.

Mango

1. Slight colour development on the shoulders.
2. When one or two ripe fruits fall from the tree naturally (tapka method).
3. Skin colour changes from dark green to olive green.
4. Counting of the days from fruit set to maturity.
5. When the specific gravity of fruits ranges between 1.01 to 1.02.
6. Flesh firmness.
7. Lenticles become more prominent and waxy bloom gradually disappears.
8. When the TSS reaches
9. The first and second methods are not of much help since these are not representative of the fruit maturity of the entire tree and the fruits harvested do not ripe uniformly.

Banana

1. The fruits are harvested at different maturity level based on distance of transport.
2. For long distance transport 75-80% maturity and for short distance transport 90-100% maturity fruits can be harvested.

3. When the pulp peel ratio reaches 1.2-1.6 and days taken from shooting, i.e., 3.0-3.5 moths.
4. Disappearance of angularity of the fingers.
5. Brittleness of floral ends.
6. Drying of leaves in some varieties.

Guava

1. Guava fruits generally take about 17-20 weeks from fruits set to reach maturity.
2. When the colour changes from dark green to light green.
3. When the specific gravity is one (1.0).

Grape

1. Grape is harvested when they reach a TSS of 16 to 24% depending on variety.
2. Bangalore blue – 12-14%.
3. Anab-e-Shahi and Selection-7 – 16-18%.
4. Thompson seedless – 20-22%.
5. Besides TSS, the following physical characteristics are also useful in judging maturity.
6. Texture of pulp (softness).
7. Peel colour (light yellow).
8. Easy separation of the berries from the bunch.
9. Development of characteristics flavor and aroma.

Papaya

1. **For local market:** When skin colour changes at the apical end of the fruit.
2. **For distance transport:** When the skin colour changes from green to yellow to the extent of 6%.
3. When the latex of the fruit becomes almost watery.

Pineapple

1. **For local market:** When 25% of surface changes to yellow colour.
2. **For long distance:** When all the eyes are still green and have no trace of yellow colour.
3. **The fruit develops a smooth surface around the eyes.**
4. The flattening of eyes.

Fig

1. When fruits become soft and wilt at neck.
2. Fruits hanging down from their own weight.
3. No milk exudation from the stem when the fruit is polled off.

Custard Apple

1. When the fruit turns to light green colour.
2. Development of yellowish white colour between the carpels.
3. Initiation of widening the gap between carpels or segments.

Ber

1. Attainment of full size of particular cultivar with softening of pulp.
2. Development of characteristics yellow or golden yellow colour.
3. Days to mature the fruits.

7

Rootstocks of Fruit Crops

Fruit crops are mostly composite plants consisting of scion and rootstock, except banana, papaya, and strawberry. Rootstock exhibits a great influence on the performance of a scion in terms of production efficiency, yields, quality, adoptability, and resistance against biotic and abiotic stresses. The primary function of a rootstock is to provide anchorage or support to the scion by growing deep into the soil and also regulating the uptake of moisture and nutrients. The rootstock cannot be changed during the life-time of a plant, so it makes it inevitable to pay more focused attention to it because its functions are long-term and complex.

Classification of Rootstocks on the Basis of Propagation

Rootstocks can be divided into two groups, viz., seedling and clonal rootstocks.

Seeding Rootstocks

These rootstocks develop from the germinated seeds and have some advantages, like, it is relatively simple and economical to raise them through seeds. Such rootstocks are usually well adapted to mass propagation. Moreover, in some cases, the root system tends to grow deeper and to be more firmly anchored than the vegetatively propagated rootstocks. Most of the seeding plants do not retain viruses dwelling in their parent plants, although some exceptions are there, like that of citrus. Seeding rootstocks have certain drawbacks, like genetic variation, which results in uniform growth and performance of scion cultivars (s) of the grafted plants.

Clonal Rootstocks

These rootstocks are multiplied through vegetative means, such as by stooling, layering, rooted cuttings, or by the aseptic tissue culture method. In polyembryonic species like citrus, seedlings of nucellar origin are clonal and, to a great extent, uniform in growth.

Characteristics of An Ideal Rootstock

1. It should be easy to propagate either vegetatively or by seeds.
2. It must have a good root system, which will provide adequate anchorage to support the tree grafted on it without staking.
3. It must yield a good, clean, upright stem that is easily budded or grafted.
4. It should have wider soil and climatic adaptation.
5. It should have better compatibility with the scion cultivar(s) or inter stock.
6. It should have a high degree of winter hardiness, tolerance to high salts, diseases, insect-pests, etc.
7. It should be easily available.
8. It must induce dwarfing, precocity in bearing and heavy cropping effects on the scion cultivar

Use of Rootstocks of Fruits Crop

Fruits	Rootstocks	Characteristics
Almond	Bitter almond, Almond Seedling, Mariana 2624, Nemaguard, Sloh (Peach X Almond) Self-fertile	
Apple	**1.Seedling Rootstock: Crab** apple or trel (Malus baccata)	Most commonly used rootstock in India
	2. Clonal Rootstocks	
	M-9	Dwarf, suitable for HDP, Chance Seedling root stock
	M-27 (M-13 x M-9)	Ultra dwarf, suitable for HDP
	MM104	Resistant to woolly apple aphid, Drought
	MM111	Resistant to drought, Vigorous
	MM106	Resistant to woolly apple aphid
	M26 (M16 X M9)	-
	Alnarp-2	Semi vigorous
	EMLA- Series	Free from virus
	M2, M6, M25	Vigorous root stock
	Jork-9	Dwarf and cold hardy
	Bemali	Dwarfing, resistant to woolly aphid
	Novole	Vigorous, cold hardy
	MAC-1	Clonal Root stock
	Mark(MAC-9)	Semi dwarf

	Merton-793	Semi dwarf, Tolerance to high temp. and high moisture
	Robusta-5	Vigorous and winter hardy
	G.141 (Malling 27 x Rubusta 5)	New dwarf rootstock (HDP), highly resistant to fire blight and phytophthora
	G.16 (Ottawa 3 x M. floribunda)	New fully dwarf rootstock, resistant to fire blight (alternative to M9)
	G.22	Winter hardy, highly resistant to fire blight and phytophthora (Alternative to M26)
	Northern spy	Resistant to wooly apple aphid
	Antonovkas	This group has widely used where winter hardiness is required. Antonovka Kamenichka is resistant to fire blight
	Ottawa-3	This was selected from Robin X M-9 AT Ottawa Research station at Canada. It is susceptible to fire blight and woolly apple aphid
Pear	**1. Seedling Roostock**	Kainth (P.pashia) commercially used in India
	2. Clonal Rootstocks	
	EMLA and Quince C	Ultra dwarf, suitable for HDP
	Oregon 211	Dwarfing, High tolerant to wet soil and collar rot
	Oregon 261, 264	Resistant to bacterial Gummosis and cankers
	Oregon 249	Nematode resistant
	OH X F 230	Semi dwarf and winter hardy
	Quince A	Most satisfaction growth
	Quince B	Semi dwarf
	Quince C	Dwarf
	(*P. betulifolia*)	It is vigorous and free from wolly aphid, leaf spot and tolerant to alkaline soils and adaptable to wide range of climatic conditions
Peach	Rubira, Hemagourd, GF 55 and Nemared	Nematode resistant
	GF 677	vigorous
	Montelar	Resistant and Fe and mg deficiency
	Myrabi and citation	Semi-vigorous
	Myrabi	Semi dwarf and tolerant water lodging

	Siberian sea	cold hardy
	Pazza	commercially used in HP
	Higma	Resistant to Fe
	Reuter's red leaf	Dwarfing
	Prenus bessiyii	Dwarfing
	Julior, Petra, Tetra	Tolerant to waterlogged soil semi-dwarfing
Plum	**1. Seedling Rootstock**	Wild apricot
	2. Clonal Rootstocks	
	Mariana 2624	Resistant nematode, crown gall, cold hardy
	Myrobal-27	Drought tolerant
	Myrobal-29 C	Tolerant to high soil moisture
	Myrobalan B	Resistant to Bacterial gummosis
	Damas (GF 1869)	Semi dwarf
	Myrabi	Semi dwarf
	Pixy	Dwarf
	Str. Julian 1, 2, 3	Dwarfing
	Brompton	Semi-vigorous and may delay fruit ripening
Cherry	Charger	Semi Dwarfing
	Gisela	Cold hardy
	Gisela-10	Dwarfing
	Weihrest-10	Semi Dwarf
	Colt	Semi dwarf resistance to Phytophthora root rot, bacterial canker.
	F 12/1	Tolerant to bacterial canker, recommended for H.P.
	Mazzard (P.avium)	Dwarf, commercial rootstock of cherry
	Mahaleb	Most drought tolerant, suited to deep, well drained loams soil
Apricot	Wild apricot *P. sibrica,* *P. mandschurica*	Commercially used in India. Suitable for colder regions
	Hybrid P 2308	Dwarfing
	Haggith	Cold Hardy
Walnut	*J. regia* (Persian walnut)	Commonly used rootstock. It is unsuitable for waterlogged, saline and oak-root fungus and nematode infected soils.

	J. hindsii	It is resistant to oak root fungus and root knot nematode but susceptible to phytophthora, but root lesion nematode and and black line disese
	J. nigra	It is tolerant to crown gall, phytophthora and oak root fungus
	Paradox (*J. hindsii x J. regia*)	It is tolerant to nematodes and phytophthora but susceptible to black line disese.
Mango	**1. Polyembryony Rootstocks**	
	Sabre, Chandrakaran, Bappakai, Olour, Goa, Salem, Pahutan, Kurukan, Nileswar dwarf, Ambalvi, Muvandan, Nekkare, Starch, Peach, Kitchner	Kurukan- Salt tolerant. Muvandan, Olour- Vigorous. Vellaikolamban- Dwarfing rootstock for Alphonso & Dashehari. 13/1 (Israel)- Tolerant to calcareous soils. Bappakai- Best for Neelum variety. Gomera (Spain)- Salt tolerant. ST-9 Vigorous. Creeping, Rumani - Dwarf
	2. Monoembryonic Rootstocks Rumani, Dashehari	Tolerant high salinity Dwarfing effect
Citrus	Sour orange	Tolerant to xyloporosis, exocortis and gummosis, standard rootstock for phytophthora root rot
	Alemow	Dwarf, Salt and cold hardy
	Rough lemon (*Citrus jambhiri*)	Highest rank citrus rootstock in India tolerant to tristeza xyloporosis, exocortis and gummosis, suitable for deep sandy soils.
	Trifoliate orange	Cold hardy, tolerant to phytophthora, tristeza and dwarfing.
	Flying Dragon (Citrus trifoliata var. monstrosa)	Potential dwarfing rootstock (HDP), resistant to tristeza , xyloporosis, exocortis and gummosis
	Troyer citrange	Most promising rootstock for Nagpur mandarin & Kinnow mandarin in India, Suitable dwarfing rootstock for kinnow in India
	Karna Katta, Jatti khatti	Suitable semi-vigorous rootstock for kinnow in India
	Gajanimma	Suitable for acid lime/lemon
	Sohsarkar	Suitable vigorous rootstock for kinnow in India

	Rangpur lime	Hardy rootstock promising rootstock for Nagpur mandarin and sweet orange (Central and South India) Tolerant to tristeza, salt, soil toxicity resistant to drought, high vigorous
	Rough lemon	Suitable for Khasi mandarin. Suitable rootstocks for sweet orange in India. Resistant to tristeza, xyloporosis, exocortis phytophthora and greening
	Citrus macrophylla	Resistant to tristeza virus
	Citrus unshiu	Tolerant freezing condition
	Citron, Kumquat	Highly resistant to citrus canker
	Cleopatra mandarin	Salt tolerant, suitable for sweet orange (North East India)
	Severinia buxifolia	Resistant to salt
Guava	Pusa Srijan (Aneuploid tetrasomic, 2n + 2)	Potential dwarfing for Allahabad Safeda, resistant to wilt
	P. friedrichsthalianum	Dwarf, resistant to wilt and nematode
	P. molle	Resistant to guava wilt
	P. pumilum	Highly dwarf, high sugar content
	Dimple	
Avocado	Duke-7	Tolerant to phytophthora root rot
	New rootstocks:	Tolerant to phytophthora root rot
	Zentmyer, Uzi, Steddom	
	Nachlat	Tolerant to lime
	Pollock	Salinity tolerant
Ber	*Zizypus nummularia* *Z. Lotus*	Drought tolerance, Dwarf stature, extensive rooting, Early fruiting
	Z. rotundifolia	Tolerant to drought
Fig	Ficus glomerata	Resistant to root knot nematode
Custart apple	Pond apple (*Annona globra*)	Tolerant to flood condition, Dwarf rootstock
Jamun	*Syzygium densiflorum*	Resistant to termite
	Syzygium fruticosum	Grown as wind break
Sapota	Ryan or Khirni (*Manilkara hexandra*)	Commercial rootstock in India
	Chrysophyllum lanceolatum	Wider soil adaptability
Grape	*Vitis berliendieri, V. riparia, V. champini*	Tolerant to soil salinity

	Salt creek (*V. champini*), 1619, 1616,	Resistant to salt and nematode
	Dogridge (*V. champini*)	Popular resistant rootstock for Phylloxera, nematode, drought and salinity prone areas
	St. George (*cv. V. riparia*)	Phylloxera resistant rootstock
	Ramsey (*V. champini*)	Preferred rootstock for warmer regions
	Temple	Multiple resistant/Pierce's disease
Olive	Olea huspidata	Nematode resistant

8

Physiological Disorders of Fruit and Plantation Crops

	Disorder	Causes	Remedy
1. Mango			
i	Black Tip	Due to deleterious effect of brick kilns fumes which contain SO_2 acetylene, and CO_2 + Boron deficiency	Borax 0.6% The mango orchard should be established 1 km ahead from the brick kilns
ii	Malformation	Caused by *Fusarium moniliforme* spp. Subglutinans + due to low temperature	Susceptible varieties: Bombay green, Chausa Resistant varieties: Ellichi, Mangeera, Bhadauran, Alib
iii	Spongy tissue	High temperature, convective heat, post-harvest exposure to sunlight	Suscebtible variety- Alphonso Resistant varieties; Ratna, Arka aruna, Arka puneet, Arka anmol.
iv	Biennial bearing	Climatic factors, C:N ratio, genetic fators etc	Soil application of paclobuttazol (cultar)
v	Jhumka (Clustering)	Due to low temperature	
vi	Internal necrosis	Due to boron deficiency	Highly susceptible- Dashehari Free- Neelum, Langa
vii	Leaf scorching	Chloride toxicity, Excess of potash	
viii	Jelly seed	Highly suscebtible variety- Tommy atkins	
ix	Softnose	Interactive effect of nitrogen and calcium in the soil	
x	Taper tip	Dashehari susceptible	
xi	Sapburn		

2. Banana			
i	Neer vazhai (Water banana)		Nendran-Susceptible
ii	Kotta vazhai (Seed banana)	Due to banana streak virus	Enlarged ovules are present in fruits Poovan- susceptible
iii	Improper bunch filling	Deficiency of potash	
iv	White leaf	Mg deficiency	Younger leaves show white discolouration
v	Degrain	Due to rotting of pedicles	Seen in ripe fruits
3. Citrus			
i	Granulation	Due to high temperature and related humidity during ripening, drying of juice vessels, faded taste	Spray of lime and Lead arsenate or Zn + Cu + K @ 0.5%
ii	Exanthema/ Ammonization / Dieback	Due to copper deficiency	
iii	Leaf mottling/ Frenching	Due to Zinc deficiency	
iv	Yellow leaf of citrus	Due to Molyndenum deficiency	
4. Guava			
i	Bronzing	Due to Zinc deficiency	L-49 is more tolerant than Allahabad Safeda
5. Grape			
i	Hen & Chicken disease	Boron deficiency	
ii	Blossom End Rot	Due to calcium deficiency	
iii	Millerandage or short berry	Due to boron deficiency	
iv	Uneven ripening	Susceptible- Bangalore blue, Beauty seedless	Application of Ethephone @ 250 ppm at colour break stage,
v	Pink berry formation	Due to high temperature, susceptible varieties- Thompson seedless, Tas-E-Ganesh	
vi	Calyx end rot	Calcium deficiency	
vii	Colour & short berries	Boron deficiency, Improper pollination	

6. Litchi			
i	Little leaf + Leaf bronzing	Zinc deficiency	
ii	Fruit cracking	Due to excess of water, high temperature	
iii	Leaf scorching	Due to K deficiency	
7. Sapota			
i	Cock's comb		
ii	Flatting of branches		
8. Pineapple			
i	Multiple crown (Fasciation)	Genetical/Nutritional factor	Susceptible variety- Kew
ii	Sun scald	Exposure of fruit to sun rays	
9. Custard apple			
i	Kelsey spot	Due to high temperature	
ii	Stone fruit	Due to high relative humidity, low temperature	Fruts turn brow and become hard
10. Avocado			
i	Dry neck		
ii	Gray pulp		
iii	Fizzles		
iv	Apoplexy		
11. Aonla			
i	Fruit necrosis	Due to boron deficiency	Francis- highly susceptiblr Chakaiya- resistant variety
12. Pomegranate			
i	Fruit cracking	Boron deficiency, moisture imbalance. Young plants due to- boron & old plants due to- Water imbalance	More n mrig bahar Tolerant varieties- Khandari, Alandi, Bedana bosec
ii	Internal break down	More in ambe bahar	
13. Persimmon			
i	Calyx end rot	Due to calcium deficiency	
14. Fig			
i	Fruit splitting	Due to sudden change in atmosphere humidity and soil moisture	
15. Apple			
i	Water core	Boron deficiency	
ii	Bitter pit + cracking	Calcium deficiency	

iii	Wither tip	Due to copper deficiency	
iv	Rossete leaves	Due to zinc drficiency	
v	Interveinal chlorosis	Due to magnesium deficiency	
vi	Endoxerosis (June drop)	Water deficiency	
vii	Early drop	Improper pollination and fertilization	
viii	Jonathan spot	Due to water deficiency	
ix	Internal browning		Susceptible variety- Yellow newton
x	Scald	Due to storage at high temperature	Immature fruits are more susceptible
xi	Storage breakdown		Susceptible varieties- Golden delicious, Yellow newton
16. Pear			
i	Black end	Water deficiency	
ii	Pear scald	Prolong handling of fruits in storage	
iii	Pink end	Due to abnormal cool season preceding harvest	
17. Strawberry			
i	Albinism	Lack of fruit colour during ripening	
18. Apricot			
i	Tip burning	Due to high temperature	
ii	Apolexy		
19. Cherry			
i	Sun scald		
ii	Abnormal reduction in fruit size	Due to Zinc deficiency	
20. Peach			
i	Woolliness		
21. Walnut			
i	Blank nut	Due to hot summer with low humidity	
22. Mangosteen			
i	Gambose	Due to heavy rainfall, high humidity	Yellow exudation of gum on fruits

23. Rambutan			
i	Alternate bearing		
24. Coconut			
i	Crown cracking	Due to Boron deficiency	
ii	Rossette or little leaf	Due to Zinc deficiency	
25. Cocoa			
i	Cheelle wilt		
26. Rubber			
i	Tapping panel dryness (TPD) or Brown blast	Due to excessive of harvesting of latex	Tapping rest
27. Tea			
i	Yellowing	Due to sulphur deficiency	
28. Coffee			
i	Black bean/Black Jollo/Normal Jollo	Depletion of carbohydrate reserve during bean development	
29. Arecanut			
i	Band disease	Improper drainage	

9

Diseases Management of Fruit and Plantation Crops

S.No.	Name of disease	Causal organism	Remarks
1	**Mango**		
i	Powdery mildew	Fungus (*Oidium mangiferae*)	Spray 0.1% Karathane or o.1% Bavistin
ii	Anthracnose	Fungus (*Collectotrichum gloesporiodes*)	Spray Blitox 0.3%
iii	Canker	Bacteria (*Xanthomonas compestris*)	
iv	Rust	Algae (*Cephaleoros mycoides)*	
2	**Banana**		
i	Panama wilt	Fungus (*Fusarium oxysporum*)	Avoid continuous cropping of banana in the same field. Basrai dwarf –Immune, Poovan and Champa-Resistant
ii	Leaf spot or Sigatoka	Fungus (*Mycosphaerella musicola*)	Spray Dithane M-45/Dithane Z-78/Captan 0.2%
iii	Bunchy top or cabbage top	Virus –Transmitted by aphids (*Pentalonia nigronervosa*)	Resistant-Virupakshi, Tetrazolium test: best for bunchy top virus detection
iv	Moko	Bacteria (*Pseudomonas solanacearum*)	Vector-Insect
v	Konkan disease	Banana bract mosaic virus (BBMV)	Nendran-highly susceptible
3	**Citrus**		
i	Gummosis	Fungus (*Phytophthora spp.)*	
ii	Triesteza	Virus-Transmitted by Aphid (*Toxoptera citrida*)	Indicator plant-Acid lime, Distinct symptoms- sweet orange, mandarins and acid lime

iii	Citrus canker	Bacteria (*Xanthomonas campestris* pv. citri) transmitted by leaf minor	Resistant varieties-Tenali and Parmalini
iv	Greening	MLO'S transmitted by Psylla (*Diphornia citri*)	Bio-control: Tatriasticus radiates, Indicator varieties-pineapple and valencia
v	Exocortis	Virpids, transmitted by contaminated tools	Indicator plant-rangpur lime and citron
vi	Xyloporosis	Bud wood transmission	
4	**Guava**		
i	Guava wilt	Fungus (*Fusarium solani, Macrophomina phaseolina*)	Apply bavistin @ 15 g in the basin of each plant, treat the soil with lime or gypsum, spraying with 0.2% zinc sulphate during March and September.
ii	Guava canker	Fungus (*Physalospora psidii*)	Paste cracked portion with Bordeaux paste or Blitox paste
iii	Anthrocnose	Fungus (*Gleosporium psidi*)	Spray the plants with 0.2% Dithane Z-78
5	**Papaya**		
i	Damping off	Fungus (*Phythium sp., Phytophthora sp.*)	Sterilize the nursery bed with 2% formaldehyde solution before sowing the seeds, treatment of seed with Bavistin @ 2g/kg seed
ii	Collar rot	Fungus (*Phythium aphanidermatum*)	Apply Bordeaux paint on affected portion, Spray 0.02 % Copper sulphate on plant
iii	Ring spot	Virus-transmitted by Aphids	Mottling of leaves
iv	Leaf curl	Vector-White fly (*Bemisia tabaci*)	Spray 0.02% Rogor to control the insect vector
v	Mosaic	Papaya mosaic virus (PMV)	Spray 0.02% Rogor to control the aphid insect-vector of the disease
6	**Grape**		
i	Powdery mildew	Fungus (*Uncinula necator*)	Spray 0.1% Karathane
ii	Downy mildew	Fungus (*Plasmopora viticola*)	Spray 1.0% Bordeaux mixture
iii	Pierce's disease	Fungus (*Xyllela fastidiosa*)	Resistant rootstock-Temple

iv	Anthracnose	Fungus (*Elsinoe ampelina*)	Spray the plant with 0.2% Copper sulphate or Bavistin
7	**Aonla**		
i	Ring rust	Fungus (*Ravenellia emblicae*)	Appearance of brown pustules on leaves, Spray Dithane Z-78 0.2% during August-Setember.
ii	Blue mould	Fungus (*Penicillium islandicum*)	Covering of fruit with bluish green pustules, treat the fruit with 0.2% Borax.
8	**Pineapple**		
i	Wilt	Virus-transmitted by lace/ mealy bug	-
9	**Fig**		
i	Rust	Fungus (*Cerotolium fici*)	-
10	**Ber**		
i	Powdery mildew	Fungus (*Oidium erysiphoides*)	Affected fruits turn brown, dry and fall down from the leaves. Spray 0.2% sulfex or Karathane
ii	Sooty mould or Black leaf spot	Fungus (*Isariopsis sp.)*	Spray Dithane Z-78 at 0.2%
11	**Pomegranate**		
i	Fruit rot	Fungus (*Phomopsis sp.)*	Spray Dithane Z-78 at 0.2%
ii	**Anthracnose leaf spot**	**Fungus** (*Collectotrichum gloesporiodes*)	-
12	**Sapota**		
i	Leaf spot	Fungus (*Phleopheospora indica*)	-
13	**Date palm**		
i	Graphiola leaf spot	Fungus (*Graphiola phoenicus*)	Premature dry of leaves. Spray the plant with 0.2% Bavistin
14	**Apple**		
i	Scab	Fungus (*Venturia inaequalis*)	Yield reduction upto 70-80%, Spray the tree with 0.5% Dithane M-45.
ii	Crown gall	Bacteria (*Agrobacterium tumefacience*)	-
iii	Pink disease	Fungus (*Pellicularia salmonicolor*)	Spray the plant with 0.2% Dithane M-45
iv	Fire blight	Bacteria (*Erwinia amylovera*)	-

v	collar rot	Fungus (*Phytophthora cactorum*)	Drench the tree basin using 0.2% Mancozeb solution
15	**Pear**		
i	Pear decline	MLO's, transmitted by pear Psylla	-
16	**Peach**		
i	Leaf curl	Fungus (*Taphrina deformans*)	Thickening, reddening, puckering and curling of newly emerging leaves.
ii	Shot hole	Fungus (*Stigmina carpophila*)	Appearance of dark brown scattered lesions on leaves. Spray the plant with 0.2% Thiram or Captan
17	**Plum**		
i	Gummosis	Bacteria (*Pseudomonas syringae* pv. *morseprunosum*)	Scrape affected portion. Apply Bordeaux paste over the infected portion
18	**Coconut**		
i	Bud rot	Fungus(*Phytophthora palmivora*)	-
ii	Cadang-cadang	Viroid (CCCCd)	viroids may be carried on mouthparts of several chewing insects.
iii	Root wilt	MLO's, transmitted by Lace bug	-
19	**Arecanut**		
i	Mahali or koleroga	Fungus	Vector-spindle bug
20	**Coffee**		
i	Leaf rust	Fungus (*Hemelia vestratrix*)	Introduced from Sri Lanka ,Resistant sp.-*Coffea liberica*

10

Insects-Pests of Fruit and Plantation Crops

S.No.	Pest Name	Nature of Damage	Control Measure
1	**Mango**		
i	Hoppers (*Amritoides atkinsoni*)	Drying of the panicle, reduced fruit setting	Spray of Monocrotophos 0.02% on the tree during panicle emergence and pea stage of fruits
ii	Mealy bug (*Drosicha mangiferae*)	Sucking of the sap from young shoots, twigs, flowers and young fruits	Digging of soil during hot summer
iii	Fruit fly (*Dacus dorsalis*)	Feeding of pulp leading to appearance of off-smelling	Collect and bury the fallen fruits deep in the soil.
iv	Stem borer (*Bactocera rufomaculata*)	Drying of branches in case of severe attack	Pour kerosene oil inside the hole and plug it using mud
v	Stone weevil (*Sternochetus mangiferae*)	Adult female weevils deposit eggs singly in the fruit skin	Use of irrigations, Alphonso is susceptible
2	**Banana**		
i	Rhizome weevil (*Cosmopolites sordidus*)	The grubs bore into the stool and feed there in initially, due to severe attack central shoot is killed.	Uproot and burn the infested plant. At initial stage of infestation, insert 3 tablets of aluminum phosphide (0.5%)/ plant in the pseudo stem
ii	Flea beetles (*Nodostoma subcostatum*)	The beetles feed on tender unfolded leaves and fruits.	Follow clean cultivation and sanitation in the orchards.
ii	Aphid (*Pentalonia nigronervosa*)	The aphid acts as a vector of bunchy top virus.	Avoid taking ratoons and intercrops in epidemic areas.

3	**Citrus**		
i	Psylla (*Diaphorina citri*)	The psyllid is a serious pest of citrus through the country. The damage is caused by the nymphs and adults which suck sap from buds and tender leaves.	Prune affected shoots and destroy.
ii	Leaf minor (*Phyllocnistis citrella*)	The larvae feed on the epidermis of tender leaves making serpentine mines which are silvery in colour.	Prune affected shoots and destroy.
iii	Lemon buttergly (*Papilio demolius*)	Nymphs and adults suck sap from ventral side of the tender leaves,	Clip-off and destroy the infested shoots at the initial stage of infestation.
iv	Aphids (*Aphis gossypii*)	Nymphs and adults suck sap from tender inflorescence and shooty and transmit the tristeza virus	Prune infested shoots at the initial stage of infestation
v	Mites (*Eutetranychus orientalis*)	The mites damage leaves and fruits.	-
4	**Custard apple**		
i	Mealy bug (*Ferrisia virgata*)	Colonies of small reddish-white mealy bugs infest leaves, shoots, buds and fruits.	Remove all infested fruits and destroy them along with mealy bugs.
ii	Fruit borer (*Anonaepestis bengalella*)	The caterpillars bore into fruits making irregular tunnels and damaging mesocarp.	-
5	**Fig**		
i	Stem borer (*Bactocera rufomaculata*)	Grubs bore into the trunk and make tunnels in branches and main trunks.	Trap the adult beetles in light traps after rains and kill them.
ii	Gall midges (*Udumbarie nainiensis*)	On hatching maggots feed on fruit pulp. Infested fruits become elongated, shrivel and drop prematurely.	Plough soil under trees to kill the pupae.

6	**Grapes**		
i	Defoliating beetles (*Adoretus lasiophygus*)	The beetles feed on leaves and in severe cases may defoliate the grapes vine completely	Spray Carbaryl (0.1%) or Malathion (0.1%).
ii	Thrips (*Rhipiphorothrips cruentatus*)	These insects cause damage by sucking sap from foliage and flowers.	Spray Malathion (0.05%) after fruit setting.
iii	Yellow and red wasps	Wasps cause damage by puncturing ripe fruits and feeding on the.	Treat all the wall and groundnessat sunset with chloropyriphos
7	**Guava**		
i	Fruit flies (Bactocera dorsalis)	The fruit flies deposit eggs in the soft skin of ripening fruits.	Collect and bury the fallen fruits deep in the soil.
ii	Fruit sucking moths (*Othreis spp*)	During dusk time the fruit sucking moths puncture ripe fruits and suck their sap.	Collect and destroy infested fallen fruits.
8	**Pomegranate**		
i	Anar butterfly (*Deudorix isocrates*)	The extent of loss vary from 20-80% depending upon the area, variety and season.	Collect all fallen, infested fruits at least twice a week.
ii	White fly (*Siphonimus phyllyreae*)	Infests severely leaves on the ventral side, causing curling, yellowing and drying	Apply little more nitrogen to infested plants and irrigate regularly.
9	**Pineapple**		
i	Mealy/Lace bug (*Pseudococus breviceps*)	Vector for pineapple wilt disease	-
10	**Papaya**		
i	Fruit flies (*Bactocera cucurbitae*)	The fruit flies damage ripe fruits of papaya. Maggots feed in the fruit pulp, causing rotting	Collect all fallen infested fruits and destroy.
ii	Red spider mite (*Tetranychus cinnabarinus*)	The red spider mite is an occasional serious pest.	Infested leaves may be clipped off and destroyed.
11	**Jackfruit**		
i	Shoot and fruit borer (*Diaphania caesalis*)	Affected shoots wilt and get dried. Bored fruits rot and drop.	Destroy affected shoots along with caterpillars, Similarly, collect and destroy fallen fruits.

12	**Litchi**		
i	Fruit borer (*Cryptophlebia illepida*)	The caterpillars bore into developing fruits and seeds.	Collect all fallen infested fruits and destroy.
13	**Ber**		
i	Fruit flies (*Bactocera zonata*)	The infested fruits become deformed,	Resistant varieties- Kaitha, Tikadi, Dodhia &Meharun.
14	**Apple**		
i	San Jose Scale (*Quadraspidiotus perniciosus*)	Polyphagous pest, small, grey or brownish grey specks with a central nipple, twigs and fruits often surrounded by reddish or pinkish rings.	-
ii	White scale (*Pseudoulacaspis sp.*)	It causes damage by sucking sap from the stems,branches leaves and even fruits.	-
iii	Woolly apple (*Eriosoma lanigerum*)	Lives in colonies both on root and aerial parts.	Use resistant rootstocks of MM series and M 25.
iv	Blossom thrips (*Thrips rhopalantennalis*)	Very small, yellow, pale-brown and black slender insects seen within the flowers and flower buds.	Spray with Chlorpyriphos (0.02%) at early pink bud stage or 7-10 days before flowerin.
v	Leopard moth (*Zeuzera multistrigata*)	White moths with attractive patterns, seen at dusk during May-June.	-
15	**Peach**		
i	Leaf curling aphid (*Brachycaudus helichrysi*)	Smooth green looking aphids suck sap from growing bud.	-
ii	Termites (*Odontotermes obesus*)	Feed on roots and stem bases near the ground level of almost all types of fruit trees in warm localities with sandy loam soil.	Add well-rotten manur. Drench tree basins with Chlorpyriphos (0.1%) spray or dust.
16	**Plum**		
i	Plum weevil (*Amblyrrhinus pricollis*)	Adult beetles feed on the epidermis of fresh and tender leaves.	Collect attacked and fallen fruits and destroy
17	**Almond**		
i	Aphids (*Brachycaudus helichrysi*)	Smooth green looking aphids suck sap from growing bud.	-

Section 2
Tropical and Sub-tropical Fruit Crops

11

Aonla

- Amla/ Indian Goose Berry/ Amritphal fruit/Scared tree/ amla, amali and nelli
- B. N: *Emblica officinalis*
- Family-Euphorbiaceae
- Chromosome No. 2n=4X=28
- Origin: Central India
- Related species: Star gooseberry or Otaheite gooseberry: *Phyllanthus acidus*
- Type of fruit: Capsular (drupaceous)
- Type of inflorescence – Racemose
- Edible portion - Pericarp

Introduction

Aonla or Indian gooseberry (*Emblica officinalis* Gaertn. Syn. *Phyllanthus emblica)* is one of the important indigenous fruits of the Indian subcontinent. Most of the xerophytes are members of the Euphorbiaceae. The tropics are evergreen, while the subtropics are deciduous. The highest concentration of vitamin C in aonla is found in mature fruit rather than immature fruit. The special attribute is its capacity to retain vitamin C even in a dried stage, which is not possible in other fruits. Aonla contains six seeds per fruit, and in it is found zygotic dormancy. Flowers are borne in the axils of the leaves on determinate shoots.

Composition and Uses

A hundred grams of fruit pulp contains 14 g of carbohydrates, 0.5 g protein, 1.2 g iron, 0.3 mg vitamin B and 600 mg of vitamin C. Because of the high content of iron (1.2mg/100g), vitamin B and vitamin C. One part or another is used in the cure of cough, bronchitis, jaundice, diabetes, dyspepsia, diarrhea and fever. This fruit extract is probably used in a number of ayurvedic and homeopathic remedies that prevent greying of hairs and falling of hairs.

Area and Distribution

Naturally growing trees of aonla have been reported from India, Sri Lanka, Cuba, Puerto Rico, Hawaii, Florida, Iran, Iraq, Java, West Indies, Trinidad, Singapore, Southern Thailand, Pakistan, Malaya and China and Panama Canal regions. However, its cultivation is more common in India, particularly in the state of Uttar Pradesh, where it is cultivated in the districts of Pratapgarh, Raibareli, Jaunpur, Banda, Kanpur, Agra and Mathura districts. The major concentration of aonla cultivation is in Pratapgarh district.

Climate Requirement

Aonla is a hardy subtropical fruit crop and is being cultivated in variable soil conditions, but its cultivation in tropical climates is quite successful. Its annual rainfall of 630-800 mm is ideal for its growth. Warm weather is conducive to the initiation of fruit growth during July-August. Aonla is a subtropical fruit. It grows very well and comes to yield in tropical humid conditions. If young plants are protected from both extremes of temperature, the mature trees can tolerate temperatures right from freezing 0 °C to as high as 46 °C. It takes 8 months from flowering to fruiting.

Soil Requirement

Well-drained, fertile loamy soil is ideal for aonla cultivation, although it can be successfully grown in light as well as heavy soils, except very sandy ones. Slightly acidic to saline soils (pH 6.5-9.5, EC 5dsm^{-1}, ESP 30-40) are suitable for its cultivation. Heavy soils with a high water table are not suitable for its cultivation.

Growth Behavior

It is supposed to be an evergreen tree but behaves like a deciduous tree due to complete defoliation of its leaves. However, before dropping of determinate shoots, side buds initiate to develop determinate shoots again in February-March. The Aonla tree is characterised by a phyllanthoid branching habit with two types of shoots. On the basis of growth characteristics, these have been characterised as long (indeterminate) and short (determinate) shoots. The indeterminate shoots are longer and continue to put new growth into the season. These shoots do not fall from the tree and also do not bear flowers, irrespective of the period of their emergence. On the other hand, determinate shoots appear on the nodes of indeterminate shoots and their number at each node may vary from 3 to 5 in different cultivars.

Flowering

Aonla bears two types of flowers, male and female, which appear on branchlets in late spring. When hot winds blow, fruit set is often a problem. It is observed that flower bud differentiation in the Banarasi cultivar takes place in the first week of March. The flowers commence opening in the last week of March, and the blooming period lasts for 3 weeks. Male flowers appear in clusters on the axil of leaves all over the branchlet, while female flowers are on the upper end of a few branchlets only. The maximum number of male flowers opens between 6 and 7 pm and the dehiscence of anthers occurs soon after anthesis, or about 10-15 minutes after anthesis. The female flowers open gradually, and it takes 72 hours to open completely. The stigma becomes receptive on the third day of anthesis. Aonla is pollinated mainly by wind and honey bees.

Fruit Growth and Development

Fruits are available from November to December. After fruit set, the embryo lies in dormant condition and does not exhibit any symptom of external growth until middle of August. The diameter and volume of the fruit increase rapidly thereafter, and the maximum growth is achieved by November, after which there is not much increase in size. The growth of the fruit is due to the enlargement of the cells of the mesocarp, while endocarp cells form the hard stone cells.

Varieties

- **NA-7** (Neelum) selection from Banarasi, Early bearing.
- **BSR-1** Selection from Banarasi, suitable for preserve and candy making, Bhavanisagar of TN
- **Laxmi- 52** selection from Francis.
- **Anand-1, Anand-2** in Gujrat.
- **Amrit (NA-6)**-Seedling selection from Chakaiya, ideal for preserve and candy.
- **NA-10 (Balwant**) Seedling selection from Banarasi.
- **Chakaiya** Alternate bearer, Ideal for making pickles, free from necrosis.
- **Banarasi** It is a shy bearer. Trees are having upright growth habit. Best for muramba
- **Goma Aishwarya**- Clonal selection of NA-7, Early variety, drought tolerant, CIAH, Bikaner.
- Amrit- Seedling selection from Francis, and free from necrosis.

- **Krishna (NA – 5)**-A chance seedling of Banarasi, it has fibreless flesh.
- **Kanchan (NA-4)**-It is supposed to be a chance seedling of Chakaiya.
- **Francis (Hathi Jhool)** The branches have dropping habit, highly susceptible to necrosis.
- Francis is highly suceptible for necrosis followed by Banarasi.

Propagation

Hence, vegetative propagation was resorted to building using 1-year-old rootstocks through the **"T"** (shield) or patch budding method is successful from June to September. The seeds are hard and take a long time to germinate. Hence, the seeds should be treated with concentrated H_2SO_4 for 3 minutes and then washed with water and soaked in 500 ppm of Gibberellic Acid for 24 hours. Such treated seeds can be sown in a nursery bed or a poly bag filled with pot mixture.

Planting

After deep and thorough ploughing, pits of 1 cubic metre can be dug and plant spacing of 9 x 9 M during May – June can be filled up with 10-15 kg of well decomposed FYM. The budded plant (especially in dry areas) can be planted at the onset of monsoon rains.

Manures and Fertilizers

For a bearing tree, 800 g N, 640 g P_2O_5, 750 g K_2O should be applied along with 30 – 40 kg of FYM every year during rainy season. This dose can be split into two equal halves, one half applied during April – May (at fruit set) and other half can be given during September – October.

Training and Pruning

In training the trees, initially all the rootstock sprouts should be nipped off. The growth of scion should be trained to develop a low-headed one. At about 75 cm to 1m height, the first two main laterals are allowed to grow out. Five to six such laterals with wide angled crotches should be encouraged and fairly well spaced on the truck so as to have a modified leaser fashion. When the crops start bearing early in the year after the harvest of fruits, dead, diseased, weak, and criss-cross branches should be pruned. The water sprouts and rootstock growth should be watched and periodically removed. Pruning is done during March–April (removal of 50% of the erect growing shoots). Top work is done to rejuvenate the inferior variety or unproductive trees.

Intercultivation and Irrigation

However, irrigation during April-June one in every 15 days will help to encourage fruit set and prevent fruit drop. During early stages, intercrops like cowpea, green gram can be grown for up to 4–5 years. Black polythene mulch is most effective in reducing the irrigation requirement in aonla.

Harvesting and Yield

Maturity stage: Change seed colour from creamy white to brown. Large sized graded used for candy and preserve. Small sized grades used for *chavanprash and trifla* preparations. Time of fruit bud differentiation in aonla is March-April. Trees remain dormant from April -June and resume growth from last week of July to first week of August.

- Harvesting time is November to December
- Yield: **100-150 kg/tree**

Insects and Diseases

- **Aphid -** *Setaphisbougainisillia*
- **White fly-** *Trialeurodesrara*
- **Bug -** *Scutelleranobilis*
- **Leaf roller-** *Caloptilaacidula*
- **Bark-eating caterpillar-** (*Indarbelatetraonis*)*:* Major pest
- **Gall Caterpillar-** *Betonsastylophora*
- **Rust - (*Raveneliaemblicae*):** Major disease of aonla

12

Avocado

- Avocado/Fruit of new world/Butter fruit/Alligator pear/21stcentury fruit/ Tasteless fruit
- B N: *Persea American*
- Family: Lauraceae
- Chromosome No.: 2n = 2X = 24
- Origin: Tropical America/ Central America /Mexico
- Inflorescence: Panicle
- Type of fruit -Fleshy single seeded berry
- Edible part: Pericarp

Introduction

It is a subtropical, evergreen fruit tree. Its trees can be short-spreading and fairly bushy or grow erect to a height of 20 m. Tree demand for this fruit has been increasing over the past few years in many countries. It has become an important fruit in international trade. Mexico, Colombia, Dominic Rep, Peru, Indonesia, South America and South Africa are the major producers of avocado. In India, it is grown as a backyard tree and is found in small pockets on the hill slopes of Tamil Nadu, Kerala, Karnataka and Maharashtra.

Composition and uses: One of the most nutritious fruits, rich in fat, protein, and minerals and low in carbohydrates, is the avocado. It can be safely eaten by diabetics. Fruit is used as a dessert in salads, ice creams, and milk shakes. Immature fruits may have a bitter flavour that may be reduced on ripening; the peel of avocado has 3 antifungal compounds. Its energy value is twice as much as banana fruit. Rich in fat 26.4% and low sugar 1% content and rich in K, Fe, Vitamin B, As the sugar content is low, it can be recommended as a high-energy food for diabetics. Fruit is a rich source of oil, ranging from 5–20% used for the cosmetics industry.

Climate and Soil

Avocados are now grown in most of the tropical and subtropical countries. Mexicans are more cold-tolerant. Guatemalan is intermediate, and West Indian is the most tropically adopted. Fruits are very sensitive to freezing temperatures. Flowering is very much influenced by temperature, with the optimum being 28–31°C. Cool, moist subtropics are best for maximum production. Subtropical dense and evergreen suited to fruit trees can be grown to about 800 feet in height. Only the Mexican race survives between the Mexican and Guatemalan races in regions where minimum winter temperatures range from -0.5°C to 3.5°C. The ideal temperature for flower induction is 25°C. well-drained and well-aerated acidic soils of sandy to clay loams. The West Indian race can tolerate a pH of 7.6 to 7.9.

Species and Cultivars

Mexican Race *(P. americana var. drymifolia*): Climate requirement semi-tropical, frost resistant rootstock, salt tolerance, and this group is characterised by small fruits weighing less than 250 g and ripening in 6-8 months after flowering. The oil content of the fruit is 30 per cent, the highest of all the three races. **Varties-** Duke, Topa, Gotfried, Pernod.

American Race (*P. americanavar. var guatemalenses*): Climate requirement subtropical, Cold tolerance, oil content ranges from 8–15%. **Varieties**-Lula, Hass, Green, Taylor, Linda, Queen, Itsamma, Benik.

The West Indian Race (*P. americana var. americana*): Requirement tropical, tolerant to salinity. The oil content of the fruit is low, ranging from 3–10%. Varieties-Pollock, Simmond, Black Prince, Peterson, Waldin, Fushsia, Purple Green (pear-shaped).

TKD 1: Early Maturing

Furete: (Calinson x Winslowson) Mexican and Guatemalan strains (Pyriform shaped). It is fairly resistant to the cold. It belongs to group B. Branches are thinned and shortened in spread. Most of the leading varieties in the world.

Paradenia, Purple Hybrid (PPH): This variety is born in clusters.

Duke seedlings: are resistant to root rot and cold hardiness

Propagation

Commonly propagation is normally propagated by seeds. Removal of the seed coat before sowing helps to speed up the germination, the viability period of avocado seed is very short (2-3 weeks). Mexican stocks are best and dwarfing in effect. Mexican race can be propagated by cuttings of young plants also. ***Persea schiediana*** is a vigorous stock. Popular Root stock selections are duke-6, Duke-7, G-6, Huntalas, Dusa and Latas.

Planting

Spacing 5-6 m Different Races with overlapping blooming periods should only be planted together. High density plantings of 800 trees/ha gave double yield than normal planting of 400 plants/ha.

Manuring and Irrigation

Bearing trees of 10 years may be given 200 g N, 45 g P and 165 g K in addition to 50 kg FYM. Graded doses can be given from early stages depending upon the growth of the plant. Fertilizer should be applied 30 cm away from the trunk only.

Irrigation

Irrigation at 10 days' interval will be optimum. Sprinkler irrigation to keep the top 60 cm soil moist is the best.

Training and Pruning

Better to train the plants to Pyramidal form. Regular pruning may be done to remove overcrowding, damaged & unproductive branches only. Better to train the plants to Pyramidal form.

Flowering and Fruit Set

A flower that is functionally female in the morning and male in the afternoon. Flowers are functionally female in the afternoon and male in the next morning. Bergh classified the varieties of avocado in groups A and B based on their flowering patterns. Flowering behaviour Protogynous diurnally synchronous dichogamy (PDSD), hence cross pollination is the rule. PDSD was first reported by Bergh (1969). The individual flowers are morphologically bisexual, having fertile male and female organs. But they exhibit dichogamy, viz., the male and female organs coming to maturity at different times, there by avoiding self-pollination of an individual flower.

Harvesting and Yield

The maturity index is oil content. The average yield is about 100-500 fruits per tree. Maturity can be known by a change in the colour of the fruit and a reduction in the glossy shine of the fruit. It is a climacteric fruit and harvesting time is August to September in South India. Mature avocados ripen in about 5–10 days at 15–210°C. Ripening can be hastened with ethrel treatments. Mature avocados ripen in about 5–10 days at 15–21°C. Ripening can be hastened with ethrel treatments. Avocados can be stored in a controlled atmosphere with 9 % CO_2 and 1 % O_2 at 10°C for 60 days. because of the polyphenol oxidase enzyme (PPO).

Pests and Diseases

Pests: mites, mealy bugs, and scales are the important insect pests.

- **Hemiberlesiarapax (greedy scale):** a major pest
- (***Colletotrichum gloeosporioides***): A significant disease is fruit rot
- **Root rot (*Phytophthora cinnamami*):** *the most* serious disease.

Physiological Disorders

- **Tip Burn**-It is due to chloride toxicity, most prevalent in the spring season.
- **Grey pulp-** is more common in warmer climates.
- **Dry Neck-** is an avocado physiological disorder.
- **Pulp spot-** Black discoloration of vascular bundles at cut ends of the fruit stalk, it occurs due to low temperature in avocado and is a post harvest disorder
- **Remedy-** Do not store at very low temperature and dry condition.

13

Bael

- Symbol of Lord Shiva fruit/Bengal Quince/Indian Quince/Golden apple/ Sirphal
- B. N.: *Aegle marmelos*
- Family: Rutaceae
- Chromosome No.: 2n=2X=36
- Origin: India
- Inflorescence: Racemose (Axillary cyme)
- Fruit type: Hard shelled berry (Amphisarica)
- Edible portion: Succulent placenta

Introduction

Bael, one of the oldest fruits cultivated in India, has a mythological significance, viz., leaves used for the worship of **Lord Shiva**. The flower contains numerous stamens, which is hypogenous. The seed is monoembryonic. It is a deciduous tree with trifoliate leaves. Aegle is a monotypic genus with deciduous foliage. Bael has dimorphic twigs.

Distribution and Uses

Bael is mainly cultivated in Orissa, Jharkhand, Madhya Pradesh, Uttra Pradesh, Bihar, West Bengal, and M P. Ripe fruit is laxative and good for the heart and brain. Mature green fruits are most suitable for making preserves. One hundred gram of pulp contains 55 μg of carotene, 0.13 mg thiamine, 1.19 mg riboflavin, 1.1 mg niacin and 8 mg vitamin C. From its pulp, sherbet and syrup can be made. It is the richest source of riboflavin. "**Marmelosin**" is present in all parts of the plants (extracted from bark), and is responsible for medicinal properties such as restorative, tonic and laxative.

Climate and Soil

A subtropical condition with a hot dry summer and a mild winter would be ideal for the cultivation of bael. It can be grown even up to an altitude of 1229 m MSL. It is not affected by temperature, even at temperatures as low as -8°C, and can withstand temperatures as high as 48°C. It is highly tolerant to sodicity and up to 30 ESP and salinity up to 9 EC. Since it is a hardy tree, it thrives well in a wide range of soil conditions, right from pH 5 to 10, where many other fruit trees fail. It can tolerate almost every alkaline soil as well as stony soils. However, well-drained sandy loam is the best. susceptible to waterlogging conditions.

Popular Varieties

NB-1, NB-2, NB 5, NB6 (Dwarf), NB7, NB9, Goma yashi, Pant Aparna. Mirzapuri, Kagzi Gonda, Kagzi Etawah, Kahzi Banarasi, Kaghli, KB 11, KB 1, Dhar Road, Ayodhya, Pant Aparna, Pant Shivani, Pant Urvashi, Pant Sujata, Darogaji, Rampuri, Azamgarh Sewan Large, Deoria, Gorakhpur, Thar Divya, CISH Bael- 1, CISH Bael- 2 and Thar Neelkanth.

Propagation

Root stocks are raised from seeds. On 6 month old seedlings, commercially propagated **patch budding** is done during **June-july**.

Planting

Pits of the size of **50 cm x 50 cm x 50** m are dug at a spacing of **10 m x 10 m**. Planting can be done during June-July or at the beginning of the monsoon. Every year, regular application of 20-30 kg FYM should be done at the beginning of the monsoon. Young plants are irrigated whenever there is a monsoon failure. Any legume or forage crop can be taken as an inter crop during early years. Young plants are trained with the help of stakes.

Harvest and Yield

Flowering is seen in May-June and fruits become ready in 8–10 months, viz., April–May. Mature fruits should be harvested individually along with the fruit stalk and they should not be allowed to fall on the ground. A well-grown tree of about 12–15 years' age gives 300–500 fruits/year. Mature green fruit is ideal for harvesting. Fruits are climacteric and need about 11 months to ripen. Ripe ones have a shelf life of one week. Bael takes the longest duration for fruit set to ripen (11 months, April-May matured).

14

Banana

- Antique fruit crop/Tree of Wisdom/Tree of Paradise /Adams fig/Plant of virtue/Apple of paradise
- Botanical name: *Musa paradisiaca*
- Family: Musaceae
- Chromosome No. 2n: 2X=22, 33, 44
- Origin: Indo-Malayan (South East Asia),
- Origin of *Musa acuminata*: Indo Malaysia,
- Origin of *Musa balbisiana:* Indo Burma
- Inflorescence: Spadix
- Fruit type: Berry
- Edible part: Mesocap and Endocarp

Introduction

The fruit is recognized as the fourth most important global food commodity. It is used as a staple fruit in most African countries and is used as ripe (table) or raw fruit (cooking). This fruit is available throughout the year. All the plant parts are being used. Banana is a tropical, herbaceous, monocotyledonous and monocarpic fruit crop.

Uses and Composition

Banana by virtue of its multiple uses is popularly known as "Kalpataru". Usually, bananas are eaten when ripe as dessert and table fruit. Plantain or cooking bananas are the staple food of people in many countries in central and western Africa. Apart from fresh consumption. The flower bud and also the central core of the pseudostem are used to prepare tasty dishes.

Banana leaves are used as plates for serving food, leaf sheaths are used as wrapping material and dry leaves as fuel, while tender pseudostems, leaves, Banana fibre can be made into attractive napkins, table mats and carry bags.

Plantain means "cooking banana. Banana leaves are used as plates for serving food (Biological plate). Cattle feed is made from tender pseudo stems, leaves, and underground rhizomes. It contains 73% moisture, 25%–30% carbohydrates, 1.4% protein, 0.3% fat, 0.5% mineral matter (Ca, Fe, P, and K), and vitamins C and K.

Origin and Distribution

The origin of banana is believed to be in the hot, tropical regions of Southeast Asia. India has the second largest diversity of indigenous bananas in the world. The largest producing countries of banana are India, China, Indonesia, Brazil, Ecuador, Uganda, Sudan, Fiji, Honduras, Hawaii, Canary Islands, Philippines etc. In India, it is the first major fruit, occupying about 20 per cent of the total area under fruit crops and 32 per cent of total fruit production, Andhra Pradesh, Madhya Pradesh, Gujarat, Tamil Nadu and Maharashtra are the leading producers. Though India is the leader in banana production, the Indian export of fresh banana is meager. T.N is a special banana grown for leaf production

Climate and Soil

Banana is a tropical fruit crop, adapted to wide range of climatic conditions. Banana is a crop of humid tropics. It is grown up to an altitude of 1200 m from mean sea level. The temperature range is 13-40°C, but the optimum is 25-30°C (26.5) for getting good yield, Temperature between 20-30°C is ideal for good plant growth. Temperature above 36-38°C causes scorching effect with increased transpiration. Low temperature 10°C leads to Choke of impeded inflorescence and bunch development. Rain fed crop in west coast and hills South India. Better quality banana: subtropical condition (Better aroma and crisp pulp). Deep well drained, loamy soil with adequate organic matter is ideal for its cultivation. It can grow well in slightly alkaline soils. Though banana requires large quantity of water, it cannot tolerate water stagnation. The optimum soil pH is 6.5-7.5.

Improvement

- **Simmonds and Shepherd (1955)**: is given scoring technique for genomic classification of banana
- ***Balbisiana*** genome in banana is associated with greater drought hardiness and disease resistance.
- All edible bananas are the result of a natural cross between two wild ancestors, *Musa acuminata* (A) and *Musa balbisiana* (B).

- The scoring technique involved the use of 15 morphological characters. One score is given to a variety for pure *acuminata* and five for the score ranges from 2 to 4 for intermediate based on intensity.
- A variety with a higher score reflects its origin from balbisiana and a lower score from *M. acuminata.*
- Edible clones from *acuminata* have a concave growth curve, while *M. balbisiana has a* convex growth curve.
- There are six classes of genomic classification: two diploids, three triploids, and one tetraploid.
- The Moongli variety, a **mutant of Nendran,** there is no male phase and plants bear one or two hands with the biggest fruit.
- The best known banana all over the world belongs to the pure acuminata (AAA) group but the clones which have both parents are associated with greater drought tolerance and resistance to diseases. eg;-AB, AAB, ABB, AA or AAA-suited for rainy conditions.
- Species whose flowers are used as vegetables: *Musa acuminata* sp. Malaccensis
- Fiber yielding species: *Musa basjoo* (used in Japanese textiles) and Musa textilis
- Contribution of parthenocarpy origin in Banana-M. acuminata sspbanksii
- Ornamental species: *Musa ornata, Musa flaviflora, Musa velutina*
- Panama wilt and nematode-musa laterita tolerance
- Tolerant to panama wilt, nematode, and sigatoka leaf Spot-Musa acuminate ssp. burmanica
- Most cultivated banana is triploid in nature.
- Hybridization work-Central Banana Research Station, CBRS, Adhuthuai(TN)
- The male flower of banana is resistant to Panama wilt disease but susceptible to bunchy top disease.
- Banana improvement work was started in 1949 in Tamil Nadu.
- Leading production: Poone, Monthan, and Karpuravalli.
- Banana is a stable food in South Africa.
- Basrai Dwarf is immune; Poovan, Moongli, and Cavendish are resistant; and Sonkel, Rasthali, Malbhog, Alpan, Mortaman, and Monthan are susceptible to panama wilt.
- Banana cultivars are screened for viruses by the ELISA test.

Species

The family musaceae has two genera viz., *Ensete* and *Musa* with about 50 species

Ensete- It is an old genus, which probably originated in Asia and spread to Africa. It has about 6-7 species, of which *Ensete* and *Ventricosa has* been reported to be grown as a food crop.

Musa – It is having 40-45 species; the genus Musa is divided into following sections.

i. *Eumusa*: *(*2n=22)- All are cultivated forms; Pseudostem usually exceed 3m in height with pendent or semi pendent inflorescence, produces 10-25 nodes of flowers and covered with dull brown colour bracts, gives the edible cultivated parthenocarpic banana and are derived from 2 wild species. *Musa accuminata* (A) and *Musa balbisiana* (B).

ii. *Callimusa*: (n=10) -It Suited as ornamental, parthenocarpy absent completely.

iii. Australimusa: (2n=20)-These fruits are parthenocarpic and predominantly female sterile; The fruiting bunch is erect and contains a red sap, the skin is orange in colour when ripe. *eg-Musa textilis* (Manila hemp).

iv. *Rhodochlamys*: (2n=22) - It is spreads from India to Indonesia, pseudostem less than 3m height with erect inflorescence, parthenocarpy absent *eg-M. ornate & M velutina*are sometimes grown as ornamental plants.

v. *Incertaesedis*: (x=7; 2n=14) - it is the largest group among the Musaceae family, *eg-M. ingens&M. beccarii* (x=9, 2n=18). *Musa ingenes* is the largest known herb which grows to the height of 10 m.

Genomic Constitution

- AA-Anaikomban, Matti, Tongat, pisanglilin, Sanna Chenkadali, Surya Kadali, Namarai, Score (16-23).
- AB-Ney poovan (Elakki bale), Kunnan, Nathu Poovan. Thaenkunnan, Adakka Kunnan. Score (46-48)
- AAB-Poovan, Rasthali (Silk), Virupakshi, Sugandhi, Rajapuri, Poovan, Pachanadan, Nendrapaditha, Nendran/Rajeli, Chinali, Score (26-46)
- AAA-Dwarf Cavendish/Basrai, Giant Cavendish, Robusta, Gross michel, William, Nagabale, Chakkarakeli, Grandnaine, Chenkadali/Red banana, Score (15-21).
- ABB-Peyan, Karpuravalli, Sugandhi, Kari Montha, Kallu Monthan, Nalla Bontha, Monthan, Sugandhi.Score(59-69)

- ABBB-KlueTeparod, Sawai (Synthetic hybrid), Score (63-69)
- AABB-Kalamagol
- AAAB-Atan, Goldfinger (FHIA).

Varieties

- Ney poovan/Elakkibale (AB): It is commercially cultivated in Kerala and Karnataka. It is tolerant to leaf spot and fusarium wilt, but susceptible to banana bract mosaic virus.
- Kunnans (AB): It is a back yard cultivar of Kerala and Karnataka. It is tolerant to leaf spot and fusarium wilt.
- Dwarf Cavendish/ Basrai/ Kadali/ Pachavazhai/Mauritius Kuzhi Vazhal (AAA)-: 58 % total production.
- Gandevi selection (Hanuman or pardase) from Basrai. It is the most important commercial cultivar of India the plant is dwarf, fruit large, and suitable for high density planting, and susceptible to leaf spot disease. Resistant to Panama wilt.
- Gross Michel (AAA): It is the main cultivar of this sub-group. Gross Michel was the leading cultivar in the world banana trade until the late 1950. The variety has lost its commercial status due to susceptibility to panama wilt.
- Red Banana (AAA): This cultivar is grown throughout the world. It is highly susceptible to bunchy top, fusarium wilt and nematode
- Rasthali (AAB): it is one of the most popular commercial choicest table cultivar of West Bengal, Tamil Nadu, Karnataka, Andhra Pradesh, Kerala and Bihar. severe susceptibility to fusarium wilt, easy dropping of fruits from bunch. Susceptible to sun injury and formulation of hard lumps in the pulp.
- FHIA: Foundacion Hondurena de Investigation Agricola, Handura released different banana hybrids.
- IC-1: The first hybrid between Gross michel x Musa acuminata sub species Malacensis developed in 1990 from IITA, Nigeria, resistant banana clone for wilt disease, which is similar to that of Gross michel, a premier cultivar in that zone. Also IC-2, S-19 - none of them are better than Gross michel.
- FHIA-01: It is popularly known as Gold finger. It is the cross between a dwarf lady finger type ie., Dwarf prata x SH-3142. and has Apple flavour & suitable for export. It is resistant to black sigatoka & fusarium wilt race.
- FHIA-03: GH-3386xSH3320-It is a Robusta, Disease resistant cooking banana.

- FHIA-04: AVP67xSH3437-A French type hybrid, resistant to black leaf streak or black sigatoka resistance.
- FHIA-17: high gate x SH-3362-Cavendish hybrid resistant to black sigatoka.
- FHIA-21: AVP-67x SH3142 resistant to black sigatoka
- The other hybrids are-FHIA-19,20,22
- CO1: (AAB) is a multiple cross between Kellar Laden (AAA) x *M. balbisiana (BB)*cv savani.x Kadali (AA). also called H-135(It is a pome hybrid).
- H-135- It is a Pome hybrid. It retains the typical virupakshi.
- H1-Agniswar x Pisang lilin – Resistant to leaf spot, fusarium wilt & burrowing nematode.
- H2-Vannan x Pisanglilin -tolerant to leaf spot & nematode.
- BRS-1: (Agniswar x Pisang lilin)- AAB-Resistant to Sigatoka leaf spot
- BRS-2: (Vannan x Pisang lilin)- AAB-Tolerant to leaf spot and Panama disease, rhizome weevil and nematodes
- TR4- Panama wilt and nematode.
- Poovan/Champa (AAB): It is resistant to panama wilt and fairly resistant to bunchy top highly susceptible to banana bract mosaic and streak virus. Pink pigmentation on ventral side, susceptible banana streak virus.
- Bodles Altafort (AAAA) is a synthetic tetraploid developed as a result of cross between Gros Michel (AAA) and *Pisanglilin* (AA) from West Indies.
- Poovan/Rasthali/ AmriTpani, Mortman – (AAB)- choicest table banana best Resistant to Panama wilt, severely affected by Banana streak virus, fairly resistant to bunchy top. Fruit cracking major physiological disorder.
- Monthan/ Bontha/KanchKela/ Ponthan(ABB)- Drough resistant, a leading culinary banana of India.
- Udhayam: (Selection from single plant "PisangAwak (ABB)" Ratoon crop, Field tolerance to Sigatoka, Suitable for long distance and processing
- ABB clone are grown under rainfed condition (MonthanKanthali, Kunnan).
- Lady finger-(Australia) AB- Resistant to bunchy top virus. diploid banana
- Grand Naine(France) AAA-Tall mutant of Dwarf Cavendish, introduction from Australia

- Harichal/ Robusta/ Bombay Green (AAA)- Semi-tall sport of Dwarf Cavendish, banana is highly susceptible to bunchy top, anthracnose, sigatoka, black tip and frost but resistant to Panama wilt
- Lal velchi (AAA)- It is growing red skin
- Grand Naine (AAA) Tall mutant of dwarf Cavendish. it required propping
- Hill banana/ Virupakshi/ Vellavazhai/ Sirumalai(AAB): Suitable for higher elevations, Perennial
- Nendran/ Ethankai/ Rajali/French/ Plantain(AAB)- good for preparation of banana chips, dual-purpose variety of banana and is known as plantain worldwide. cooking variety used in Kerala
- Kunnan(AB)- Popular dessert variety, for making baby food
- Peykunnan(ABB)- Karpuravalli, Kanthali- Good for making baby food, juice, wine
- AAB, AAA clone are grown under irrigated condition.
- Amritsagar, Dudhisagar, Chakia, Monohar
- Karupuravalli(ABB)- Tolerant to drought, salt, wind and suitable for juice, wine
- *Musa acuminata* spp. *Burmonica*- Panama wilt and nematode and Sigatoka leaf spot
- Pisang lilin, Tongat, Anaikomban, *Musa laterita*- Panama wilt and nematode
- While all tripoids (AAA) are susceptible to sigatoka leaf spot, ABB clones are resistant.
- Gandevi Selection (also known as Hanuman or Padarse) it heavy bunch weight is so popular.
- FATOM-1: Early flowering mutant from Grand Naine (G9)
- Dwarf Cavendish and Robusta are widely adopted commercial bananas.
- Cooking banana varieties: Monthan, Nendran
- Varieties suitable for multi-storey system: Poovan and Ney Poovan (Semi-tall bananas)
- Resistant to biotic and abiotic stress: Poovan, For excellent quality banana variety: Rasthali
- Smoke treatment is the commonest method to induce ripening in Tamil Nadu and Maharashtra
- Frost is the major limiting factor of banana cultivation.
- Weterloggig condition in banana cultivation leads to the more incidence of Panama wilt.

Propagation: Tissue cultured plants 500-750 g were also used as commercial scale and advisable ratooning is 2-3 times. Popular tissue culture variety in India-Grand Nine. Banana is traditionally propagated vegetatively through suckers or rhizome or tissue culture plants. Sexual propagation is not possible due to Parthenocarpic nature of fruits. Banana produces two types of suckers 1- Water suckers. 2- Sword suckers.

Sword Suckers

- Leaves are pointed, narrow & upright
- Growth is vigorous & fast.
- Bearing is early (11 months.)
- Sword suckers are best.

Water Suckers

- Leaves are Broader, spreading. with roundish tip.
- Rhizome flat with non-sound heart.
- Growth slow.
- Bearing late (15 months.)

Main Criteria for Selection Planting Materials (sucker)

- The orchard/mother block should be disease free. The weight of the suckers should be 1.0- 1.5 kg.
- The mother plant should be heavy yielder.
- Always select sword suckers for planting.
- Select the suckers free from rhizome weevils.
- Age of suckers: 3-4 months
- The whole or split rhizomes can also be used when suckers are not available.
- Bits of rhizomes may also be used as a planting material (cut of Rhizome) and peeper are also used for propagation
- Tissue cultured plants were also used as planting material on commercial scale.

Pairing & Prolinage

It is the removal of older leaves, roots, adhered soil and other particles on the surface of the rhizome and the top portion of the suckers, leaving 15cm from the rhizome should be removed and immersed in cow dung slurry. Sprinkle phorate granules @ 10-15 g/rhizome in order to avoid soil pathogens and rhizome weevils. Also, rhizomes are dipped in fungicide solution by giving a slant cut.

Planting: Except during severe winters and heavy rains, bananas can be planted all year. In general, June - July is the most common season of planting. In Gujarat and Maharashtra, the furrow method and in Tamil Nadu trench method of planting are followed. The trench method is especially followed in the wetland system of cultivation. Strong winds are a threat to successful banana production. For long distance transportation, harvesting is done at 75-80 maturity. Taller varieties 2.54x2.54m spacing is suitable for leaf purposes. Semi-tall varieties 2.51 m x 2.51 m spacing and dwarf varieties spacing 1.8x1.8m. A multi-storey system is commonly followed in coastal Karnataka and Kerala. High density planting in banana accommodates 3000–4500 plants/ha.

Pit Method: The pit method and furrow methods are commonly followed. A 60 cm^3 pit should be opened at a distance of 1.8 x 1.8m (dwarf varieties) or 2 x 2m (Tall varieties) using a square system. These pits are filled with top soil weighing 20–30 kg. FYM should be applied at least 15–30 days prior to planting. During planting, each pit will be supplied with 250gm of neem cake and 50 gm of trichoderma to prevent nematode and rhizome rot problems. Plant suckers at the centre of the pit and irrigate immediately after planting.

Furrow Method: This is the most common method of planting. Furrows of 15-20 cm deep are opened at a regular distance and rhizomes are planted in the furrows. Furrow planting is practised in Gujarat and Maharashtra. Trench planting is practised in the Cauvery delta of Tamil Nadu.

Tissue Culture Plants: Banana are also grown commercially by using tissue-cultured plants. These plants require more care throughout the growth period compared to suckers and yield about 10-20 per cent more than suckers. In recent years the concept of HDP is being practiced where suckers are planted at closer spacing or by planting two suckers per pit by accommodating more number of plants at a specified spacing to get a higher yield and reduce the cost of production. The cultivar Robusta and Dwarf Cavendish spaced at 1.5x1.5m accommodate 4444 plants/ha and are recommended by IIHR. They also recorded the highest yield.

Irrigation: Banana requires a high amount of water, ranging from 1800–2500 mm annually. From planting to harvest, approximately 40-45 irrigations are required at 4-5 day intervals. Drip irrigation promotes early emergence, saves water (40–45%), and increases yield. Banana are moisture-loving plants. The water requirement for banana through drip systems is 10–25 litres/plant.

Manures and Fertilizers: 50-75 cartloads of FYM or compost/ha. Preparatory tillage: 100 g/N, 40 g/P_2O_5, 100 g/K_2O. full dose of FYM, P_2O_5 and K_2O before planting. While N gives 3 split doses 3rd (40g), 4th (30g) and 5th (30g) month after planting. The addition of 50g N is after fruit set for better development of set fruit. Application of N during the reproductive phase delays senescence of leaves and improves bunch weight. Application of K_2O (potassium) at the shooting stage improves finger filling.

Weed Control: Integrated weed management programme should include growing of cover crops. Use of herbicides, inter cropping and hand weeding where ever necessary. Pre emergence application of Diuron at the rate of 4kg/ha controlled grasses and broad leaved weeds without affecting the yield and quality of banana. Application of Glyphosate 2kg/ha followed by gramoxone1.8kg/ha proved effective in controlling weed growth. Vegetative phase (1-6 months) is crucial for weed control in banana, cause 60-70% yield loss.

Important Cultural Practices

Desuckering: Desuckering is done by cutting the pseudostem of a sucker at the ground level followed by the application of kerosene/2-4, D @ 0.5 percent. The growth of suckers can be inhibited by damaging the cut end to prevent further growth. If allowed, banana suckers compete for nutrients with the mother plant; one or two healthy sword suckers may be allowed to grow for a ratoon crop. Desuckering once in 45 days is common practice in banana cultivation.

- **Earthing up:** To prevent uprooting of plants by wind, soil is mounded around the pseudostem during the rainy season.
- **Propping:** Providing support to the plant when it is at a bunching stage. If a dwarf variety and closed spacing are used, no propping is required, but for all tall varieties, it is required by using bamboo poles or any other support.
- **Denaveling:** removal of male buds after the last set of fruit. It increases the bunch weight/fruit weight and quality of fruits also.
- **Thrashing:** A process of removal of old, dry, diseased, and senescent leaves. This could reduce the disease and facilitate better light, temperature, and air. However, if leaves are pruned before bunch initiation, flowering is delayed and cycle time is increased. A minimum of 12 leaves are required to be retained for maximum yields.

- **Bunch covering:** Bunch bagging with perforated polythene cover or dried leaves to protect against cold sun scorching, thrips and other scraping insects, and pesticides during the bunch maturity stage. The bags may be coated with pesticides.
- **Mattocking** is the process of cutting the pseudostem after harvesting the bunches. After harvesting, the pseudostem should be cut, leaving a stump of about 0.6m high. The left over stump, with its stored food material, continues to nourish the daughter sucker (follower) till it withers and dries up.

Growth and Development of Plant: Depending on the variety, the plant produces 30-40 leaves per month during its life cycle. The last leaf produced at shooting, which is small in size, is called the flag leaf. A minimum of 10-12 leaves must be retained on the mother plant to achieve maximum yield. The first distinguishing feature between the vegetative and reproductive phases is the production of bract primordium. The basal (proximal) nodes of the inflorescence bear female flowers, and the upper (distal) nodes contain male flowers. In between male and female buds, hermaphrodite flowers have stunted ovaries and do not develop into edible fruit. Banana fruit, botanically known as berries, the edible bananas are vegetative Parthenocarpy, the female sterility gene and lack of pollen due to triploid causes seedless nature. While pollination is essential for fruit development in wild-seeded bananas.

Flowering: Pollination in banana is mediated mainly by bees and birds. Crop improvement in banana is hampered due to Parthenocarpy, polyploidy, Inflorescence type: Spadix (Female and Hermaphrodite flower). Banana initiate flowering 9-12 months after planting, about 40 leaves are formed till flowering. Bunch covering with polythene and 2% sprays of KH_2PO_4 at fruit development stage increases fruit yield in banana. Banana fruit follows double sigmoid growth curve (seeded-single sigmoid, parthenocarpic-double sigmoid).

Use of Plant Growth Regulators

Spraying of NAA at 100 ppm after 5 and 7 months of planting markedly increases fruit size and yield. Spraying of 2-4 D at 20 ppm increased the quality of fruits. If it is poured into the growing apex, then the bunch will have more female flowers/fingers. Application of GA3 at 50mg/L resulted in maximum yield and required a lower number of days for fruit maturity in giant governor banana. **Banana seedlessness is controlled by spraying 2,4-D at a concentration of 25 ppm.** Growth regulator 2,4-D at 25 ppm (25 mg/lit) may be sprayed after the last hand has opened. This also helps to remove the seediness in the poovan variety. The dual factor hypothesis proves flowering in banana: GA3 induces stem growth and elongation, and anthesin acts as a flowering hormone.

Fruit Maturity and Harvest

Under favourable conditions, banana start flowering in 9–12 months and fruit matures in about 4-5 months, depending upon varieties, climate etc. Maturity 90-150 days, banana are harvested at 3/4th of the maturity stage for distant markets or for chip making purposes, while those for local markets are harvested at full maturity. Drying of the top leaves. The fruit's colour changes from green to light green; the fruit becomes plump and the angles disappear; the fruit's starch content (22-25%).

Yield

Tall cultivars usually yield 15-20 tones/ha. Cavendish Group varieties yield about 40 t/ha. Hill banana/cooking varieties yield about 11–15 tons/ha.

Post Harvest Management

The fruits should not be stored / shifted under refrigerated conditions. The storage life can be increased by keeping the fruits in a high concentration of carbon dioxide and a low concentration of oxygen. Store them in polythene bags that have been sealed and contain ethylene absorbents such as potassium. Shrink film wrapping or Waxol (12 per cent) treatment can extend shelf life up to 3 weeks. Bananas are not usually allowed to ripen on the tree. Smoking is done with straw, leaves, and cow dung in a closed chamber for 18–24 hours in summer and 48 hours in winter and later shifted to a ventilated room for uniform ripening. The exogenous application of 100 ppm ethylene gas in an enclosed chamber for 24 hrs will produce uniform colour and ripening. Banana can be stored at about 13°C with the Relative Humidity of 85-95 per cent for 3 weeks and is ripened in a week at 16.5-21°C.

Rhizome and Sucker Production

Production of banana suckers in large quantities is currently receiving attention in the wake of great demand for elite planting material. Besides, rapid multiplication of suckers in successful hybrids will enable their quick spread in a short period. Sucker production in banana is influenced by a complex of factors. Some authors reported that the diploids ranked first in sucker production, followed by triploid and tetroploid in the order. Pueraccuminata diploids and triploids produced a larger number of suckers than the balbisiana derivatives. The nutritional status of the mother plant has an overwhelming influence on sucker production. Among the cultivars, greater uptake of nutrients by a cultivar of the same ploidy level resulted in more sucker production. For instance, Anaikomban (AA) among the diploids and Monthan (ABB) among the triploids showed higher uptake of nitrogen with a resultant increase in the production of suckers.

Pests

Pseudostem borer): *Most* of the commercial cultivars are attacked by the borer. Exudation of plant shape is the initial symptom, and a blackened mass comes out of the holes bored by the larvae.

Banana aphid- Vector of the virus disease bunchy top

Fruit and leaf scarring beetle-The beetle feeds on young leaves and skin of young fruits, occurrence is maximum in rainy season.

Banana weevils *(Cosmopolites sordidus)* are serious pests of banana.

Virus Dieaseas

- Banana streak virus-transmitted by citrus mealy bug –***Planococcus citri*** foliar symptoms resemble those of banana mosaic.
- Banana bract mosoic virus

Diseases

- **Panama wilt** - ***Fusarium oxysporium F. sp*cubens** It is the most severe and important disease of banana. Rasthali is a highly susceptible cultivar. It is serious in poorly drained soil. Resistant varieties are **Robusta and Dwarf Cavendish.** The leaf blades yellow, buckle, and hang around the pseudostem. Severe in acid soil and wet soil –Soil borne. Gross Michel is susceptible, while Bassarai is immune and Poovan is resistant.
- **Finger tip disease is serious in high density planting (HDP).**
- **Tip over or heart rot *(Erwinia carotovora)*** Seen mostly in tissue cultured plants.
- Leaf spot/sigatoka
- **Sigatoka Leaf Spot Disease** is a serious fungal disease in banana. It was first observed in the Sigatoka Valley in 'Fiji' in 1913. While *Mycosphaerella musicola* is a sexual stage of the fungus causing sigatoka leaf spot, *Cercospora musae is an* asexual stage. It is a fungal disease, initially, the presence of light yellowish spots which leaves under sever conditions causes the formation of brown spots and later die, turning light grey surrounded by a brown ring. The Gros Michel and Cavendish groups are all (AAA) highly susceptible to sigatoka. All ABB clones are resistant.

Banana bunchy top virus (bbtv)/Bunchy top (Cabbage top): Bunchy top of banana is a viral disease *(Pentalonia nigronervosa). It was first* reported in Fiji (1891) and infected plants show short and narrow leaves together at the top of the pseudostem to form a bunch. Hence this disease is known as "bunchy top' A Tetrazolium test is used for the diagnosis of presence. The dwarf banana cultivars are very susceptible. The leaves are bunched together like a rosette at the top, the margins are wavy and slightly rolled upward. Dark green streaks of the lamina or midrib.

Other Diseases

- *Pseudostem* heart rot
- Diamond spot
- Anthracnose
- Cigar end tip rot
- Crown rot
- Bacterial soft rot
- Moko disease or Bacterial wilt *(pseudomonas solanacerum)*
- **Burrowing nematode (*Radopholus similis*)** is commonly present in banana plantations.

15

Barbadas Cherry

- Barbados Cherry/West Indian Cherry/Puerto Rican Cherry/Garden Cherry/ Native Cherry
- B N: *Malpighia punicifolia*
- Family: Malphigiaceae
- Chromosome No. 2n = 20
- Origin: Tropical America
- Inflorescence: Crymbos
- Fruit type: Drupe
- Edible part: Mesocarp and epicarp

Introduction

The West Indian Cherry is a very important fruit because of its high ascorbic acid content (10-33g/kg edible pulp). It is widely distributed throughout tropical America, Africa and the West Indies. It is also called Acerola, Cereza in Puerto Rico, Barbados cherry in Florida and Texas. Leading producer's countries of cherry are Turkey, USA, Chile, Uzbekistan and Iran. The plant is ornamental and when laden with clusters of its crimson fruits, presents a very attractive appearance. It is a medium-sized shrub and can serve as an ornamental tree on account of its fruits, flowers, foliage and drooping branches. Eating one or two fruits every morning will meet the daily requirement of vitamin C. This is a small shrub or small-sized tree growing up to 8.0 m in height and doing well in a tropical climate. The flowering seasons are May and November.

Composition

It is an excellent source of Vitamin C (1000–4000 mg/100g edible pulp), Malic acid, Glucose, Fructose, Sucrose, Calcium, P, Fe, Na, and K in trace quantities. Because of its high vit-C content, the fruit of the West Indian cherry can be rightly called the "pill of nature" vit-C. Eating one or two fruits every morning will meet the daily requirement of vitamin C. The fruit also contains protein, carbohydrates, fiber, carotene, thiamine, riboflavin, etc.

Uses

There is good potential in the form of processed produce and also for the food and pharmaceutical industries. The fruits are eaten as fresh; they can be juiced for punch, jellies and in the preparation of gelatin desserts, salads or for fortifying other juices with low in Vitamin C. The fruit is used to cure throat pain, diarrhea. The bark extract (**26 per cent Tannin**) used for leather conditioning and for colouring silk and woollen clothes. Another fruit of *Barbados cherry – M. glabra,* are useful against diarrhea, dysentery and liver disorders. Green fruits are the richest source of vitamin C, while half-ripe and ripe fruits contain much lower quantities. Green fruits are the richest source of vitamin C, while half-ripe and ripe fruits contain much lower quantities. Tannin is found in tree bark (26%). It also yields a dye having a coffee-brown colour. Tropical evergreen shrub, Ideal fruit plant for kitchen garden.

Climate and Soil

It is a tropical plant that thrives well in warm lowland climates. The number of flowering cycles depends upon temperature and the rainfall pattern. A well distributed average rainfall of about 175 cm is good for growth and fruiting. It cannot withstand low temperatures (less than 5°C) for a long period of time. It can be grown in a wide range of soils with good drainage. The soil pH should be 5.0–6.5.

Cultivars and Species

The genus Malphigia mainly having 2 related species – ***Malphigia coccijera*** and ***Malphigia urense*** produced edible fruits.

Cultivated types are grouped into:

A- Acid types

These are generally better yielders, with a more desirable form with more vitamin C.

Varieties – Maunawik, J.H. Beanmont, C.F. Rehnbong, F. Haley, and Red Jumbo.

B. Sweet Type

These cultivars were developed for general home planting and for potential use in baby foods and products requiring minimum acidity.

Varieties: Manoa sweet, Tropical Ruby, Hawaiian Queen, Florida sweet

Propagation

It is propagated by seeds. It takes about 10–12 days for germination, and 6–8-month old seedlings are used for transplanting. Vegetatively, commercially Air and ground layering are followed during the rainy season. Hardwood cuttings treated with IBA at 2500ppm give better rooting.

Planting: Spacing of 7x7m for general cultivation and at 3x3m for hedge purpose.

Manuring: 40-50 kg compost/plant/year should be applied in 2 splits.

Training and Pruning

The plants are trained to a single stem with scaffold branches produced at a height of about 60-90 cm above ground, to facilitate easy cultural operations. Since the fruit bears only on current season growth, light pruning is necessary to encourage new growth.

Flowering

The flowering starts after two years and generally commences with the onset of the mansoon (May to August). Flower-bud production appears to occur between 15 and 18 days after pruning.

Harvesting and Yield

The fruiting starts in the second year after planting. The fruits are harvested at different stages of maturity based on their intended use. Half-ripe or fully ripe fruits can be utilised for various purposes. Mature green fruits are suitable for vitamin C extraction. All fruits may not ripen at the same time. Hence, repeated harvesting is required. The fruit is thin-skinned, delicate and highly susceptible to bruising. An average yield of 2 kg/plant could be obtained from four-year-old trees.

16

Ber

- Ber/King of Arid fruits/Poor man's fruit/Summer deciduous fruit/Chinese fig/Chinese date/Apple of desert
- B. N.: *Zyzyphus mauritiana*
- Family: Rhamnaceae
- Chromosome No.: 2n = 48. x = 12.
- Origin: Indo-China or India
- Inflorescence: Fasicle
- Fruit type: Drupe
- Edible part: Pericarp

Introduction

The ber is a vigorous-growing, small-spreading tree with almost vine-like drooping branches. The round to oval reddish brown has 5.4–8.0% sugar and 85–95 mg of ascorbic acid (Vit. C) per 100g. The tree is a host plant for rearing lac insect (*Tachardia lacca*). Lac insect rearing helps in the production of lac. The powder of ber roots has very many medicinal properties, such as cure for ulcer, fevers and wounds. The stem bark powder is a remedy for diarrhea. Ber also shows strong self-incompatibility (gametophytic). Madhya Pradesh, Gujarat and Chhattisgarh is a leading state in "ber" production.

Climate and Soil

India ber ***Zizyphus mauritiana*** is to grow in tropical and subtropical regions, while **Zizyphus *jujuba***, the Chinese ber is a deciduous tree, found in temperate regions. Ber is an ideal fruit tree for arid and semiarid regions where copious irrigation is impossible. It can tolerate a high temperature of even 40°C. But below freezing temperatures are injurious. It can be grown in extreme moisture stress and in a wide range of soils, up to an infertile one, which is unsuitable for major fruits and other crops. In alkaline soils with high ph (even up to 9.5) and sodic soil, five kg of gypsum has to be applied per pit, mixed with top soil and flooded the pits one week before planting has to be done.

Flowering and Fruiting

Irrigation during October causes flower shedding and that during March-April causes fruit spoilage and delays ripening. **Sylleptic** branching pattern is observed in ber. flower colour, greenish to yellow. Index tissue for nutrient analysis in ber is **the 5th–6th** mature leaf collected at the flowering stage (during September and middle of November) from the secondary shoots. It is salt tolerant and extremely drought tolerant. Flowering in North Indian conditions is from **August, September to November.** Fruits are harvested in **November–December (South India).** Flowers in ber are **hermaphrdite.** Pollination is mediated by insects. The type of **placentation** in ber is basal. Fruit growth follows a **double sigmoid** curve and the current season's growth habit. Ber fruits are higher calorific value and Vitamin-C than orange.

Species

- Indian ber (***Zyzyphus mauritiana***), Evergreen, Spreading habit.
- **Chinese ber (*Zyzyphus jujuba*)**-Deciduous, Upright, Highly resistant to frost.
- ***Z. rotundifolia is*** (Commercially used rootstock for arid regions).
- *Z. rotundifilia*is a better rootstock as compared to *Z. nummularia;* because
- ***Z. nummularia-*** Dwarfing rootstock suitable for high density planting.

Cultivars

- **Kaithali:** It is a variety with straight thorns but not so pronounced.
- **Seo (Sanaur No.2):** Resistant to powdery mildew.
- **Seb:** It is an early variety. It acts as a good pollinizer for a number of varieties.
- **Banarsi Karaka:** It is a mid-season variety. It has performed well under Tamil Nadu condition.
- **Chhuhara:** Fruits ovate-oblong. Fruits are suitable to be heated and made into dry fruits like dates.
- **Sandhura Narnaul (Sanaur No.1):** Fruits are oval-oblong to longish, style end slightly pointed.
- **Elaichi:** 90% pollen sterility, Trees spreading with fruits having the characteristic shape of cardamom hence called **'Elaichi'**.

- **Gola, Seb, and Mundia-** are suitable for extremely dry areas.
- **Gola** is tolerant against **salinity and alkalinity.**
- Darakhi, Dandan
- Umran and Illaichi are **Late maturing** cultivars.
- Sanaur 2 and Mehrum are for comparatively humid region.
- **Umran** is an early bearing selection made at Godhra, Regional Research Stataion,
- are early maturing than Rashmi, Kathali, mundia
- Late maturing varieties: **Umran and Illaichi**
- **Mid-season varieties:** Rashmi, Mundia and Banarasi Karaka.
- **Thar Sevika:** Seb x Katha, Early maturing
- **Thar Bhubraj,** Selection from Bhusavar area of Bharatpur
- **Goma Kirti (Ganesh kirti):** Early maturing selection from Umran
- **Kaithali, Tikdi, Sanaur No. 1, Mehrum** and **Dodia** are resistant to **fruit fly.**
- **Gola, Seb, Illaichi, Safed Rohtak**and **Mehrum**are resistant against **powdery meldew.**
- **Safed Rohtak, Banarasi Karaka** and **Mundia** are resistant against **leaf spot.**

Propagation

Ber is propagated by Ring and T budding. Highly suitable for arid conditions: *In-situ* budding (July to September).

Use of seedling rootstock

Z. rotundifolia- Drought tolorence

Z. nummularia- Dwarf and salt tolorence

Nursery Raising

Fruits of wild species are taken; seeds are extracted and soaked in a 17% salt solution to remove the ill-filled ones which float. The seeds which sink are taken and soaked in concentration H_2SO_4 for 5 minutes, washed and soaked for 48 hours in cold water. Then the seeds can be sown in the poly bags (25 x 15 cm) of 300-gauge thickness. The seeds take 10–15 days for germination. If we want to bud the seedlings raised in polybags, large-sized polybags have to be used since the tap root grows very fast. June to august is the best period for budding for getting maximum bud-take. The scion sticks with 0.9 cm diameter, about 1-year-old maturity with plumpy buds should be selected to take buds.

The buds will take about 7–10 days to sprout. The treated seeds can also be sown directly into the pits @ 2-3 seeds per pit at a depth of 3 cm. Normally, the required varieties are budded in situ on this rootstock seedling after 90 days.

Planting

Pits of size 1x1x1 m filled with 20 kg of FYM + top soil and irrigated.

Best time of planting: July to september and spacing for: 5m x 5m.

Training and pruning: For young plant, a support should be provided by bamboo stakes. In the first year, all the branches arising up to 75–100 cm should be removed so that a straight trunk can be developed. Above this, 3-5 primary branches should be developed so as to have a balanced frame work in all directions. In the second year, on each primary branch, 3–4 well-distributed secondary branches may be allowed, and during the third year, final frame work should be decided. In ber, the fruit buds are borne on the current season's growth in the leaf axils. Therefore, the plants should be pruned every year to induce new growth to provide maximum fruit bearing area. After the harvest of fruits, the plants have to be pruned by heading back 25% of their one-year-old shoots. If severe pruning is attempted, it will adversely affect the growth, leading to a poor yield. The type of training system adopted system of modified leaders. A spray of 3 % thiourea or KNO_3 once in 2 days before pruning induces bud sprouting from the maximum number of nodes.

Interculture: The inter space can be utilised to raise a crop like pulses such as green gram, black gram, cowpea etc, for about 3 years. If sufficient irrigation facilities are available, papaya or phalsa can be grown as an intercrop for the first three years.

Irrigation: For young trees, irrigation should be done once a day. As the trees grow older, it can be restricted once in 15 to 20 days. At the time of flowering and fruiting, there should not be any moisture stress.

Manures and Fertilizers

The manorial dose can be split into two equal halves and applied once during June and another after pruning.

Year	FYM(kg/tree)	Ng./Tree	Pg/Tree	Kg/Tree
1 Year after planting	10	125	40	75
2 year after planting	15	250	80	150
3 year after planting	20	375	120	225
4 year after planting and then onwards	25	500	160	300

Pests

Fruit fly: (***Carpomyia vesuviana***) -The flies puncture the young developing fruits by inserting them into the oviposit or and lays eggs singly. Hatching of eggs is completed in two to three days. The larvae (maggots) feed inside the fruit pulp, make small holes in the rind and come out of the fruit when fully grown. The affected fruits become misshapen and their growth is retarded.

Management

Collect infested fruits and destroy them. Spray monocrotophos (0.04%) or Rogar 30EC (0.06%). Spray with a mixture of 100 ml Malathion 50EC and one kg of jaggery or sugar in 100 litres of water twice starting from September to October at a 7–10 day's interval (3 sprays). Cultivate ber orchard soil during April-May and apply 10% BHC to destroy pupa Grow fruitfly tolerant varieties like Umran, Sanaur.

Bark eating caterpillar: *(Inderbela quadrinotata)*

- **Hairy caterpillar: (*Euproctis freterna*)-** caterpillar feed on leaves and cause damage.
- **Ber beetle or leaf chafer:** (***Adorentus pallers***)
- **Ber fruit borer *(Meridarchess cyrodes)***

Diseases

- **Powdery mildew: (*Oidium emysiphoides*)** This is a serious problem, affected fruits show white powdery spots which later cover the whole area of fruit. The white powdery mass also spreads on flowers and leaves. Later, white spots turn brown and fruits drop.
- **Management:** Spray dinocep 0.1% at 15 days' interval after fruit set preferably at pea stage.

Harvest and Yield

Fruits harvested at the correct stage of maturity alone will ripen properly. After attainment of full size of a particular cultivar and turning of colour to yellow or golden yellow. The normal harvesting season is **October-November,** while in **North India** it varies from place to place, viz., **December to April.** The average yield from a 10-to-20-year-old tree would be **100-200 kg/year**. If the fruits have to be stored, they can be stored at **3**°C and **85-90% for 30 to 40** days.

17

Bread Fruit

- B. N.: *Artocarpus altilis*
- Family: Moraceae
- Chromosome No. 2n = 56
- Origin: Tropical Asia
- Fruit type: Sorosis

Introduction

Breadfruit has long been an important staple crop and a primary component of traditional agroforestry systems in the Pacific, where numerous varieties are grown. Breadfruit was first domesticated in the western Pacific and has been spread by humans throughout the region over the past 3,000–4,000 years. From Melanesia to Micronesia to Polynesia, many islanders grow and use breadfruit as part of their daily diet. Breadfruit has a significant and often unappreciated legacy in Hawai. It is one of the canoe plants brought from the society Islands centuries ago, and the variety they brought, known simply as 'Ulu (same as the Hawaii a word for breadfruit), was widely grown throughout the archipelago.

Climate and Soil

A hot, humid, tropical low land is suitable for its cultivation. It is ideal to have an annual rainfall of 1500-2500 mm and a temperature of 21.10-32.2 °C. Young plants grow better under shade but later require full exposure to sunlight. It can be grown on a variety of soils, provided they have sufficient depth and drainage.

Propagation

Since bread fruit has no seeds, it is propagated vegetatively. Root cuttings about 2.5 cm in diameter and 20 cm in length are planted horizontally or at an angle in shade beds. Root suckers can also be used as propagules. Airlayers made from offshoots can also be used for planting.

Planting

June–December is planting time. Pits of 60 cm x 60 cm x 60 cm are dug at a spacing of 10–12 m. There is no exclusive fertilizer recommendation for breadfruit. Organic manure @ 25 kg/tree can be applied. Depending upon the age of the plant, an NPK mixture (7: 10: 5) @ 1–2 kg/plant can be applied.

Irrigation

Irrigation should be given during the summer, which helps to check fruit drop. Spraying of Bordeaux mixture (1 %) controls fruit rot, especially during the rainy season.

Intercropping

Inter planted with small fruit. intercropping trees or short-term fruit and vegetable crops. It is a monoecious tree, growing to a height of 15-20 m. The inflorescence is known as a catkin, and the fruit as a syncarp, consisting of the fused ovaries of numerous female flowers.

Harvesting and Yield

Fruits require about 3-4 months for attaining maturity after fruit set and they will turn from green to yellow wish green colour. For culinary purposes, they are to be harvested at the maturity stage but before ripening. The tree produces regular crops and bears about 50–150 fruits (25 to 75 kg) per year.

18

Carambola

- Carambola/Star fruit/Kamrakh/Five Corner Fruit
- B N*:* *Averrhoa carambola*
- Family*:* Oxalidaceace
- Chromosome No.: 2n = 2x = 24
- Origin place = Indonesia/ Malaysia
- Inflorescence: Panicles or Cymes
- Fruit type: Capsule
- Edible part: Pericarp

Introduction

Sweet form sugar: Glucose, 0.16% oxalic acid 0.06% malic acid. The fruits are attractive, golden-yellow in colour. **Bilimbi (*Averrhoa bilimbi* L.)** 'Cucumber tree' and 'Tree Sorrel' are synonyms of bilimbi. Its produced **gherkin.** It is also known as five-corner fruit or five-edge fruit. Malaysia, Popular in China, Taiwan, Malaysia, Thailand, Pakistan, Indonesia and India. Carambola is a good source of vitamin A, B, and C, minerals, especially Fe, K, Mg and P. The composition of fruit is 75 to 80% pulp, 5 to 7% total soluble solids, 9.4% carbohydrates, 0.4% protein, 0.5 to 1.0% acidity, 0.9% fibre, 570 to 700 IU of vitamin A and 0.3 to 0.5% minerals. It is used as a dessert fruit or for preparing juice, jam, jelly and pickles. It has medicinal value also. Ground leaves and shoots for chicken pox, root extract as an antidote for poisoning and pulp of immature fruits for cleaning brassware. It is used as an alternative to tamarind in cooking.

Climate and Soil

Carambola grows well in both tropical and subtropical climates free from frost. But it prefers warm and moist conditions optimum temperature range is 21-32°C, 1800-2500 mm annual rainfall is best. It can be grown up to 1200 m. It can be grown in any type of soil. But well-drained, deep clay soils are preferable optimum pH is 5.2 to 6.2.

Species and Cultivars

No specific cultivars but 2 types namely sour and sweet are grown in India.

A) **Sour type**: 1% acid Var: Goldstar, Icambola.

B) **Sweet type:** 0.4% acid with 5% sugars Var: Fuang Tung (Chinese sweet type), Goldstar.

Florida:	Goldenstar, Fwangtung, Dahpon. Teanma, Artin, Maha, Starking. Thaiknight, wheeler, New comb.
Singalopre:	Lengbak, Turong
Taiwan:	Intho, Cheng, Tyeyi, Dahpon, Erlin soft sih, Tean Ma and Min Tao
Malaysia:	B1, B2, BA, Bb, BB, B102, Bill and B6 (Orange colour)
Australia:	Giant siam, Jungle gold, BCP-1 Hosie, Jungle Gold. Chujuba, Giant Siam
Thailand:	Fwangtung, Thai Knag
(Hawaii): **Golden Star**	Developed by Florida, Queensland, Singapore, Taiwan and Japan.
(Columbia):	Icambola

A few of them varaties are: Ma, Mih Tao, Dah Pon, New Comb, Theyer, Arkin

Propagation: Seed **(Recalcitrant)** should be sown immediately as they loose viability at a faster rate.

Manures and Fertilizers

One-ton of Carambola fruit is reported to remove 1.28 kg N; 0.12 kg P, 1.58 kg K. No recommended fertilizer dose is available. The addition of 'N' fertilizer improves the yield. Normally, organic manures are only applied. Irrigation is given during fruit development only if the weather is dry. Trees may be provided 200 litre/ tree/ weak in dry periods. Mulching will be very helpful in reducing weed growth and irrigation needs.

Planting: Spacing 7-9mx 5-7m or high density planting with 7x3m or 4.2x 3.6m is also.

Pruning and Training

The plant should be trained to have four to six main scaffold limbs at a suitable height on the trunk. Pruning back old branches will improve the quantity and quality of the fruit.

Flowering and Pollination

Fruit blooms in June-July and bears fruit in December-January. Carambola trees' three flush are found First (April-May), 2nd (July –August), 3rd (Sep-oct). Short Style are found Self-incompatible and long style self-fertile, flower –heterodistylous. A cauliflorous type of flowering is noted as in other *Averrhoa* species. Trees are small-sized with drooping branches.

Harvesting

Carambola produces fruit on both young and old branches, and even on trunks in clusters. Carambola produces fruit throughout the year, but the peak seasons of harvest are **Jan-Feb and Sept-Oct**. Carambola takes about 3–4 months from flower to maturity. Carambola, being non-climacteric, should be harvested at the proper stage of maturity only, i.e., when 50% of the fruit becomes yellow. On ripening, the fruit develops an orange tinge. Yield varies from 20-50kg/tree in the early stages and a grown up tree may give 400-500 kg/tree. Carambola is excellent in its keeping quality. Even under room temperature, they store well for many weeks. They can be stored at 5 °C for 12 weeks.

Pests and diseases: Fruit fly, Aphids, Mealy bugs, Scales, Leaf spot and Anthracnose

19

Chironji

- Charoli/Chironji/Cuddapah Almond/Achar/Chawar
- B. N.: *Buchanania lanzan.* Spreng.
- Family: Anacardiaceae
- Chromosome No.: 2n = 22
- Origin: India
- Inflorescence: Panicle
- Fruit type: Drupe
- Edible part: Mesocarp & Kernel

Introduction

An herb known as priyal is a drug of the Ayurveda and the Unani system of medicine. It is known to have tonic, cardiotonic, and astringent properties and is also used in the treatment of skin diseases. It is commonly known as Chironji. It is a commercially useful tropical plant. The Chironji tree is a medium-sized evergreen deciduous tree, growing 50 ft tall. It bears fruits, each containing a single seed, which is a popular edible nut, known as chironji. It is common in India, mostly in eroded land. This tree is distributed in the states of Madhya Pradesh, Bihar, Orissa, Andhra Pradesh, Chhattisgarh, Jharkhand, Gujarat, Rajasthan, Maharashtra and also commonly found in Uttar Pradesh.

Climate and Soil

Chironji is very hardy plant and thrives well on rocky and gravelly red soils. Though it is a very hardy tree but plants do not survive under waterlogged conditions. It can grow even in pockets of soil between crevices of barren rock. Trees are grown on degraded rocky area including salt affected soils. However, for its better growth and productivity, well drained deep loam soil is ideal. It prefers tropical and subtropical climate and can with stand drought admirably.

Propagation

Chironji plants are generally propagated by seed, giving a long gestation period (10–15 years) and large variability. Vegetative propagation through soft wood grafting uses one-year-old seedlings raised for rootstock to commence putting on new growth and the leaves are of light green colour. These are ready for soft wood grafting. Shoots of 3–4 months, which have a prominent apical bud, are taken as the scion material. Defoliation of such shoots is done 8–12 days before detaching them from the parent tree for grafting operations. At the time of the removal of these shoots, the apical buds should remain intact. The top of the new growth of the stock is cut and the scion is fitted by cleft grafting. The union is tied with a 200-gauge polythene strip. If the selection of scion and rootstock is proper, 60–70% success may be obtained during the months of July and August.

Planting

The seedling plants should be planted 10 m apart, whereas grafted ones 8 m apart. The pits of 1 m x 1 m x 1 m size are dug and filled with a mixture of top soil + 25 kg of farmyard manure up to a level of 30 cm from the ground level. After onset of the monsoon, soils are settled and planting is done in the centre of the pit during July–August.

Training and Pruning

Training is essential to develop the framework of chironji plants. The framework should be developed by encouraging the growth of 4–6 well-spaced branches on the trunk at a height of 60–90 cm from the ground level. Pruning is not performed in chironji plants except for the removal of dead, diseased, and crossing branches.

Cultivars

Thar Priya Variety of Chironji from CIAH, Bikaner.

- Grown up to elevation of 1200 m.
- Rich source of fat 35 to 47.2 %
- Kernel is highly nutritious and rich in protein 25 – 30 %

Intercropping: Compatible crop combinations are necessary with regard to species, cultivars, planting method, and sequence pea, chickpea, lentil, black gram, cowpea, cluster bean may be grown as inter crops in the orchard.

Manures and fertilizers: A dose of 100g N, 50g P, and 75g K/plant should be given to a one-year-old plant. It should be increased every year in the same proportion up to the age of 10 years. In addition, 5 kg FYM should be added to the 1-year plant. It should be increased to @ 5 kg up to 10 years. Thus, fully grown up to trees requires 50 kg farmyard manure, 1 kg N, 0.5 kg P and 0.75 kg K. Farmyard manure should be applied during July–August. Half dose of N and a full dose of P and K should be applied in the month of July. The remaining dose of N may be applied by the end of August under the rainfed conditions.

Irrigation: Plants can survive under rain-fed conditions. If the irrigation facilities are available, a basin system of irrigation should be used for young plantations. Irrigation should be given during the summer season at an interval of 15 days. Proper moisture after fruit set is important for the retention and development of fruits.

Flowers and fruits set: In Chironji, flowers appear in the months of January–February in different agro-climatic zones of the country. The time taken for the complete development of a flower varies from 16–20 days. New plantations of grafted plants start flowering after the 4th year.

Harvesting of chironji: The fruit of chironji matures in 4 to 5 months and is harvested manually in the months of April to May.

20

Citrus Fruits

- Fruit of godly tree/Buddha's hand fruit
- B. N.: *Citrus* spp.
- Family: Rutaceae
- Chromosome No.: 2n = 18
- Origin: Tropical and subtropical South East Asia
- Inflorescence; Cymose or Solitary
- Fruit type: Hesperidium
- Edible part: Juicy placental hair

Groups of Citrus

Sr. No.	Citrus Species	Botanical Name	Origin
1	Mandarin (Orange)	*Citrus reticulate*	China
2	Sweet Orange (Malta or Mosambi)	*Citrus sinensis*	China
3	Acid lime/Kagzi lime	*Citrus aurantifolia*	India
4	Grape fruit	*Citrus paradise*	West Indies
5	Lemon	*Citrus limon*	South-East Asia
6	Pummelo	*Citrus grandis or Citrus maxima*	South-East Asia
7	Rangpur lime	*Citrus limonia*	India
8	Tahiti lime	*Citrus latifolia*	
9	Sweet lime	*Citrus limetta*	India
10	Trifoliate orange	*Poncirus trifoliata*	

Introduction

Citrus is the leading tree fruit crop in the world. In India, citrus ranks second in area, sharing 12.8% of the total area under fruit crops with a 9.6% share in production. In the citrus group, mandarins are the most important, occupying 50% of the total area under citrus, followed by sweet oranges and limes. **"Limolin":** the glycoside which is responsible for the bitter taste of citrus fruit juice.

Origin & Distribution

South East Asia, tropical and sub-tropical regions of Indio- china. Major citrus-growing countries are the USA, Spain, India, Italy, Japan, Argentina, Citrus fruits are grown mainly States Maharashtra, Andhra Pradesh, Punjab, Karnataka and the North-Eastern region.

Climate and Soil

It grows well in dry semi-arid climates, producing good quality fruits. It can tolerate occasional light frosts. The optimum temperature range is 16–20°C within a range of 17–40°C. An annual rainfall of 500–775 mm is optimum. Sweet orange can be grown in any well-drained soil, sandy or clay loams are preferable.

Classification of Citrus

The genus Citrus contains many kinds or types that differ as to their fruits, flowers, leaves, and twigs. The arranging of the kinds of citrus into groups (as indicated by features such as looseness of peel, size, shape, and color) is termed classification. The naming of these groups with valid names is termed nomenclature.

There are definite international rules of nomenclature but not of classification. Several per-sons have placed the various kinds of citrus into groups (classified them) and given them valid names. The classification one accepts is one of personal choice, based on utility, common usage, and natural relationships.

Currently, there are 2 outstanding systems of classification for citrus. They are those of Walter T. Swingle, a USDA scientist who did much of his work in Florida, and Tyosaburo Tanaka of Japan.

Swingle's system is relatively simple, containing 16 species. He is commonly referred to as a **"lumper"** because he lumps a large number of kinds into a relatively small number of groups. Tanaka's system initially included 145 species, and he is known as a **"splitter"** because he has split the genus Citrus into many small groups. He has continued to add to this list.

From the standpoint of the grower, most horticulturists and other plant scientists, Swingle's system appears the most useable. Tanaka's system has some features that are more reasonable than some of Swingle's; however, it is not, in total, a very convenient or botanically sound system. Until someone develops a better system, horticulturists will probably use Swingle's with some modification. Some portions of Tanaka's system have wide acceptance.

It is well to understand that neither Swingle nor Tanaka classified horticultural varieties, such as "Duncan" grapefruit and "Pineapple" orange. These names have been established by usage in the industry and do not always agree. For example, the 'Key' lime of Florida is the 'Mexican' lime of Texas and the 'West Indian' lime of the West Indies.

Outlined below is Swingle's classification of citrus species, with some comparisons with Tanaka's system, some points of interest concerning them, and examples of horticultural varieties contained in the various species.

This summary is brief and designed only to acquaint the reader with the kinds of citrus and how they are classified and named.

Citrus (Genus)

A. ***Eucitrus*** (Subgenus)includes all those species with edible fruits, pulp vesicles not containing drops of acrid orbitter oil. Contains the following species:

1. ***C. medica*** (Citrons) *medica* var. *sarcodactylis*. The fingered citron. *medica* var. *ethrog*. The citron of the Mediterranean and the 'Etrog'of the Hebrews. The citrons are used mainly for their peel, which is candied. Thecitron isalso used in certain religious ceremonies by Hebrews. More recently, the citron has been shown to have use as an indicator plant for exocortis (symptoms expressed in a few months versus several years when trifoliate eorange is used).
2. ***C. limon*** (Lemons, including rough lemons). Includes the lemons of commerce, suchas **Lisbon** and **Eureka**. The **Meyer** is considered by Swingle to be a lemon hybrid of unknown parentage. of significance to Florida growers is the rough lemon, which is considered a separate species C. jambhiri, by Tanaka. It is frequently referred to by the latter name. *C. limonia* or the 'Rangpur'lime is considered close to the lemon by Tanaka. It is used as a rootstock but is susceptible to both exocortis and xyloporosis. *volkameriana* is a lemon type be ingusedin citrus rootstock breeding programs. The **Ponderosa leirion** is considered a lemon hybrid by Swingle but is classified as *C. pyriformis* by Tanaka.

- Sweet lemon—see sweet lime.
- The lemon has been hybridized to form such kinds as:
  - i. Lemonage (lemon × sweet orange)
  - ii. Lemonimes (lemon × limes)
  - iii. Lemandarins (lemon × mandarins)
  - iv. Eremolemon (*Eremocitrus*×lemon)

3. ***C. aurantifolia*** (Sour limes).

- Includes the '**Key** or **Mexican**and **Tahiti** or **Persian** horticultural varieties.
- Tanaka classifies the 'Tahiti'limeas (*C. latifolia).*
- Tanakalistsalime-like fruit called *C. macrophylla* that is not treated ass uch by Swingle.*C. macrophylla* has shown considerable promise in California as arootstock.
- The limes have been hybridized with other species to form suchkinds as:
  i. Lemonimes (lime × lemon)
  ii. Limequats (lime × kumquat)

  C. limetta (sweet limes).

  There is much confusion among writers regarding this species, many calling these fruits sweet lemons. Tanaka lists C. limetta as the sweet lime or lemon of the Mediterranean. Sweet limes are used as a rootstock in Palestine. It has much promise as a rootstock in Florida but its susceptibility to the xyloporosis virus has reduced interest in it for rootstock purposes. Tanaka lists C. limettiodes as Palestine sweet lime. This is the sweet lime of India and not the same as the sweet lime (C. limetta) used as a rootstock in Palestine.

4. ***C. aurantium*** (Sour orange): This includes the common sour orange and many forms and variants of unknown origin, such as: Bittersweet, Oklawaha, Vermillion Globe, Paraguay, Bergamot, Trabut, var. myrtifolia (Myrtle Leaf or Chinnoto), Daidai (Japanese name for sour orange). Swingle would include Tanaka's *C. taiwanica* in this species. *C. taiwanica* is a sour orange type originally reported as not susceptible to tresteza. However, recent evidence indicates it may be susceptible to some strains of tristeza. It survived the 1962–63 freeze in Gainesville, Florida with negligible damage (less than 10% leaf drop), even though adjacent 'Owari' satsumas were badly damaged. *C. anrantium* has been hybridised with the trifoliate orange to form citradias. The USDA has a citradia that is reportedly edible and extremely cold-hardy.

5. ***C. sinensis*** (Sweet orange): This includes the many cultivated varieties, such as the **Navels, Ham-lin**, Parson Brown, Pineapple, Jaffa, Queen, and Valencia. The term **orange** has been used to denote many fruits that are not sweet oranges, such as:
   - Temple orange
   - Satsuma orange

- Trifoliate orange
- Poormans orange
- Murcott orange

Unshiu orange (Japanese name for Satsuma orange) Recently the '**Page**' orange has been released by the USDA not because it has any sweet orange parentage but because it reportedly resembles a sweet orange. As citrus breeding programs develop this practice may become more common.

The sweet oranges have been hybridized with many species and other genera to form:

- Citranges (sweet orange × trifoliate orange)
- Citrangors (sweet orange × citrange
- Citrangequats (citrange × kumquat)
- Citrangeremos (citrange × Eremocitrus)

6. ***C. reticulata*** (the mandarins, satsumas and tangerines).This speciesincludes the commonly grown Florida varieties:
   - Dancy tangerine
   - Ponkan or Chinese Honey
   - Clementine or Algerian
   - Owari satsuma
   - Cleopatra mandarin
   - King orange
   - Temple orange
   - Murcott or Murcott Honey orange

- It should also be noted that **King**, **Murcot**, and **Temple** are often classified as hybrids. However, the origins of all are unknown. They are also loosely called oranges. From the standpoint of flavor, color, and looseness of peel, they closely resemble *reticulata*. It is here that Swingle and Tanaka differ widely and examples are worth noting.

Common Name	Swingle	Tanaka
Satsumas	*C. reticulata*	*C. unshiu*
Ponkan	*C. reticulata*	*C. reticulata*
Dancy	*C. reticulata*	*C. tangerina*
Cleopatra	*C. reticulata*	*C. reshni*
Clementine	*C. reticulata*	*C. clementina*
Willowleaf	*C. reticulata*	*C. deliciosa*

King	tangor	*C. nobilis*
Temple	tangor	*C. temple*
Calamondin	*C. reticulata var. austera? × C. ichangensis*	*C. madurensis*
Yuzu	*(tangerine × C. ichangensis)*	*C. junos*

- The **Clementine** and **Temple** are used widely in breeding programs because they are monoembryonic and of high quality. All crosses (using these as mother plants) give hybrid seed and not nucellars. There are several hybrids of tangerines with grapefruit that are good commercial varieties in Florida. Most of these are called tangalos.
  - Orlando tangelo
  - Minneola tangelo
  - Page orange
  - Robinson tangerine
  - Nova tangelo
  - K-Early tangelo.

  There are many more of these hybrids of little importance or promise of importance. The **Ortanique** is a very famous Jamaican cultivar that is much like the **Temple** in flavor and appearance but often seedless. Moreover, it is nucellar rather than monoembryonic. reticulata has also been hybridized with other species to form such hybrids as:
  - Tangors (tangerine × sweet orange)
  - Tangelo (tangerine × grapefruit)
  - Citrandarins (tangerine × trifoliate orange)

7. ***C. grandis*** (Pummelo or Shaddock): Native to southeast Asia where it is widely used and distributed. It is very similar to grapefruit but the fruits are much larger and thicker peeled than grapefruit. The pummelo is not as juicy as grapefruit and its pulp vesicles are larger. Typical grapefruit flavor is lacking. The leaves of the pummelo are generally larger than grapefruit leaves and the twigs are pubescent. Both contain naringin and both have white and red fleshed varieties. Seeds of pummelos are monoembryonic and those of grapefruit poly- embryonic. It is thought by some that the grapefruit originated as a mutant of *C. grandis*. This is largely on the similarity of their gross morphological features. grandis hybridizes easily with other species of citrus. Two of the better varieties grown in Florida (as dooryard fruits) are 'Thong Dee' and 'Tresca'.

8. ***C. paradisi* (Grapefruit):** The origin of this species is unknown. It is not known in the wild state in either the Monsoon Region of the Orient or Africa. It includes several horticultural varieties- **Duncan**, **Marsh**, **Red-blush** and **Thompson** are important varieties in Florida. Hybridized with the tangerine it has given some very fine quality fruits, some of which resemble sweet oranges more than either of the parents. See C**. reticulata**. It has been crossed with trifoliate orange to form citrumelos that are of little value. It is highly nucellar. An attempt was made to give grapefruit the common name of pomelo but it failed. There are some varieties of unknown parentage (**Triumph**, **Royal**, **Isle of Pine** and **Silver** Cluster') that do not have typical grapefruit flavor. They are probably not true grapefruit.
9. ***C. indica* (Indian wild orange):** An unpalatable loose-skinned fruit included in the Metacitrus subgenus by Tanaka. Found in the truly wild state northeastern India.
10. ***Tachibana* (the Tachibana orange):** A cold hardy type of loose-skinned fruit. Swingle thought the **Shekwesha** to he a hybrid of this species but Tanaka calls the Shekwasha *C. depressa*. tachibana is described by Tanaka as a very primitive type of native citrus of Japan.

Papeda (subgenus). A group of Citrus species having inedible fruit with acrid oil drop- lets in the juice vesicles. The flowers and fruits are small and the petioles are very long and broad as compared with those in Eucitrus

1. ***C. ichangensis*:** Probably the coldest hardy evergreen citrus. It is monoembryonic and hybridizes readily with other species of Citrus. Swinqle thought the Yuzu and Ichang lemons were hybrids of C. ichangensis but Tanaka gave them species status of *C. junos* and *C. willsonii*, respectively. Included by Tanaka in the Metacitrus subgenus and the Acrumen sec-tion.

2. ***C. latipes*:** Another cold hardy species similar to *C. ichangensis* but with a thicker peel and more variable leaves. Tanaka placed this species in his Archicitrus subgenus and Papada section.

3. ***C. micrantha*:** (small fruited types found in the Philippines): Includes a botanical variety C. micrantha var microcarpa has the smallest fruit and flowers in Citrus. These are given separate species status by Tanaka in his Papeda section (*C. micrantha* and *C. westerii*, respectively).

4. ***C. celebica*** (a small thick-peeled fruit of the Papeda type found in northeastern Celebes). Contains a larger thin-skinned botanical variety, *C. celebica* var. south-wikii. Tanaka gives both species status (*C. celebica*

and *C. southwikii*). Swingle also lists a number of possible hybrids that Tanaka considers species.

5. ***C. macroptera***: (inedible fruits with leaves 10–12 inches long and fruits as large as sweet oranges). Here too, Swingle lists several botanical varieties and possible hybrids that Tanaka calls species.

Swingle	Tanaka
Webber's Philippine hybrid Kansi	*C. weberri*
C. macroptera var. kerrii	*C. kerrii*
C. macroptera var. annomensis	C. combara

6. ***C. hystrix***: a very bumpy or warty fruit of the Papeda type.

Mandarin Group-1

(Tangerines or Satsuma, Kinnow)

1. ***Citrus reticulata***: This group of oranges is otherwise called the Kamala orange. Nagpur santra of Maharashtra, Coorg of Karnataka and Kodai orange of Tamil Nadu. This group is characterised by the loose skin of fruits of Chinese origin. Polyembryonic cultivars are Nagpur, Coorg, Khasi, of India and Ponkan of China.
2. ***C. unshu:*** Japanese origin-seedless, cultivars are **Satsuma mandarins** of Japan and Owari, Kara and Silver hill.
3. ***C. deliciosa***: **Willow leaf mandarin**.
4. ***C. nobilis***: natural Tangor, Kunemboof Japan, **King Mandarin** of USA.
5. **C. reshmi (Coleopatra)-** It is originated in China, fruit produce in clusters.
6. **C. temple- Temple (**tangerine x C. sinensis
7. Kinnow mandarin: ***King x willow*** leaf
8. Wilking- (Willow leaf x king)

Varieties of Mandarin-1

Ponkan or Nagpur Mandarin or Nagpur Santra, Coorg Mandarin, Satsuma mandarin,King Mandarin,Willow Leaf Mandarin, Dancy Tangerine, KodaiOrange, Sastuma Mandarin, Willow Leaf Mandarin, Khasi Mandrin, Clementine, Beaury, Campeona, Ellendale, Sutwal, Desi, Darjeeling orange, Campeona, Emperor, Fuetrelles and Laddu.

Kinnow:King orange (Citrus nobilis) x Willow leaf mandarin (Citrus deliciosa),

Orange Group-2

1. *C. **sinensis***: sweet orange Polyembryonic cultivars are Mosambi, Malta blood Red, Sathgudi, Valencia, Pineapple, Washington Navel Orange, Shamouti of Israel, Succari of Egypt, DoblaFina of Spain, Mudkhed (bud mutant of Nagpur mandarin), Hamlin, jaffa, Valencia Late from China.
2. ***C. aurantium***- sour orange. Narthankaai in Tamil, used for pickling).
3. ***C. multifolia*-Multiple leaf orange**
4. C. natsudaidai-Japanese summer grape fruit:

Forbidden Fruit/ Grapefruit/ Breakfast Fruit-3

1. **Pummelo (*C. grandis***–Monoembryonic. Malaysia and Polynesia origin. Leaves pubescent in lower surface, fruits in clusters. Cultivars are kaopan of Thailand and Buntan of Formosa.
2. **Grape fruit C. paradisi-** Grape fruit- south china origin, polyembryonic. Leaves non-pubescent, fruits solitary. Cultivars are Ruby, Marsh, Duncan seedless, Triumph, Sharanpur special. Duncan, Marsh Seedless, Foster, Red Blush, Triumph, Davis, Foster and Burgundy, Whitefleshed, Marsh, Walters, Hudson, Thompson (Pink fleshed), Flame, Star Ruby (Red fleshed).

Acid Group-Lime and Lemon-4

1. ***C. limon:*** Lemon: weakly polyembryonic, cotyledons white. Cultivars are Eureka, Lisbon of USA, Feminello and Monactiello of Italy, Bernia of spain. Lemon oil is very important.
2. ***C. jambheri*:** rough lemon- polyembryonic Indian origin, cotyledons light green, popular rootstock, fairly tolerant to virus diseases.
3. ***C. aurantifolia:*** Acid lime or sour lime- Polyembryonic, cotyledons whitish- popular cultivar is Kagzi lime-susceptible to tristeza and canker.
4. ***C. medica**:* citron – Indian origin – Monoembryonic, Persisting style. Citron- Persian apple
5. ***C. karna:*** KharnaKhatta- Popular rootstock, cotyledons white.
6. ***C. limonica*:** Rangpur lime-hardy- popular rootstock, tolerant to tristeza and also salt.

Other Related Wild Species are

7. ***C. indica-*** Indian wild orange with inedible fruits
8. ***C. latipes-*** Khasi papeda- cold tolerant.

C. macroptera– Melanasianpapeda- has medicinal value

9. ***C. ichangensis-*** Ichangpapeda- cold hardy, fruits inedible.
10. ***C.assamensis***-Admajor/Gajanimbe*(C.pennivesiculata var assamiensis)* is resistant to greening.
11. **Tahiti or Persean lime**: ***Citrus latifolia***is a seedless triploid.
12. **Sweet lime: *Citrus limettoides***

Varieties of Lime and Lemo

Mithachikna, Mithotra, Anamalous, Baramasi, Gandhraj, Seville, Pramalini, Villafranka, Nepali round, Nepali oblong, Pusa Udit and Pusa Abhinav, Lucknow seedless, Assam lemon, Eureka, Lisbon Kagzikalan, Pant lemon-1, Kagji lime, Pramalini, Chakradhar seedless, PKM-1, Sai Sarbati, Jai Devi, Vikram, Saisarbat, Chakradhar. Selection-49, Seedless lime, Balaji, NRCC-7, NRCC-8, Phule sarbati and Ganganagar lime.

Related Genera of Citrus

1. **Ponicirus**-Trifoliate orange, fruits inedible, Polyembryonic rootstock.
2. **Fortunella**: Kumquat- the species are margarita, japonica, errasiflora, hindsii- Polyembryonic plants ornamental with small oval fruits.

Intergeneric Hybrids

1. **Citrange-** Trifoliate orange x *C. sinensis*, cultivars are Troyer, Carriyo, Morton, Stonia, Rusk, Coleman.
2. **Citrange quat-** (*P. Trifoliate x Kumquat*), Citrange xfortunella- trigeneric hybrid (kumquat)
3. **Citrangedin:** (*Ponicirus trifoliate x C. sinensis*) x *C. mitis* (calamondin) – bigeric hybrid.
4. **Citrangor**- citrange x *C. sinensis*
5. **Citrange**- citrange x ponicirus trifoliate
6. **Citrandarin**- *P. trifoliata x C. reticulata* (mandarin).
7. **Citrumelo**- *P. trifoliata x C. paradisi* (grape fruit).
8. **Citermon-***P. sp x C. aurantium*
9. **Citrumquat-***P. sp x C. japonicum x F. margarita* (kumquat)

Hybrids of Fortunella (Kumquat)

1. **Procimequat:** *F. japonica x C. aurantifolia (acid lime) x F. hindisi*
2. **Limequat:** *C. reticulata x F. japonica x F. margarita*.

Intrageneric Hybrids

1. **Tangor:** *C. reticulata x C. sinensis*, cultivars are temple, clamentine, montreal, Umatilla, monoembryonic
2. **Tangelo:** *C. reticulata x C. paradise*, cultivars are Orlando, Sampson, Seminole.
3. **Lemonima:** *C. limon x C. aurantifolia*
4. **Lemmonnage:** *C. limon x C. reticulata.*

Sweet Orange

1. Second largest citrus fruit in cultivation and commercially grown in India.
2. **AndhraPradesh** (Ananthpur, Cudappah, Nalagonda, Mahaboobnagar and Chittor district)
3. **Maharasthra:** Marathwada, Ahmednagar, Pune and Nasik.
4. Karnataka, Punjab, Haryana, Rajasthan

Propogation of Sweet Orange

1. Budding (T or Patch) during Jan-March or Sept-Oct.
2. Jatti Khatti (*C. jambheri* x Karnakatta (*C. karna*) are popular stocks for blood red in Punjab.
3. Rangpur lime is popular rootstock for mandarins and sweet oranges.

Planting

Pits of 60 cm cube are dug and filled with soil + FYM + Carbofuron and planting is normally done during monsoon season, spacing is 6m x 6m.

Manure and Fertilizers

The bearing tree should be given the fertilizer dose at the ratio of 550gm: 370gm: 550gm NPK/plant/year, depending upon its performance. A graded dose of fertilizers can be applied from the 1st year to the 10th year. Fertilizers should be applied in a ring 30-40 cm wide just below the canopy of the tree at a distance of at least 1-2 m from the trunk. **Citrus is a micronutrient-loving plant.**

Application of Micronutrient

- Zinc sulphate - 2.25 kg
- Copper sulphate – 1.45 kg
- Magnesium sulphate – 0.90 kg
- Manganese sulphate - 0.90 kg
- Ferrous sulphate – 0.90 kg
- Boric acid – 0.45 kg
- Slaked lime – 4.0 kg
- Urea – 4.50 kg
- Applied in 450 liters of water 2-4 sprays every year on the new flush of fully grown leaves.
- Two sprays minimum during the new flush is recommended
- 1st spray-on new flush
- 2nd spray-after the leaves on new flush fully expand.

Irrigation

Irrigation should be regular during fruit development. Water should never come into direct contact with the trunk of the tree. For this reason, the double ring or check bund method is best suited for this crop. Presently, drip irrigation is becoming popular, which helps in saving irrigation water. Stopping irrigation 1 or 2 months prior to flowering is beneficial to the crop, until the tree withers and drops half of its leaves.

Interculture

During the pre-bearing stage of the plants, vegetables other than solanaceous crops can be grown, taking care not to waterlog the soil around the trunks of the plants. Leguminous crops are the best intercropping with 'pea' was found to improve the yield of sweet orange. Cucurbits can also be grown successfully.

Weed Management

Weeds can be controlled with a pre-emergence spray of diuron @ 3 kg/ha twice at 120 day intervals. Other weedicides used are simazine, atrazine, bromacil, 2, 4-D. etc.

Pruning and Training

No regular pruning except removing dead, diseased and overcrowding branches after harvesting of the fruit. Plants should be trained during the first 3 years to have a well distributed frame work at 1 metre height on a single trunk.

Harvesting and Yield

Sweet oranges take 9–12 months for maturity. It is non-climacteric, and should be harvested only after full maturity of the fruits. The yield varies from 500-2000 fruits/tree depending upon the variety.

Handling and Storage

Fruits can be stored for 20 days at room temperature by dipping them in 500 ppm Benlate or 0.1 % carbendazim (Bavistin). Malta fruits can be stored for 2-3 months at 4.4°C Sathgudi for 4 months at 2°C and Mosambi for 3 months at 5°C and 85–90% RH. Post-harvest handling consists mainly of washing, drying, storing, grading and wrapping in tissue paper and packing in corrugated boxes.

Physiological Disorders

Fruit Drop

Citrus drops flower and fruits at different stages as a natural means of adjusting to its resources. These are normal and may not affect the yield. Pre-harvest drop of well-grown fruits prior to maturity is a serious problem, reducing the yield.

The main causes may be

I. Climatic factors

II. Improper water management

III. Lack of nutrition

IV. Relation of seed to fruit drop.

Control: Spraying 10 ppm 2, 4-D, 2 months before harvesting. Fruit drop may be due to presence of pests and diseases also which can be controlled by spraying 1 % Bordeaux Mixture.

Mandarin

Mandarin occupies 50 % of the citrus area in India. Nagpur santras are grown in the Satpur hills of central India, the hilly slopes of Darjeeling and Coorg. Nagpur mandarin was introduced in India in 1894 by Shriji Raja Bhosle. Kinnows-Rajasthan, Punjab, Haryana, H.P, J & K and U.P, Wynad, Nilgiri, Palni and Shevoroy hills of South India, hills of Meghalaya, Mizoram, Tripura, Sikkim, Arunachal Pradesh and Bramhaputra Valleys.

Climate and Soil

Mandarins grow well in frost-free tropical and sub-tropical regions with an elevation of 370-1500 m, average temperatures of 10–35°C and an annual rainfall of 100–1200 cm. Hot winds and severe heat during flowering and fruit set result in a heavy loss of yields. (The Kinnow cultivar requires light chilling for good yields). Mandarins grow in a wide variety of well-drained soils, but light loamy soils are best. It is grown in a soil pH range of 4.5-9.0 depending upon region and type of soil, but 6.0-8.0 is an ideal pH.

Cultivars of Mandarin

- **Coorg:** Fruits medium to large, bright orange, obvate globose, base depressed, rind thin (matures during Feb-March) - Coorg and Wynad
- **Khasi:** fruits depressed, globose to ovate, orange yellow to orange, smooth surface even based or short necked- grow in Assam, Meghalaya, and North Eastern states.
- **Nagpur-** fruits medium, subglobose, coloured, smooth surface, base slightly drawn out (matures Jan-Feb) Nagpur is the most popular variety.
- **Kinnow:** (King x William leaf) fruit medium obvate, base flattened, deep orange yellow, thin rind adhered to segments- irregular bearing, matures in mid-January.

Propagation

By seeds using nuclear seedlings and by T budding on Rangpur lime, rough lemon, Cleopatra, Troyer Citrange, and Kharna Khatta rootstocks.

Planting

The monsoon is the best season for planting. Pits can be 50 cm cube. Spacing is 4-6 m. At Kinnow on Troyer Citrange can be planted at 1.8 m spacing (3000 plants/ha). Kinnow can be grown in high density planting by using the cultivar "Troyer Citrange" as a root stock and spacing the plants at 1.8 x 1.8 m.

Manuring

A fertilizer dose of 450 g of nitrogen, 450 g of phosphorous and 900 g of potassium per plant in two split doses The 1st dose in May and the 2nd after harvest may be given from the beginning.

Micronutrient are Recommended

- Zinc sulphate - 2.5 kg
- Copper sulphate – 1.5 kg
- Magnesium sulphate – 1.0 kg

- Manganese sulphate - 1.0 kg
- Ferrous sulphate – 1.0 kg
- Boric acid – 1.0 kg
- Slaked lime – 1.0 kg
- Urea – 4.5 kg
- Dissolved in 450 liters of water and may be sprayed on new flush of growth once in a year.

Irrigation

Irrigation may be 10–15 days in winter and 5-7 days in summer. Drip irrigation was found to be more beneficial. Water stress during growth, flowering, fruit set and development should be avoided. In south India, mandarins are grown under rainfed conditions in heavy rainfall areas.

Training and Pruning

Mandarins are better trained as low-headed plants (50 cm) with a single stem. Pruning in bearded trees is mainly to remove dead, diseased and overcrowding branches; water shoots; and stock sprouts after the harvest of fruits.

Crop Regulation (Bahar Treatment)

Under south and central Indian conditions, mandarin produces 3 flowerings in a year during Feb. (Ambe bahar), June (Mrig bahar), and October (Hasta bahar) of which Mrigbahar is preferred. So, flower regulation is done by withholding water a month or two before flowering, till the plants wither and drop some of their leaves. Then they are manured and irrigated, which results in profuse flowering.

Harvesting

Mandarin, being non-climacteric, should be harvested at the right stage of maturity only. Mandarin starts bearing from 4 years but commercially yields can be obtained from 10-12 years. The harvesting season starts in August and extends up to April, depending upon the locality and the variety grown.

Yield

Yield may vary from 500-1000 fruits/plant. Dipping the fruits in 50 ppm ethrel develops a golden yellow colour in 5 days. Mandarins can be stored for many months at 8–10°C and 85–90% RH. Under room temperature, they can be stored well for 3–4 weeks.

Citrus Decline

- Serious problem of citrus industry of the world.
- Citrus decline is not a disease but it is a symptom of expression of many diseases and disorders.

There are Number of Reasons which are Responsible to Tree Decline

- Wrong selection of site heavy soils, soils having hardpan, high salts, water logging.
- Nutritional deficiencies: Nitrogen and or micro nutrient.
- Fungal Diseases-Gummosis.
- Improper orchard management practices- lack of fertilization, improper use of irrigation, heavy cropping etc.
- Insects and pests.
- Viruses-Tristoza and Greening.
- Nematodes. (citrus nematode & burrowing nematode)

Control for Citrus Decline

- Removing all dead wood before new growth starts
- Spraying Bordeaux Mixture and
- Covering all cut surfaces with Bordeaux paste.
- Applying more organic manures to the plants.
- Regularly following the recommended plant protection
- Spraying 0.3 % zinc sulphate and B. M. whenever new growth appears.
- New plantations should be with plants grafted on Rangpur lime which is reported to be highly tolerant to tristeza.

Granulation

Most of the Mosambi, Hamlin, and blood red cultivars are mostly affected by this disorder. Granulation is drying up, becoming hard with a greyish colour and enlargement of the juice vesicles with an increase in pectin, lignin, etc., resulting in a reduction of the juice content. This occurs more in young, vigorous trees than in old trees.

Factors Affecting Granulation are

Climate (humid) Cultivars Rootstocks Mineral Nutrition Enzymes and plant growth regulators Crop load.

Control

1. No successful method, but the following can be tried as remedies.
2. Spraying lime, $ZnSo_4$, Bordeaux mixture or $ZnSo_4$ + $CuSo_4$ (0.5%).
3. Early picking reduced irrigations spraying 200-500 ppm lead arsenate

Fruit Drop

Mandarins also drop the flowers and fruits at different stages, adjusting to their resources. But preharvest fruit drop reduces the yield. It can be controlled by proper cultural management and treating with 2,4-D (10 ppm) or NAA (5ppm) or 2,4,5-T (5 ppm) and Aureofungin (20 ppm) to check any fungal diseases.

Limes and Lemons

Origin: Limes -India, Lemons –Malaya

India is the largest producer of acid limes in the world but ranks fifth in the production of limes and lemons. Acid lime is the third important citrus fruit after oranges and mandarins. It is cultivated in almost all states; A.P., Maharasthra, T.N., Karnataka, Gujarat, Bihar and H.P, are the major ones. Commercial lemons are grown in Punjab, Rajasthan and Uttar Pradesh.

Climate and Soil

Acid lime grows well in areas which are free from frost, strong winds, and are warm and moderately humid. Can grow upto 1000 m elevation with annual rainfall not exceeding 750 mm. Sweet limes can tolerate frost better than acid limes. Rangpur lime grows well in areas with 20-30°C. Lemons can be grown in both humid and semi-humid areas up to 1200m elevation and tolerate frost better than acid lime. Lime grows well in deep (2–2.5 m) well-drained soils rich in organic matter. Optimum pH is 6.5-7.0. It cannot grow in waterlogged soils, alkaline soils with high lime content. Lemons grow better in shallow soils and if well drained.

Lime Varieties

1. **Kagzi lime** is the best most popular one.
2. **Pramalini, Vikram, Saisarbati-** supposed to be canker free and prolific bearers.
3. **Chakradhar:** seedless strain of acid lime
4. **PKM-1**: heavy yielder.
5. **Selection 49:** tolerant to canker, tristiza and leaf miner. It is also a prolific bearer.
6. **Seedless lime:** fruits are oblong-prolific bearer.
7. **Tahiti lime-** plants nearly thornless, fruits also much larger- triploid.

8. **Mithachikna (Thin rind) and Mithotra (Thick rind):** Two varieties of sweet lime with less acid, sweet fruits and are mostly used as stocks.
9. Rangapur lime is grown for root stock purpose.

Lemon Varieties

1. **Eureka:** Fruit elliptic to oblong, base necked.
2. **Lisbon:** Fruits oblong, apex nippled
3. **Villafranka:** Fruits oval-oblong, apex pointed, bear more seeds.
4. **Lucknow seedless (Assam lemon):** Fruit oblong, apex nippled. seeds few or absent.
5. **Nepali round:** nipple has a scar.
6. **Kagzikalan:** fruit spherical, apex slightly nippled.
7. **Pant lemon-1**: Selection of Kagzi kalan tolerant to canker tristiza and die-back, selfcompatible.

Propagation

1. Acid lime is propagated by seeds
2. Sweet limes can be propagated by sub-terminal stem cuttings or layerings or budding.
3. Lemons by budding on trifoliate orange or Rangpur lime.
4. Trifoliate orange: Resistant to Phytophthora and nematodes.

Planting

It can be done before or after rains during the rainy season depending upon amount of rainfall. Spacing should be 4-6 m for limes, and 6-8m for lemons. Pits can be of 1 m cube and filled with mixture of soil and manure.

Training and Pruning

Limes may be trained as low headed plants. Pruning is mainly to remove dried, diseased and overcrowding branches. Grow up lemons may be given light pruning every year to remove already fruited shoots.

Manuring

Fully grown lime may be given 5 kg FYM, 900 g N, 250 g P and 500 g K per plant per year. Total P, half of N and total K after rains and remaining dose should be given after flowering. Organic manures are more preferable than chemical fertilizers. Lemons may be given 500g N/plant/year only. Composite spray of micronutrients as given for sweet orange may be given once a year on the new flush.

Irrigation

Limes require more water; irrigation may be given if top 25 cm soil becomes dry. Check bund system is best as it will not allow the contact of water with the stem. Drip irrigation improves yield.

Interculture

Mulching basins will be very beneficial. Weeds can be controlled by using weedicides like Monouron, Diuron, Gramaxone.

Harvesting and Yield

Harvesting season extends from **June to September** depending upon the locality for acid lime and May for lemons, June-Aug for Rangpur limes. Lime yields **2000-5000** fruits/plant. Acid limes can be stored for 6-8 weeks at 8.3 to 10°C and 85.9 % RH and lemons for 8-12 weeks at 7.2-8.6°C and 85-90 % RH.

Diseases

1. Greening

Caused by *mycoplasma Chlorosis* of leaves, thickening of leaf, shortening of internodes, off season blooming. & die back. Fruits small. Spread by vector *Citrus psylla*.

Controlled by

Removing infected trees

1. Control *Citrus psylla* by pesticide.
2. ***Phytophthora Rot*** (Root rot, collar rot, crown rot, gummosis, etc) Fungal **Most serious, soil inhabitant**

   Controlled by raising nursery in disease free condition, avoid water logging; paste bordeaux paste 60 cm on the trunk. Use tolerant root stocks like Sour Orange, Trifoliate Orange, etc. Apply qurofunsin (1 g in 10 lit water)
3. **Citrus canker**. Bacterial: Wide spread in India. Serious in Acid lime. Leaves, twigs, fruit & thorns affected. Fruits of poor quality. Controlled by pruning & burning of diseased parts. Spray Bordeaux mixture (10:5:100) to new flushes. Spray 500 ppm streptomycin sulphate.

Pests

1. **Citrus leaf miner:** Serious pest in all species. Spray Monocrotophos (0.03%) weekly during mansoon.
2. **Lemon Butterfly:** Larvae feed on leaves, fruit drops. Dusting of BHC (5 %) or spraying carbaryl (0.1%).
3. **Fruit sucking moth:** Suck juice from fruit, fruit drops, unmarketable. Spray monocrotophos (0.03%)

21

Custard Apple

- Custard Apple/Sugar apple fruit /Fruit of poor people
- B. N.: *Annona squamosa*
- Family: Annonaceae
- Chromosome No.: 2n = 2x = 14
- Origin: Tropical America
- Inflorescence: Solitary
- Fruit type: Aggregate (Etario of berries)
- Edible part: Pericarp

Introduction

It is found growing wild in the sub-tropical and tropical parts of India. The edible part of annona is the areoles. It has about 120 species, five of them having pomological significance, and four of the five species are important in commerce. Custard apples are cultivated mainly in Maharashtra, Madhya Pradesh, Gujarat, Karnataka, Andhra Pradesh, Uttar Pradesh, Bihar, Assam, and Orissa.

Climate Requirement

The custard apple is a hardy plant that prefers dry climates with mild winters and is very sensitive to frost. The tree goes into dormancy in **November-January** after harvesting. The tree can be grown successfully at 1000 m above mean sea level. A moist climate is favourable for proper growth and development of fruits. The plant can tolerate extremes of heat, but the yield goes down as shedding of flowers occurs above 39°C.

Soil Requirement

Custard apple grows well on sandy, rocky gravel, even on heavy soils. It does not require deep soil, but drainage should be proper. Soils rich in organic matter with a pH of 6.0 – 6.5

Annona Species

- ***Annona cherimoya***- Cherimoya/Lakshmanphal - Requires a lower temperature and a subtropical mild climate, considered to be best fruit of annonaceace
- ***A. squamosa***- Custard apple/Sitaphal/Sweet Sop/Sugar apple
- ***A. senegalensis***- wild soursop / African species
- **Atemoya/Hanumanphal-** Atemoya is a cross between *A. squamosa x A. cherimola.* (hybrids between custard apple and cherimoya) have the superior tastes of cherimoyer can tolerate fairly higher temperature just like A squamonsa.
- ***Annona Montana***- Mountain Soursop, Native to the West Indies, Peru and Brazil cultivated in South America and Philippines, Propagation is exclusively by seed.
- ***Annona reticulata***- Bullock's Heart (West Indian Custard apple, Ramphal, 'anta', 'lavani' and 'Krishnabeejam' are more common grown in South Indian than North Iindia, does not tolerate severe summer when compared to sitaphal, and is a good rootstock for *Annona squamosa and Annona cherimoya*
- ***Annona glabra***- Pond apple /Monkey apple/ is known as drought tolerance.
- ***A. muricata*** - Soursop/Prickly custard apple/MundlaSitaphal/Mullu Ramphal/Mamphal: (evergreen tree, heart shaped fruit, Among the annonaceous largest annona in the world), incompatible with *A. squamosa and A. cherimoya.*
- **Illama (*Annona diversifolia*)** Fruit pulp like apricot flavor.

Varieties

- **Hybrids Atemoya:** Hybrid between *A. squamosa x A. cherimoya*– Natural hybrids and manmade fruit, Shy bearing habit
- **Balanagar:** It is a cultivar of *A. squamosa.* It is a popular variety of Andhra Pradesh.
- **Red sitaphal:** It probably originated as chance seedling. Fruits of this cultivar through belong to *A. squamosa* are pinkish dark with red pulp.
- **Mammoth:** It is a cultivar of *A. squamosa.*
- **African Pride:** It is a cultivar of *A atemoya* (Cherimoya x Sugar apple). It is a popular variety of subtropical region of Australia.
- **Pink Mannoth:** It is another popular cultivar of atemoya. The pulp is similar to cherimoya.
- **Arka Sahan**: Island Gem (Atemoya) x Mammoth (Custard apple) Artificial interspecific hybrid
- **APK (Ca)-1**: It is a clonal selection, Tamil Nadu.

- Islander, Israel hybrids, Washington, Mahaboobnagar, Kakarlapahad, Local Sitaphal, Saharanpur Local, Barbados seeding, British Guinea.

Propagation

It is commercially propagated by grafting and also propagated by seed, the seed should be sown, after removal from ripe fruits. The seeds of cherimoya treated with 10000 ppm GA3 significantly increased germination by 70%. Fresh seeds germinate in **20-30** days.

Planting

Spacing

- Cherimoya – 6.0 x 4.0 m
- Custard apple – 3.0 x 3.0 m
- Sour soup – 4.0 x 4.0 m
- Bullock's heart – 5.0 x 5.0 m

Manures and Fertilizers

Application of castor cake and bone meal, in 2:1 ratio is found beneficial. Application of 250:125:125 g NPK in two splits along with 50 kg FYM/Plant/ Year.

Irrigation

Custard apples are mainly grown as a rainfed fruit crop in India. This crop benefits from the rainfall that is received during the fruiting season. One or two irrigations during fruit development improve the size and quality of fruits.

Training and Pruning

Newly planted seedlings/grafts are supported with wooden sticks. Generally, pruning is not recommended in custard apples. Fruits are borne on **old as well as current-season** growth. Light pruning helps in inducing better branching. Pruning can be done to avoid overcrowding or to remove dead/diseased wood stems. Attend a single stem training system.

Flowering and Fruiting

Annona plants start fruit bearing after 3 years of planting and economic yields start from 5-6 years after planting. Flowering periods extend from spring to the end of the rainy season.

Factors Affecting Fruit Set

Many flowers failure due to lack of pollination. It is desirable to set fruits early in the season, but in India the early flowers produce no pollen and hence natural pollination is difficult. Hand pollination gives about 85 percent fruit set compared to natural pollination's 30 percent.

Dichogamy

The stigmas are not receptive when the pollen is shed and the receptivity also remains for a short period. Some custard apple trees shed pollen on the morning and others in the afternoon. Protogyny is observed in Annona except in *Annona muricata* (Protandry). Annona seedlessness is due to **ovule sterility**. An increase in fruit set is possible by the application of NAA at 20 ppm during flowering for 3-4 times at an 8-10 day interval and spraying with GA_3 at 50 ppm.

Maturity Indices

Custard apple fruits are climacteric fruit, therefore they are harvested when they are mature, firm and plump. On maturity, the fruits turn light green. The inter aereolar space widens, the fruits turn creamy white, the skin between the segments or tubercles turns a light yellow colour.

Harvesting and Yield

In dry regions, custard apples mature and ripen during October-November and December, whereas in humid areas, harvesting is done during August. A good bearing tree gives about 100-150 fruits, *i.e.,* about 8-10 tonnes/ha. A centrifugal process was developed by (CFTRI) for the separation of the gritty portion from pulp.

Pests and Diseases

- **Collar rot and Root rot:** *Pithium* sp., *Phytophthora* sp.
- **Leaf spot:** *Cercospora anonae*
- **Fruit rot**: *Phomophsis anonacearum*
- **Fruit rot: Phytophthora** *citrophthora and p. nicotiana*-Major disease of custard.
- **Anthracnose**: ***(Glomerella cingulata)*** Infection begins at blossom-end of the fruit.
- **Mealy bugs**: (*Ferrisia virgata, Maconellicoccushirstus*). These cause blemishes on fruits

Physiological Disorders

- **Stone Fruits-** Remain fruits turn brown, become hard is major physiological disorder.
- **Woodiness-** Presence of seed pockets and gritty lumps in flesh.
- **Brown pulp-** Discoloration of pulp are the physiological disorder of custard apple
- **Tree decline-** Due to water **stagnation**
- **Poor fruit-** Set is problem in custard apple due to **dichogamy**.

22

Datepalm

- Datepalm/Century plant/Oldest plant/Tree of desert oases
- B. N.: *Phoenix dactylifera*
- Family: Palmae/Arecaceae,
- Chromosome No.: 2n = 36
- Origin: Mediterranean region (Iraq).
- Inflorescence: Spadix
- Type of fruit: Single seeded berry
- Edible part: Pericarp

Introduction

The flesh contains 60–65 percent sugar, 2 percent protein, 2.5 per cent fiber, 0.4 per cent fat and 2 per cent mineral matter. Date palm contains 75-80% carbohydrates in the form of invert sugar. It is a good source of easily assimilable iron (7.3 mg/100g) and is hence prescribed to anaemic patients along with honey. It fed on running water and its head on the fire of the sky. Monocotyledonous, unbranched stem growing plant. Fruit crops are dioecious, and fruit is a highly nutritious fruit. One Kg of fully ripe fresh dates provides approximately 3150 calories. source for jaggery and a drink-toddy. Suitable places in India: Northern-Western districts. The drink of date palm is known as Dibbis. Liquor prepared from date palm, is Arrack popular in Iraq.

Climate and Soil

Dates can be grown successfully in areas with a long hot summer and mild temperatures during flowering (February to April) and fruit ripening (May to August), with temperatures ranging from 25 to 29°C. It suffers greatly from early monsoon showers. There should not be any rain or high humidity during fruit maturity and ripening. Otherwise, the fruits will be spoiled. The crop is susceptible to frost. Deep sandy loam is the best suited, though it is very hardy and can be grown in a wide range of soil conditions. Better water holding capacity with good drainage is desirable. It can grow in alkaline and saline soil. However, the growth and productivity are affected. Date palms tolerate high soil salinity (pH 9–10).

Species

1. Closed related species found in India ***Phoenix sylvestris***
2. Wild species found in Western and Eastern ghats hills: *Phoenix humilis*
3. *Phoenix humilis* and *P. acaulis* can be used in dwarfness.

Varieties

- **Exotic cultivars:** Halawy, Barhee, Khalsa Khunezi
- Sevi are best for fresh eating (dessert).
- Medjool, Khadrawy and Sharan for preparation of Chhuharah (boiled and dehydrated doka stage fruits).
- **Zahidi** is for the preparation of **pind khajoor** (soft date).
- **Sharman** cultivar to date exhibit uneven ripening.
- Cane sugar dates (semi-dry dates): Dayari, Deglet Noor and Zahidi.
- Invert sugar date (soft date): Halawy, Khadrawy, Bahee and Medjool
- Date juice suitable varieties include: Zagloul and Hayani.
- Dry dates (*Chhuhara)* and soft dates (*Khajoor)*
- Chhuhara making: Khadrawy, Medjool and Hsaran
- Fresh eating: Halawy, Barhee, Khalas and Khunezi.
- **Medjool-Orange** yellow with reddish brown stippling at doka stage.
- **Khadrawy:** small to medium-sized fruits; light yellow at doka stage.
- **Barhee:** Mid –season to slightly late cultivar yellow colour and low astringency at doka stage.
- **Sharan-**For preparation of Chhuhara (boiled and dehydrated doka stage fruits)
- **Halawy:** At full maturity before ripening (doka stage), the fruits are yellow in colour and free from astringency.
- **Thoory-**Bread type/dry date
- Khadrawy and Medjool were developed by somatic embryogenesis.
- Shamran (Sayer), Khadraway, Medjool, Hayany, Barhee, Khalas, Zahidi (Mid-season variety).
- **Early ripening varieties:** Halawy and Khunezi
- **Soft varieties:** Barhee, Halawy (tolerant to rain, early soft variety)

Propagation

Vegetative propagation through off-shoots produced by a female mother plant of a particular cultivar can be used for planting. After about 4–6 years, 2–3 offshoots can be obtained from a mother palm for a period of 8–10 years. Offshoots separation time: February to March and August to September. To encourage rooting, the bases of off-shoots arising from the mother plant are applied with moist soil by putting in soil boxes at least for a period of one year before separation. Commercial propagation: offshoots (ideal weight: 10–25 kg is preferred)

Planting

Planting of off-shoots can be done during late spring or early summer by cutting with the help of a specially designed chisel. Most of the basal leaves are removed and only the terminal bud with 10-12 leaves around it is retained. commercially as spacing of 6 to 8 m is adopted. In India, a spacing of 3 m – 4 m is adopted. To effect pollination under commercial cultivation, 2-3 male plants are planted for every 100 female plants. The rainy season is the best time to plant.

Irrigation

Light and frequent irrigations have to be given immediately after planting and the basins can be mulched. Though date palm is drought tolerant, 2-3 m around the root zone needs to be kept moist for maximum growth. The irrigation frequency varies with the season. Date palms are highly tolerant to saline water (even up to 2500 ppm). But to overcome the drainage problem arising out of saline water usage, periodic leaching with food water should be done.

Manure and Fertilizers

Each palm should be applied with 50 kg of FYM, 600 g N, 100 g p and 700 g K.

Pruning

In date palms, older leaves that are more than needed are typically pruned. About 75-100 leaves are found to have an optimum good yield. The leaf pruning is done during June. By this pruning, the bunches will be better ventilated during July and early August, which will be otherwise spoiled. The spines from the leaves around the bunches are also cut during spring to facilitate pollination. Ethephon is an effective fruit thinning.

Flowering

Bearing in date palm starts from 4-5th years and reaches maximum by 10-12 years. Date palms have a tendency to alternate bearings. The date palm needs 3300 units of heat. (A unit of heat means a degree above a daily mean of 180°C between flowering, fruit development and the ripening period).

Pollination

Date palm is a highly cross-by hand pollination. Hand pollination is done by inserting 2-3 strands of male flowers between strands of female flowers when the female spathes have cracked open. 10 % of males should be raised in orchard to provide adequate pollen grains. Though the dried pollen can be stored at 4–5 C until next season, the fresh pollen produces the best fruit. The spathes emerge during February–March and flower opening starts during March–April. Immediately, the flowers should be pollinated (2–3 days after spathes open). Pollen grains of certain varieties can advance ripening of certain varieties. A trailer-mounted palm duster is used for pollination. Specific pollinizer varieties should be identified for specific varieties of female plantation.

Fruits Set

Fruit thinning has to be resorted to so as to retain 1300–1600 fruits in 8–10 bunches per palm. This will be optimum. Bunch thinning can be done either by the removal of an entire strand or by shortening of strands. Spraying ethephon @ 200 ppm 10–30 days after fruit set will help to thin fruits effectively. It also helps to overcome biennial bearing, encourage earlier ripening and get better fruit weight and soluble sugar. Metaxenia (pollens have a direct influence of somatic tissue of fruit outside the embryo and endosperm) results in early or late ripening, change in size and colour of fruits and seed. Exploitation of "Metaxenia" from pollen of these species has great scope to induce earliness and improvement in quality.

Stages of Fruit Development

Sr No.	Indian Name	Arabic Name	Stage	Quality
1.	Gandara	Chimiri	4-13 Weeks after pollination	Hard, Green Colour
2.	Doka	Khalal	13-17 weeks after pollination	Hard, yellow pink or red may be sweet or astringent, Edible stage.
3.	Dang	Rutab	17-21 weeks after pollination	Fruits soften at tip, edible stage.
4.	Pind	Tamer	21-22 weeks after pollination	Fully ripe 60-84% TSS, edible stage.

Harvesting

In India, dates are harvested at the Doka stage, whereas in other countries they are harvested at the Dang stage. A pind stage is never attained in India. In India, dates are harvested at the Doka stage (70-80 % moisture). It is useful for the processing of chhuhara. For fresh consumption, dates are harvested at the peak of the doka (colour turns green to yellow) stage. For fresh eating, dates are harvested or preferred at the dang (translucent and starting softening) stage. For storage purpose dates are harvested at tamar or pind (attain full mature) stage. Chhuhara' which is the commercial date or produce of commerce. The Chhuhara recovery would be 33–35%. Fruits harvested at Doka stage can be softened and ripened within 24 hours by spraying 1-4% NaCl alone or in combination with 0.05 % acetic acid. Fruit ripening in date palm can be advanced by the removal of one third of strands at the pea stage of fruits from the centre of branches. Under Indian conditions, since the fruit ripening period is not free of rain, the fruits have to be harvested at the doka stage during June–August. Soft varieties of date palm are also referred to as invert sugar dates, while semi-dry and dry are cane sugar dates. It is based on the conversion of sugar from cane sugar to inverted sugar.

1. For early ripening: Ethrel @ 500-100ppm
2. Date harvested at stage have 70-80 moisture

Yield and Processing

The fruits harvested at doka stage have to be dipped in boiling water for 5 minutes followed by dehydration in an electric oven at 50 °C for 160 hours or in a solar drier to obtain the best quality "Chhuhara", which is the commercial date or produce of commerce.

1. The Chhuhara recovery would be 33 – 35%.
2. Average yield: Rain fed condition 40-50 kg/palm and Irrigated condition-200kg/palm.

23

Dragon Fruit

- Dragon fruit/Cactus plant/Night blooming cactus/Night blooming cereus/Pitaya/Strawberry pear/Belle of the night
- B. N.: *Hylocereus undatus*
- Family: Cactaceae
- Chromosome No.: 2n = 2x = 22
- Native Place: Central America

Introduction

Pitaya also known as Dragon Fruit is a superfruit indigenous to central and South America that is grown all over the world. It has a faint, sweet taste much like an earthy watermelon. It's a fast growing crop and the plant as can have produced for more than seven to eight years. The fruit of this plant having red skin covered with large scale and white or pinkflesh contains some black seeds. It can be used in jams, ice creams, jelly production, fruit juice, and wine. This fruit is famous in Thailand, Vietnam, Israel and Sri Lanka. This fruit cultivation is excellent in regions where less rainfall is expected. This fruit plant is treated as an ornamental plant as well as a fruit producing plant. Dragon fruit contains 73.9 kcal, 1 g protein, 0 g fat, 0.5 g fiber, 23 g carbohydrates, 23 g sugar, calcium 30 mg. 10.9 mg sodium and 1.79 mg vitamin C. Dragon fruit is consumed as a fresh fruit or can be used in jams, ice cream, jelly production, fruit juice and wine. This fruit is also used in face packs.

Climate and Soil

This fruit plant survives in poor soil conditions and temperature variations. However, the tropical climate region is best for its cultivation. This plant requires a minimum annual rainfall of 50 cm and a temperature of about 20°C to 30°C. Too much sunlight is not good for its cultivation. In high sunlight areas, shading can be provided for a better yield. Its fruit can be grown on a wide range of soils, from sandy loam to clay loam, but sandy soil with good organic matter and internal drainage is best for its cultivation. A soil pH of 5.5 to 7 is best for dragon fruit cultivation.

Propagation

The dragon fruits are cultivated in two ways, viz., through seeds or cuttings, let us look into it. If you use seeds for propagation, it will consume too much time. Stem cuttings are the best preference for dragon fruit cultivation. To propagate stem cuttings, a 20–25 cm long stem cutting is suitable. Make these cuttings to get dry just 2 days before planting. These stem cuttings should be treated with a fungicide to prevent diseases. These stem cuttings will be treated with root hormones and then directly planted in the field or plant these cuttings in a 12 x 30 cm polythene bag with 1: 1: 1 ratio soil. Farmyard manure, sand, then keep it dry, shaded area. These cutting roots are preferred for planting after 5–6 months. Cuttings will grow rapidly and may produce fruit 6–9 months from propagation.

Spacing

Keep the plant to plant space of 2 m x 2 m.

Training and Pruning

To get the proper growth and development of the plant, these should be supported by concrete or wooden columns. Immature plant stems are required to tie to these columns. Lateral shoots should be limited to 2 to 3 main stems should be allowed to grow. It is recommended to have a round or circular metal frame to maintain the balanced dragon shrub.

Manures and Fertilizers

Adding manure and fertilizer to dragon fruit has shown a good effect. At the beginning, the plant needs more nitrogen and at the time of flowering, it needs phosphorus and potash. Application of urea can cause stem rot disease in the plant. Therefore, 5-10 kg of good rotted manure of cow dung keep adding on a per plant basis at an interval of every 45-60 days. Apart from this, apply NPK and zinc manure at the time of flowering and fruiting according to the soil test. Biofertilizer or liquid organic manure or fruit stimulator plant-hormones can also be used.

Irrigation

Since the plant requires less water, irrigation is recommended once a week and drip irrigation should be used for better efficiency.

Harvesting and Yield

Generally, these plants start flowering in May -June and bear fruit from August to December month. Its fruits become ready for harvesting after 1 month of flowering. Fruiting time continues till December. Picking up these fruits can be done up to 6 times within this period. Identifying the fruit harvesting stage is very simple as the immature fruit colour is a bright green colour and will turn into a red colour once it has ripened. The exact time for harvesting is after 3–4 days of colour change. But in the case of export, they should be harvested 1 after colour change. Use the sickle or hand to pick the fruits. An average yield of 5–6 tonnes per acre can be expected.

24

Durian Fruit

- Durian/Civat fruit/King fruit of Malaysia and Indonesia
- B N: *Durio zibethinus*
- Family- Bombaceae
- Chromosome No.: 2n = 56.
- Origin: Malaysian region (Borneo)
- Type of inflorescence: Cyme
- Fruit type: Capsule
- Edible part: Aril

Introduction

The durian is an important fruit in several parts of the tropics and is the famous fruit of the far East, where it is widely cultivated throughout Malaysia. It is indigenous to Malaysia, *i.e.*, South East Asia, and is distributed in Malaysia, Indonesia, Philippines, Thailand, Singapore, Kampuchia, Myanmar, India, and Sri Lanka. The term durio is the Malaysian word duri-thorn, and zibethinus is the Italian word for "zibetho," meaning "strong-smelling. Thus, this is a fruit characterised by a greenish, hard, thick rind on the outside and a yellowish, sweet flesh with a strong odour. The fruits are nothing but seeds covered with a cotton-like fibre.

Composition and Uses

The fruits are a rich source of food energy (124 cal), Carbohydrates (28.89 g), Ca (20.0 mg), P(63mg), Fe, Vit. A, ascorbic acid protein (2.5g), fat, etc. Self-incompatibility and heterostyly are observed in durian. Each five-lobed fruit has 2-3 seeds in each locule. The fresh fruits are used for table purposes. The rind is slit longitudinally along the grooves and the inner segments of flesh are taken out to be served as dessert, preserved with salt or made into a sauce. Also used in making durian paste, sweet durian candy or confectionary, preparation of ice- cream. The seeds are roasted and eaten.

Medicinal Importance

As an aphrodisiac, the boiled decoction of roots is used to relieve fever in Malaysia. Leaves are used for curing jaundice. Durian shell placed under beds to discourage bed bugs.

Climate and Soil

Durian trees can grow in deep, well-drained, sandy clay or clay loam soils rich in organic matter with a pH of 5.0-6.5. It requires abundant rainfall (200–300 cm), equally distributed throughout the rainy season. High humidity and high temperature with only a short dry season (24-30 °C is ideal). It grows up to 800 m. near the equator, where 30-50% shade is required for young trees.

Species- The genus durian has 27 spp. 6 spp produce edible fruit, they are.

- Durian- ***Durio-zibethinus***- widely cultivated
- Lai- ***D. kutejensis***- very good fruits with good flavour.
- Karatogan- ***D. oxleyanus***.
- Tabelak- ***D. graveolens***
- Lahong- ***D. dulais***
- Munjit- ***D. Grandiflorus***.

Cultivars

- The selection programme in Thailand and Malaysia involved in evolving > 100 cultivars and few cv. are recommended
- Chance, deception, frog, Longstem, Gibbeon, Cha-Nee, Golden Pillow, Long stem, Mawn-Tawng grown in Thailand.
- Monthong, Kan yao, Luang, kradum thong- grown in Malaysian.
- **Spinless variety:** Davao
- **Long stem-** Mawn-Tawng
- The **chance** root stock resistance to root rot is commonly used in Thailand.

Propagation

Mostly propagated vegetatively by patch budding but also seed.

Planting

Plants are spaced at 10-12 m, apart in orchards and more than one clone or cultivars should be planted for cross pollination.

Manures and Fertilizers

1. 1st year- 1.7:0.4:0.4kg NPK/ha
2. 2-4 years- 23:6:13kg NPK/ha
3. >5 years- 38:8:21kg NPK/ha.

Irrigation

Irrigation is essential if the dry period extends up to 3 months or during flowering and fruit development. In the case of chance cv mild stress is required to initiate flowering and is reported to require a low amount of irrigation.

Pruning

Pruning is done to remove laterals & water shoots, leaving a central leader. It is done after harvesting and propping is provided for heavily cropped trees.

Flowering & Fruit Set

Flower clusters arise from the lateral main branches and occasionally on tree trunk. It produces about 20,000-40,000 flowers per tree during the dry period **(Mar-Apr/Sept-Oct)** Self-incompatibility is common & cross-pollination is required. Flowering habit: cauliflorous major pollinator, bats' long living tree: 8–150 years. Self-incompatibility is noticed in Durian. Application of GA_3 to durian pedicels at 6 weeks after a thesis reduces fruit drop & increased fruit size. The largest producer in the world: Thailand.

Harvesting and Yield

Climacteric to prevent natural fruit abscission, fruit may be tied to the limb or harvested at maturity, *i.e.*, a hollow sound when tapped with a finger. Harvest the fruit by picking. Durian fruits are not normally harvested. Fruits that drop naturally have better flavour and taste. Fruit ripen during May to September. Harvesting time: August to September. Climacteric fruit Commercial yield starts at 12-15 years old tree about 500-1000 fruits/tree/year. Fruits stored at 15°C. The chemicals that are responsible for the flavour are hydrogen sulphide, ethyl hydrosulfide, and dialkyl polysulfide.

25

Fig

- Forbidden fruit/Wholesome/Nutritious and delicious fruit/Peace and prosperity/Anjir
- B.N.: *Ficus carica*
- Family: Moraceae
- Chromosome No.: 2n = 26
- Origin: Western Asia (Arabian)
- Inflorescence: Hypanthodium
- Fruit type: Syconus
- Edible part: Fleshy receptacles
- Fig is a climacteric fruit

Introduction

Ficus carica is widely distributed in tropical and sub-tropical countries. The related genera with edible fruits are Artocarpus, Cudrania and Morus of the family Moraceae. Some important species are *Ficus glomerata, Ficus bengalensis, Ficus religiosa, Ficus elastica, Ficus hispida and Ficus roxburghii.* A fig is a large shrub or low-growing deciduous tree with a short and twisted trunk. Fruits are solitary, axillary green or yellow, pear-shaped.

Area and Distribution

Fig is an important and oldest fruit extensively used in fresh and dried form. It is extensively cultivated in countries around the Mediterranean, especially Turkey, Egypt, Morocco, Algeria, Iran, Italy, Spain, Greece, Portugal, California, and Afghanistan. The main areas of fig cultivation in India are Pune (Maharashtra), Srirangapatna, Raichur, Gulbarga and Chitradurga (Karnataka), Lucknow (Uttar Pradesh) and parts of Gujarat and Andhra Pradesh.

Composition and Uses

The fresh fruit contains 11.5 percent total sugars, traces of iron, Vit. A, Vit. C, Protein, Fat, Calcium, Riboflavin, Thiamine, etc. Dry fruits are very delicious. It is also used for the preparation of jam and jelly. Figs boiled in milk are repeatedly packed against swollen gums. The fruits are also used against tumor and other abnormal growth. Leaves are used as fodder. In southern France, fig leaves are used as a source of perfume extraction called fig-leaf absolute. The edible fruit of fig is botanically called **"Synconium"**, which consists of a hollow receptacle with a narrow aperture called "Ostiole" at tip and numerous small flowers living on the inner surface. The true fruit are tiny drupelets inside the cavity of the fused peduncle.

Climate and Soil

Figs are a subtropical fruit. It behaves as deciduous in temperate and subtropical regions and evergreen in the tropics. Where temperature rarely goes below 4.5°C (South India). It withstands low temperature when it goes into dormancy. Best quality fruits are produced in the region with dry climate during fruit development and maturation. Fruits ripen prematurely if the temperature exceeds 38°C and such fruits have tough skin. Fruit skin may show sunburns, less pulp, and insipid taste. Good vegetative growth is observed at 15.5–20.0 °C. Fig prefers deep soil, clay loam, non-alkaline, medium black soil with well drained and good water holding capacity. A good quality of fruit is obtained on heavy soils with a ph of 5.0-7.5.

Types and Cultivers

The figs are classified into four types based on the nature of flowers and the methods of pollination.

1. Common fig
2. Capri fig
3. Smyrna fig
4. Sanpedro fig

Common fig / Edible / Adriatic fig (mostly grown in india)

The flowers are long pistillate, Fruits develop by parthenocarpy, viz., without the stimulation of pollination and fertilization. Kadota, Mission, Adriatic, Brown Turkey, Poona fig, Deann', 'Conadria, Excel, Celeste, Black Itchia, Turkish White, Kabul, Marseilles, Bangalore, Lucknow and Yercaud Timla.

Wild/capri fig/ Male/ Goat fig (pollinizer)

This type has short-styled pistillate and functional staminate flowers. Most caprifigs are not edible but are grown because they harbour a small wasp, viz., ***Blastophaqa psenes,*** which is necessary for pollination and fruit set in other types like Smyrna fig by transferring the pollen grains from caprifig.

Varieties **Samson, Stanford, Brawley**.

Smyrna fig

It is the most important one commercially. However, the fruits develop only when the flowers are pollinated with pollen from the male flowers of the caprifig, transmitted by the *Blastophaga wasp.*

Varieties -**Calimyrna, Taranimt and Zidi.**

Sanpedro fig

In this type, the first crop is completely parthenocrpic, but the second crop develops only if the flowers are pollinated. Varieties Sanpedro, King Gentile, Dauphine and Lampeiria.

Propagation

Hard wood cuttings are the common method of propagation in figs. Rooting was the best in cuttings from 3-year-old wood, with 30–40 cm in length and 1.5 cm in diameter. Cuttings from the base of the shoot and lower part of the crown have to be used as they root better. Cuttings are taken during January-February at the time of pruning in North India, whereas the cuttings are taken during the rainy season in South India. **Poona fig** is believed to be a hybrid between *Ficus carica and F. glomerata* and is a nematode resistant rootstock of fig.

Planting

A spacing of 5-7 m is recommended depending upon the fertility status of the soil for maximum yield. Planting season varies from place to place viz.

1. South India – August – September
2. Western India – June – July
3. North India – **January – February**.

Training and Pruning

Fig can be trained on a single stem or multiple stems. "Multi-stem" is most commonly followed. Pruning of fig is important because fig generally gives two crops in a year, the first borne on previous year's growth is **breba** and the second crop is on current season growth. Therefore, pruning intensity and its type will depend on bearing habit and type of crop desired. The trees are headed back to about 2 buds of previous growth regularly to keep them dwarf. This reduces the production but improves the quality of fruits.

Notching

It is also practiced to force new fruit-bearing shoots from the lower portions of the branches; for this purpose, a 0.5 cm wide notch of bark is removed above two buds on the middle portion of the branch in July. Application of HCN (Hydrogen Cyanide) at 1.5–2.0% advanced the date of bud burst.

Manure and Fertilizer

Fig responds to heavy manuring, and 50–75 kg. A FYM for the bearing plant is recommended. For 1-2-year-old plants, about 75:50:50 g NPK is sufficient and should be doubled 5th year onwards. A dosage of 300:200:200 g NPK/plant gives good yield and quality fruits.

Flowering

Three Types of Buds Occur

Flower buds, which are spherical, have 3-5 scales, and these buds produce parthenocarpic fruits only. Mixed buds are conical, made up of 5-8 scales and these buds also produce fruits. Vegetative buds are made up of 3-5 scales which usually remain dormant and rarely produce shoots. Flowers appear on the plants even at two years' age, but considerable yield is obtained from the 5th year onwards. In central and southern India, figs bear fruit twice a year, viz., once in July–September and in February–May. Indian figs develop fruit without pollination. Figs are high in the digestive enzyme **"ficin."**

Pollination

Parthenocarpy is favoured or inhibited in a given type by the climatic conditions of the place where it is growing. In Coeur, cultivars like Black Ischia, Brown turkey and Pune are parthenocarpic. While the cultivars like Turkish White have failed to set fruits without caprification in the said place, In Allahabad, the cultivated varieties, Pune and Black Ischia, do not set fruit without caprification.

Pests and Diseases

1. Leaf eating caterpillar – (*Oeinaria varians)*
2. Stem borer (*Bactocera rufomaculata*)
3. Leaf rust - *Cerotelium fici* – Affected plant show small, round brown spots on the leaves, small rusty, raised spots on the underside of the leaves. In severe case defoliation of leaves will be observed.
4. Leaf spot – *Cylindrocladium scopartium*
5. Anthracnose – *sphaceloma ficicuricae*
6. Fig mosaic – Viral transmitted by fig mites
7. Aceria ficus – Affected leaves show yellow- green spots, white mottle on leaves.

26

Grapes

- Sophisticate fruit crop/Symbol of charity
- B. N.: *Vitis vinifera.*
- Family: Vitaceae.
- Chromosome No.: 2n = 2x = 38
- Origin: Black Sea and Caspian Sea
- Inflorescence: Panicle
- Fruit type: berry
- Edible part: Pericarp and Placenta

Introduction

Grapes are a delicious, refreshing, and nourishing fruit. It was introduced into India by the in values of Iran and Afghanistan. 50% of the total production of fruits in the world is contributed by grapes. Cultivation of grapes is called **"viticulture".** Easily digestible and rich in **energy-giving sugars**, certain minerals and vitamins. Grape juice is a refreshing drink a **stimulant to kidenys and laxactive.** In India, **80%** of the production is used for **table purposes, 10% for raisins and 10% for wine production**, while in western countries 80% is used for winery, 10% for raisins, and 10% for table purposes. The area of the same is increasing fast in Maharashtra, Karnataka, Mizorum, Andhra Pradesh, Tamil Nadu, Punjab, Haryana, Madhya Pradesh and Uttar Pradesh are grape growing.

Different Genus

- True grapes belong to sub genera *Euvitis*
- Wild species of *Vitis* are polygamous-dioecious
- Progenitor of grapes- *(Vitis vinifera ssp. Sylvestris)*
- Edible grape species- *(Vitis and Muscadinia)*
- American grapes/Fox grape (*Vitis labrusca)*

- European grape/Wine grape/table grape (*Vitis vinifera-)* Nutural origin from *V. labrusca x V. vulpina,* susceptible to pest phylloxera, World leading grapes 90%
- Muscadine grapes (*Vitis rotundifolia)*- Dioecious species, resistant to powdery mildew
- Summer grape (*Vitis aestivalis)*
- Winter grape (*Vitis berlandieri):* Tolerant to high pH.
- Sand grape (*Vitis rupestris)* Suitable for sweet preparation (jam, jelly and preserve).
- River bank Grape/Frost Grape
- *Muscadinia rotundifolia, M. munsoniana* and *M. popenoi*are three species of American grape

Climate and Soil

It is a fruit of semi-arid subtropical regions, requiring a warm, dry summer and a cool, rainy winter. During winter, the grapes shed their leaves and take rest. In spring, they put forth new leaves and flowers. The fruits mature during the summer when there is no rain. A long, warm-to-hot, dry summer is needed for proper maturity and ripening. Grapes do not thrive in humid summer climates because of fungal diseases. In North India, heavy rains during July–September cause low production. The plant takes a rest during winter since the winter is very severe and puts forth new growth in summer. When the crop reaches its ripening stage during June, there is heavy rain, resulting in poor production. In Western India, grapes do not rest because of the warm winter. Hence the grape puts forth new growth twice a year. In Bombay – Deccan region and in Hyderabad the crop produced by the new lush in April maturing during the rainy season (July–August) has lower market value due to less sugar content. Here the maximum temperature goes up to 35°C and the minimum temperature does not fall below 12°C due to the warm winter conditions. There is practically no rest period. An almost rainless period during November–June favours heavy as well as sweet crops. Here the vines are pruned twice. Early December pruning yields a sweet crop during April and summer pruning (May) yields a slightly sour crop during September. The climatic conditions are favourable in parts of Tamil Nadu so that five crops are taken in two years by staggered pruning techniques. Well-drained, rich loamy soil with a pH of 6.5-7.0. The soil depth should be almost 1 m.

Propagation

Propagated by hard wood cuttings prepared from matured canes (one-year-old shoots). Cuttings of 25–30 cm in length are prepared by making the lower cut just below a bud and the upper cut slightly above the bud. Cuttings should be tied and stored in moist sand for a month for callusing. The callused cuttings start well in the nursery. While planting, only one bud is left above ground level and the remaining portion is buried in soil. At the end of winter, the sprouted and rooted cuttings can be lifted and planted in the main yield. Treating cuttings with IBA improves rooting. Single bud cuttings can also be rooted by keeping the base of the cutting at a constant temperature of 18–20°C for 2-3 weeks.

Rootstocks

Phylloxera Resistant Rootstocks are

- ***Riparia gloire-*** Selection from *V. riparia.*
- **St. George** - Cultivar of *V. rupestris*
- **4 x 2 (Ganzian)-** (*V. vinifera* (Aromon) x *V. rupestris* (Gonzin).
- **1202-** (*V. rupestries* and *V. mataro*).
- **99-** (*V. berlanderi x V. rupestris*).
- Nemotode resistant rootstocks are:
- **Dogride**- cultivars of *V. champin.*
- **Salt creek**, 1613- 1616- (*V. solonis x V. paria*).
- **Telecki 5A-** (*V. berlandarix* x *V. riparia* 5A 1616).
- **Solonis 1616-** suitable for saline & wet soils.
- **Temple rootstock** is used for resistance against **Pierce's disease.**

Varieties

Table Purpose

Table grapes are meant for use as fresh fruit for consumption. These grapes are attractive in appearance and eating quality, with good shipping and keeping qualities. Most of the varieties grown in India are table fruits. The important table grape varieties are Muscat Humburg, Cardinal, Perlette, Thompson Seedless (Sultanina), Tokay, Concord, Anab-e-shahi, Pusa Seedless, Delware, Catawba, Ohanez, Red Malaga, Emperor, Italia, Muscat of Alexandria, Bangalore Blue, Beauty Seedless, Arka Shweta, Arka Majestic and Arka Chitra.

Raisin Making

These are the grapes intended for making dried grapes. The raisin variety of grapes should have a soft texture, seedlessness with good sugar content, a marked pleasing flavour, a large or very small size; and little tendency to become sticky in storage. The varieties most extensively used in the commercial production of raisins include Thompson seedless (Sultania, Oal Kishmish), Seedless Sultana, Red Corinth, Cape Currant, Black Monukha, Black Corianth, Gold, Kishmish Beli-Seedless and Arka Vati.

Juice Making

The varieties of sweet juice grapes produce juice of acceptable beverage. The juice should retain the natural fresh grape flavour throughout clarification and preservation. In the United States of America, Concord grapes are generally used for juice. The varieties White Riesling and Chasslas Dore are used for juice in the central Europe. The varieties Aramon and Carignan are utilized for sweet juice in France. Juice grapes: Bangalore Blue, Beauty Seedless, Arka Krishna, Early Muscat, Champion.

Wine Making

Most of the vineyards in Europe, North Africa, South Africa, South America, Australia and the United States of America produce wine grapes. Wines are classified as table wines and dessert wines. Table wines contain less than 14 per cent alcohol, while dessert wines have more than 14 per cent alcohol. Usually, 17 to 20 percent sugar acid ratio, total acidity, and tannin content, etc., will determine the wine's quality. The varieties such as White Rieslin, Chardonnay, Cabernet Sauvignon, Tinta Maderi and Muscat Blanc (Muscat canelli) produce wines of high quality, outstanding in bouquet, flavour and general balance (Winker, *et al.*, 1974). Wine grapes: Muscat Canelli, Black Champa, Arka Soma, Arka Thishna-High sugar content and low acid.

Canning Grapes

Seedless varieties like Thompson Seedless and canner are generally canned in combination with other fruits. The varieties grown in Tamil Nadu are classified as "table grapes."Pachadraksha, Muscat ("Panneer"), Anab-e-shahi & Bangalore Blue, Pusa Seedless, Thompson Seedless, Kishmish Charni, Kishmish Beli, Seedless White.

Special Characters

- **Gulabi, Abyad-** Table, juice- early type.
- **Banaqui Abyad-** Table, wine- early,
- **Kishmish charni, Black champa-** Processing & for hybridization purpose.
- **Coloured Seedless:** Beauty seedless (blue black) mid-season,
- **Krishmish Charni (brick red)** -Mid-season (sharad seedless).
- **Flame seedless-** (Purple) late.
- **White seeded- Anab-e-shahi**-Late
- **Dilkush**-clone of AES
- **Cheema sahebi** (Selection from **Pandari sahebi**) late.
- **White seedless:** Perlette (Scolokertekhirolaynoje X Sultania marble)
- **Pusa Seedless:** Superior than perlette- Mid.
- **Thompson Seedless** - Tas-e-Ganesh, ManikChaman, Sonaka are clones.
- **Arkavati**-Black Champa X Thompson Seedless-late.
- **Delight**-Sister seedling of perlette- early.
- **Homorod**- Contario X Sultania.
- **Coloured seeded:** - Banagalore Blue- a cross between *V. vinifera* x *V. labrusca*- table, juice, wine.

Seedless Cultivars

- Crimson seedless- Introduction from USA, Colored seedless.
- Thompson Seedless- Australian Seedless, Sultania, Kishmish.
- Arka Vati (Black Champa x Thompson Seedless), Suitable for table and raisin purpose Seedless.
- Arka Neelamai (Black Champa x Thompson Seedless) Suitable for table purpose and red wine.
- Sharad Seedless (Selection from Beauty Seedless).
- Sonaka Seedless- Bud sport of Thompson Seedless.
- Tas-e-Ganesh- Bud sport of Thompson Seedless.
- Arka Kanchan (Anab-e-Shahi x Queen of Vineyard) Suitable for table and wine purpose, Late maturing.
- Arka Shyam (Bangalore Blue x Black Champa) Suitable for juice and wine making

- Arka Hans (Banagalore Blue x Anab-e-Shahi)-Suitable for white wine purpose
- Anab-e-Shahi- Popular cultivar throughout India
- Pachadrakhsha (Syn. Bhokri or Green Grape*)* Best adopted to tropical climate of TN.
- Muscat- *Syn. Gulabi, Black Prince, Panner*
- Banagalore Blue (*Vitis vinifera x Vitis labrusca)*- Suitable for juice and wine making
- Pusa Urvashi (Hur x Beauty Seedless)- Tolerant to anghracnose
- Punjab Purple –Cattawa x Beauty Seedless
- Pusa Navrang (Madeline Angavine x Ruby Red)- Tenturier
- Arka Shweta (Anab-e-Shahi x Thompson Seedless) Suitable for table and raisin purpose
- Arka Majestic (Anab-e-Shahi x Black Champa)- Suitable for head system of training
- Arka Chitra (Angur Kalan x Anab-e-Shahi)
- Arka Soma (Anab-e-Shahi x Queen of Vineyards)- Suitable for wine making
- Arka Trishna (Bangalore Blue x Convent Large Black)- Suitable for wine making
- Arka Krishana (Black Champa x Queen of vineyard)- Suitable for Beverage industry

Specific Features Varieties

- Self-thinning variety; New Perlette
- Loose Perlette from Perlette, Red Niagra from Niagra and Robin Cardinal from Cardinal are grape mutants.
- Perlette is a Polyploid variety of grape produced through colchicine's treatment commercially grown in Germany and is cold hardy
- Arka Trishar – (Hur x Beauty Early x Beauty Seedless)
- TSS of Thompson Seedless is 22-24% and juice 69 %
- Kishmish, Black Hamburg, Black prince, Foster seedlings, Delight, Perllete, Bhokri, Champion
- Delight and Perlette varieties were evolved by H. P.

- Grape guard crat paper coated with KMS and plastic polymer.
- Short berries- Perlette – cause – B deficiency and poor pollination
- Pusa Adity – Bangui Abyad x Perlette.
- Pusa Swarnika - Hur x Cordinal
- Raisin- Dried seedless grapes they should not have more than 17% moisture Cane Pruned varieties- Gulab, Pusa Seedless, Kishmischarni.
- Thompson Seedless ia a spur pruned varieties.
- Delight and Perlette are sister seedlings of a cross between Queen of Vineyard and Sultanina, Marble-26.
- Pusa seedless, Tash-A- Ganesh, Manik champa and Sonnaka parent is Thompson seedless
- Dilkhush parent is Anab- A- Shahi, Rao sahebi –Cheema Shahebi, Cheema Shahebi-Pandari Shahebi, Sharad Seedless-Kishmis Chorni.
- **Cane pruned varieties:** Pusa Seedless, Kishmish Charni.
- **Spur pruned varieties**: Perlette, Beauty Seedless, Bangalore Blue, Early Muscat
- **Tetraploids:** Thompson Seedless, Pusa Seedless, Perlette, Black Champa.
- **Induced mutants table:** New Perlette, Niagara, Robin, Cardinal.
- **Coloure Seeded**-Bangalore Blue, Gulabi (Muscat), Kishmish chorni
- **Coloured Seedl**ess- Beauty seedless, sharad seedless. Fantasy Seedless, and Crimson Seedless
- **White seeded**: Anab-a-shahi, Dilkhush.
- **White seedless**: Perlette, Pusa Seedless, Tas-A- Ganesh, SonnakaManik Chaman.
- **Juice variety in Tamil Nadu**: Gulabhi, Bangalore Blue.
- **Green Seedless varieties:** Thompson Seedless, Tas-A-Ganesh, Sonaka.
- **Teinturier (French language)** means wine term applied for variety having a red colour skin and flesh
- **Teinturier variety in India-** Pusa Navrang

Planting

Trenches of 0.6 m width and 0.6 m depth. Spacing 3 x 2 m for Muscat and 4 x 3 m for other varieties. The pandal system requires maximum spacing. The best time for planting is Jan-Feb. for rooted cuttings.

Irrigation

Immediately after planting on the 3rd day and once a week. Irrigation is withheld 15 days before pruning and also 15 days before harvest. Grapes require less water during fruit bud formation and more water during berry growth. Reduced irrigation during ripening improves the quality. Drip irrigation is becoming more popular.

Manures and Fertilizers

N. India: N-250 Kg Feb+ 350 Kg March + 100 Kg Oct/ha.

P-800 Kg April + 400 Kg May.

K-400 Kg May + 400 Kg June+200 Kg October.

60 g urea + 125g/Plant are recommended for newly planted plants in April & June.

Foliar application of micronutrients, particularly iron, zinc, boron, and manganese, has been shown to improve yield quality and quantity.

Advantages of Pandal System

The climate prevailing in India affords full scope for the vine to grow as luxuriantly as possible. The crop gets enough sun shine uniformly for its proper development. Plant protection measures are more effective in this system, especially against.

Kniffin System

In this method, the vine is allowed to put forth a single shoot which is trained erect and tipped at a height of about 45 cm from the ground level. Only three shoots are allowed to grow from this point, all others being removed. Two of the retained shoots are trained horizontally, and the remaining is trained vertically to a height of another 60 cm. when it is tipped to a height of another 60 cm. When it is tipped again to produce two more shoots, which are trained horizontally, all four horizontal arms are supported by bamboo poles or wire tied horizontally to posts fixed at regular intervals of 3 meters. Primary laterals are allowed to develop from these four main arms at fairly regular intervals and they are pruned to produce the crop every season. For every fruiting cane, a renewal spur of 2 to 3 buds is left. Normally only four canes are allowed in the 4-cane kniffin system and 6 canes in the 6-cane system. Under tropical conditions it is possible to have doubled the number.

Single-stake System

A single shoot is allowed to develop from the vine of rooted cuttings and is trained vertically by staking to a support. When this shoot reaches a height of 120 cm, it is tipped and allowed to produce 4 to 5 secondary branches or canes, which are pruned after every bearing season. The main stem and the primary laterals are supported by a bamboo post planted nearby.

Compare Different Systems of Training

From the point of view of fruit production, the pandal system was found to be superior both in vine yield and acre yield to the kniffin and single-stake systems and the single-stake system was slightly more productive than the kniffin system. It was observed that the vines trained in the pandal system of training recorded the maximum yield (3164 bunches weighing 675.4 kg per plot), followed by the double horizontal cordon system (2,432 bunches weighing 526.5 kg). The head system of training recorded the minimum yield. With regard to the quality of the fruits, the bunches from the pandal system of training recorded the highest total solid content. Coimbatore with Muscat Hamburg on pandal (bower), telephone, kniffin and head systems of training revealed that the pandal system was more economical with the highest cost benefit ratio of 1: 2.09, followed by 1:1.71 in telephone, 1:1.42 in kniffin and the lowest being in the head system with 1: 0.05 (Anon. 1982).

Wire Trellis System

In this system, two or three wires are strung in rows from vertical posts. A single stem in between the posts is trained as far as the top of the wire (2.1 m) and two arms are allowed to develop along the wire on either side. Each arm will meet the arm from the adjacent vine and will have two arms with six tertilaries. The spurs on the fruit-bearing shoots growing on these canes are seasonally pruned for fruit every year. Because of its resemblance to the telephone pole with its flat-topped mast bearing the supporting wires, this is called the "Telephone Trellis System". In this system, there is a large flat-topped T-trellis above ground level. The horizontal cross is 120 to 150 cm wide, with wires strung on the top 30 to 45 cm apart. The vines are trained along the wire in the direction of the row. Steel angle iron or waste structural steel is used when available and the cross arm is welded to the upright. In a few vineyards, the flat top is modified by bending each half of the mast upwards about 30° so that a trough-shaped or "H"-shaped form is obtained. The disadvantage to this system is that there are no cross-supports between the rows to withstand strong winds. Cultivation and movement are limited tone direction. But they are provided with better exposure to light.

Pruning

The grapevine is a vigorous climber. If it is not properly trained and pruned, it does not bear fruit properly. Pruning is one of the most important operations in grape culture.

Objectives of Pruning

1. To reduce the amount of old wood in order to keep the vine within manageable limits.
2. To secure fruit bearing branches in predetermined places.
3. To expose the fruiting branches to sufficient sunshine.
4. To reduce the excessive vegetative growth.

It is essential that one should be familiar with the various parts of the vine and their functions for successful pruning. The following are some of the technical terms used:

Term of Pruning

1. **Shoot:** Young growth of green stem of the current season, which bears the grape cluster.
2. **Cane:** A well mature and ripened shoot of the past season or that of the previous year which gives rise to shoots.
3. **Spur:** A portion of the cane or ripened shoot left behind on the plant after pruning.
4. **Fruiting spur:** A cane or well ripened shoot leaving 304 buds, producing a bunch after pruning.
5. **Foundation spur or Renewal spur:** A well-ripened shoot or cane bearing bud. This normally remains after the shoots are pruned in March-April or summer in Hyderabad. It is called a foundation spur as it forms the base of the foundation wood on which next year's canes and fruiting spurs are formed or on which both growth of the year is borne.
6. **Trunk:** Main stem of the plant.
7. **Long spur:** A ripe shoot, carrying more than five buds. Normally it is 25-30 cm long with about 5-10 buds on it.
8. **Medium spur:** It is a cane cut back keeping 3-5 buds.
9. **Spur:** It is cane pruned to 1-2 buds.

Special Practices

1. **Tipping** of shoots and tying of clusters in the pandal after the fruit set. Remove tendrils.
2. **Nipping** the growing shoots of axillary buds and terminal buds at 12 to 15 buds.
3. **Thinning** the compact bunches by removing 20% of the berries at pea stage.
4. The clusters are dipped in a solution containing brassinosteriod 0.5 ppm and GA_3 25 ppm at 10-12 days after fruit set to maintain vigour, yield and quality parameters

Harvesting

A grape is a non-climacteric fruit and has to be harvested at the correct stage of maturity. The degree of days from full bloom gives a correct indication of maturity. Early cultivars need 1600–2000 degree days, while late cultivars need 3,000 or more. A cluster having underdeveloped seedless berries known as "shot barriers" or mummies. These will be very sweet in taste. Treating clusters with AVG (2 Aminoethyl Vinyl Glycine) 50-300 ppm 1-3 weeks before anthesis improves berry set. Cluster thinning also helps in improving the quality. 60-70 clusters/ vine spaced 3m x 3m is best. Girdling the shoot one week after bloom improves berry set, after berry set improves berry size and before veraison advances ripening and uniform coloration. GA (40 ppm) Ethrel can also be used to improve yield and quality. Ethrel (250 ppm) can be used for uniform colour development (5 weeks after anthesis, 4 weeks after berry set for colour).

Yield

The grape starts yielding in 2-3 years and continues for more than 20–25 years. Average yield will be about 25-30 tonnes/ha but higher yields of 60-75 ton/ha also possible with good management. Grapes are packed in corrugated fibre board boxes having grape guard (craft paper coated within layer of mixture of sodium bisulphate and a plastic polymer) improve their storage life. Grapes can be stored grape guard or for 7-12 weeks under a controlled atmosphere with 15-25% CO_2 and at 0-10°C.

Fruiting Buds of Different Varieties

1. Bangalore Blue-3-5 buds.
2. Bhokri-3 to 4 buds.
3. Anab-e-Shahi-6 - 8 buds.
4. Cheema sahebi- 5 to 8 buds.
5. Pusa seedless, Kishmish charmi, Gulabi-10 to 12 buds.

Physiological Disorders

- **Blossom End Rot (BER)-** develops at the blossom end of the berry which later on spreads + water suncans and due to rot Calcium deficiency.
- **Interveinal chlorosis:** Mn, Zn or Fe deficiency-0.2%.
- **Stalk necrosis: Calcium** deficiency.
- **Bud, flower and berry drop:** Girdling 10 days before full bloom, 500 ppm ethrel at ripening NAA 100ppm at 10 days before ripening, Reducing irrigation during bloom, Benzyl adenine 200ppm, 4-CPA 20ppm-for thinning.
- **Bud killing:** Excessive nitrogen.
- **Hen & Chicken-** Due to boron deficiency.
- **Uneven Ripening** (Bangalore Blue, Beauty Seedless)- Ethephone @ 250 ppm at colour stage.
- **Post-harvest berry drops** (Anab-e-Shahi, Beauty Seedless)- NAA @ 50ppm.
- **Flower or Bud drop** (Thompson Seedless, Tas-E-Ganesh)- Stem girdling.
- **Millerandae or Shot berries** (Bad weather and heavy crop load) Ethephone 25ppm+Seven 2000 ppm for 1 minute
- **Pink berry formation** is common problem in Thompson seedless and it clone
- **Berry drop in grape** due to improper pollination and fertilization
- Major nutrient deficiency in grapes growing area in the world: Mg

Mumification

- Berries during development become soft, loose and water.
- Berries may be at the tip of rachis or scattered.
- Cluster thinning gradlings, manuring helps control problem.

Short Berries

- Small, seedless berries in cluster is short berries.
- Unfertilized flowers may remain cluster and develop small seedless berry.
- Heavy cropping and lack of poor pollination are some reasons for development of short berries.

Pests

- **Thrips-** spray 1ml monocrotophos per litre of water.
- **Chaffer beetle-** dusting any insecticide in evenings.

- **Nematodes-** Application of neem cake 200 g/vine also control.
- **Flea beetles Phosalone** – 2 ml/lit after pruning,
- **Mealy bug:** Monocrotophos – 2 ml/lit
- **Scales**

Diseases

- **Anthracnose-** Removing & burning, spraying Bordeaux Mixture, Benlate or Bavistin regularly or Ziram
- **Downy mildew**- Spray Bordeaux Mixture, or Mancozeb (2.5 gm/1 litre.)
- **Powdery mildew**- Spraying or dusting sulphur or Neem products.

27

Guava

- Apple of the tropics/Poor man's apple
- *B. N.: Psidium guajava*
- Family: Myrtaceae
- Chromosome No.: 2n = 2x = 22
- Origin: Tropical America
- Inflorescence: Solitary or Cyme
- Fruit type: Berry
- Edible part: Thalamus and pericarp

Introduction

Guava ranks 4th in area and production after mango, banana, and citrus. It is a very hardy sub-tropical plant, prolific bearer. It is known as the **"poor man's apple"** in India and has spread widely throughout the country. Allahabad has the reputation of growing the best guava in the country as well as in the world. A native of **tropical America**, guava was spread rapidly throughout the world's tropics by the Spanish and Portuguese. The major guava producing countries are India, China, Thailand, Indonesia and Pakistan. In India, it is grown in Uttar Pradesh (largest area and production), Bihar, M.P, Maharashtra and Andhra Pradesh. The fresh fruits are very rich in vitamin C (100-260 mg/100 g pulp). Vit-A, B and minerals like Calcium, P, acidity 2.4 per cent, carbohydrates 9-10 per cent, TSS-13 per cent, pantothenic acid, riboflavin, thiamin and niacin, also a rich source of pectin. The fresh ripe fruits are used as table/salad fruit. Fruits are rich in pectin. The best quality jelly can be prepared.

Climate and Soil

It is the fruit crop of a sub-tropical region but is basically a tropical fruit; the young plants are susceptible to drought and cold. The trees are very hardy and can withstand heat and prolonged drought. A cool winter induces heavy fruiting and produces good quality fruits. The rainfall of 100 cm with uniform distribution is ideal for production. The optimum temperature lies between 23-28 °C and can be reached up to 1500 m above mean sea level. Being very hardy, is grown successfully in a wide range of soil types like light sandy loam, clayey, deep, rich alluvial with the pH of 4.5–8.2.

Important Species

- ***Psidium guajava-*** It is the commercially cultivated species, rest of them do produce fruits.
- ***P. guajava var. aromaticum-*** small scented fruits.
- ***P. guineense-*** (called Brazilian or Guinea guava) which bears small fruits of poor quality.
- ***P. cattleianum- (strawberry or cattley guava)*** small fruits, red colour and use as a roots stock.
- ***P. friedrichsthalianum*** - (Costa Rican guava or china guava) dwarfing rootstock, fruits having high acid content and resistant to wilt and nematode, Small fruits globose in shape.
- ***Feijoa sellowiana-***Pineapple guava/feijoa guava
- ***P. montanum*** - Mountain guava, is a shrub, about 1.5 m high with flat branches.
- ***P. pomiferum*** -Apple shaped.
- ***P. pyriferum*** - Pear shaped.
- ***Guisaro guava- Psidium molle.***
- ***Psidium cujavillis-*** *(Highest* Vitamin- C)
- ***P. pumilum-*** Highest sugar content species.
- ***P. pumium*** -Dwarfing
- ***P. cattleianum var. lucidum-*** Seedless against guava wilt for breeding.

Varieties

- **L-49 (Sardar Guava)** is a chance seedling selection from 'Allahabad Safeda' in 1927.
- **Lucknow-46**- Pear-Pyriform shape.
- **Allahabad Safeda:** The most famous variety grown in Uttar Pradesh for table purpose.
- **Allahabad round-** Parthenocarpic variety.
- **Arka Mridula** – Seedling selection from Allahabad safeda and soft seeded.

- **Pant Prabhat, Dhawal and Lalima, CISH-G-4.**
- **CISH-G-1**- Seedling selection from Allahabad Safeda, Red colour and longer shelf life.
- **Lali**t- Seedling selection from Allahabad Safeda, Red colour pulp.
- **Red Fleshed:** Which has yellowish skin with pink colour flesh.
- **Harijha**- Popular variety in Bihar.
- **Allahabad Surkha-** Pink flesh and large fruited variety.
- **Chittidar**- Highest TSS guava variety and red dot skin.
- **Shweta**- Selection from haf-sib population of Apple colour, Colour variety. High TSS (14^0 Brix).
- **TRY(G)-1-** Selection, Off-season, drought, soditicity tolerant.
- **Apple Colour-** Red skinned guava variety, Tolerant to anthracnose and fruit canker.
- **Hafshi**- Red fleshed guava.
- **Behat Coconut-** Sedless guava.
- **Saharanpur Seedless, Nagpur seedless** -Seedless varieties.
- **Pusa Srijan-** Potential dwarfing rootsock for HDP, Tolerance to wilt.
- **Aneuploid-82 (2n+2=24**) tetrasomic type of aneuploidy, Recommended for dwarf rootstock from IARI.

Hybrid Varieties

- **Punjab Pink** – (Porttugal x L- 49) x Apple Colour.
- **Hisar Safeda-** Allahabad Safeda x Seedless.
- **Hisar Surkha-** Apple colour x Banarasi Surkha, Pink fleshed.
- **Kohir Safed** - (Kohir x Allahabad Safeda).
- **Safed Jam** (Allahabad Safeda x Kohir).
- **Arka Amulya-** Allahabad Safeda x Triploid-Seedless.
- **Arka Reshmi-** Kamsari x Purple local.
- **Arka Kiran-** Kamsari X Purple local, Pink fleshed soft seeded variety.

Propagation

In India, guava is propagated from seed, which germinates in about three weeks. Boiling the seeds for five minutes, soaking them in water for weeks prior to sowing or treating them in strong sulphuric acid for five minutes facilitates their germination. Vegetative propagation through **layering** is therefore recommended. Both air layering and simple layering have been found to be successful. In about 45 days, layers can be separated from the mother plants. These separate layers should be planted in full size pots and they are hardened by gradually exposing them to direct sunlight. Such hardened layers are ready for planting in about six months.

Planting

Pits of 0:5 m x 0.5 m x 0.5 m size are dug at a spacing of 6 m x 6 m. The layers with the ball of earth are planted in the centre of the pit.

Manures and Fertilizers

Therefore, for the bearing trees, 50 kg of FYM and one kg each of N, P, and K are applied per tree in two equal split doses, once during March and again during October. The manure and fertilizers are spread in the entire basin of the tree, 15 cm away from the trunk up to the leaf drip and incorporated by shallow digging. It also responds to foliar spray of nutrients and spraying of urea 1 % + zinc 0.5% twice a year during March and October to increase the yield. Guava sometimes suffers from micronutrient deficiencies. As a result, a mixed spray containing $ZnSO_4$, $MgSO_4$, $MnSO_4$ @ 0.5%, $CuSo_4$ and $FeSo_4$ @ 0.25% plus a wetting agent @ 1 ml per 5 litre of solution is recommended at various stages, including new flush, first month after first spray at flowering, and fruit set.

Bronzing

Bronzing (Fatio disease) The leaves turn a pale or purple color; they are small in size; there is little flowering; the growing tips die; and the plant eventually dies. It could be due to a lack of NPK, Zn, Mg, and low pH, as well as weekly spraying of 0.5 percent DAP and $ZnSo_4$ during October and November.

Training and Pruning

Training is done primarily to give form to the tree. For the development of a strong framework, the first 60 to 90 cm from the base of the trunk should be cleaned, followed by 4 to 5 scaffold branches at an interval of 20–25 cm. When the plants have attained a height of about 1.5 m to 1.8 m, they are headed back to make the centre open. In some parts of India (Maharashtra and West Bengal), the branches are bent downward and tied to each other. Thus forcing the dormant buds to grow and this results in increased yields. Trees are rarely pruned in North India, but light annual pruning after harvesting to promote vegetative growth and flowering is desirable. All dead, diseased, crowded growth and suckers sprouting from the base and sides of the framework are pruned back annually. Pruned trees give larger fruits and early ripening.

Flowering and Fruiting

The guava tree flowers throughout the year, but the peak flowering is observed in 2 seasons: the rainy crop (April-May) and the winter crop (Aug-Sept). Flowers are produced in leaf axils or in cymes. The period of flowering varies from 25-45 days. Honey bees are the pollinators.

Flower Regulation

Bahar treatment expands root exposure, with holding water and also deblosoming the rainy season crop due to the insipid taste of the fruits and inferior quality. So winter crop is generally desirable.

Fruit Set

Only 35-50% of the fruits are carried to maturity, though initially 80-86% of the fruit sets. In seedless varieties, it is as low as 6 per cent to improve fruit set, GA_3 at 200 ppm.

Crop Regulation. (Bahar Treatment)

Only for **tropical region**. In **west & south India**, guava flowers **thrice**. June, October. & February (Mrig Bahar, Hast Bahar &Ambe Bahar). Mrig Bahar is taken in all over India, become of quality & rich in Vitamin C.

Why only one Bahar?

Flowers throughout year without rest. It is not desirable. So only one best bahar is taken.

Type of Bahar

1. ***AmbeBahar:*** Flowering in February. Fruit ripening in **June-Sept**. Quality poor, insipid & watery. No keeping quality.
2. ***Mrig Bahar***: **Flowering in June**. Fruit ripening in **Nov -Jan**. quality is best as no infection of fruit fly. (Gujarat prefer Mrig bahar)
3. ***Hast Bahar:*** Flowering in **October**. Fruit ripening in **Feb.- April**. Taken if late rains fail & clear weather in Sept.- Oct. Gives good price.

Methods of Crop Regulation (Bahar Treatment)

1. **To restrict Irrigation-**Withholding irrigation from December to June or until the beginning of monsoon.
2. **Expose of Roots-**Root exposure followed by root pruning is recommend to suppress rainy season crop so

3. **Hand thinning**- Removal of some fruits they reach maturity.
4. **Bending** – Trees in which shoots are growing erect and bear very poorly. Such shoots may be bent and tied on the pegs driven on the ground. Doing so, dormant buds are activated which bear flowers and fruits.

Harvesting

The guava is a climacteric fruit; it ripens after harvesting; the fruits are harvested throughout the year (except during May and June). However, peak harvesting periods in north India are August for rainy season crops, November-December for winter season crops and March-April for spring season crops. The change in colour of fruits from dark green to pale green is an indication of maturity. The fruits are harvested at their full yellow but firm for the local market, whereas half-yellow fruits should be picked for distant markets. The fruits are harvested selectively by hand, along with the stalks and leaves. The tree reaches its peak bearing stage within fifteen to sixteen years after planting. Amature tree yields about 90-150 kg of fruit, or 10-15 t/ha. The Sardar variety gives about 25 t/hectare.

Storage

Guavas are perishable in nature. They are disposed of immediately after harvesting in the local market, and a very small quantity is sent to a distant market. They are not kept in cold storage. However, shelf-life of guava can be extended up to 20 days by keeping them at low temperature of 5°C and 75-85% relative humidity. It can also be stored in polybags for about 10 days at room temperature (18-23°C) with 0.25% ventilation.

Pests

1. **Fruit fly- *Chaetodacusspp-*** Rainy season crop more affected. Dark greenish patchess on fruit surface. Fruits drop. Spray Malathion, Demecron 10 ml/10 lit water. Destroy affected fruits. Avoid Ambe Bahar crop.
2. **Mealy bugs- *Cryptolems spp*.** Small white tiny bugs suck sap from tender parts. Treat soil with aldrin, Thimet etc.
3. **Scale insects-**
4. **Tea mosquito-**
5. **Shoot & bark borer-**

Diseases

1. **Guava wilt- *Fusarium spp*. Serious during monsoon.** Occurs where soil pH is above 7.0. Branches weather & die. Uproot & remove affected trees. Drench soil with 0.3 % coercion.

2. **Anthracnose- (*Colletotrichum psidii*)-** It is also severe during rainy season crops. Pin head **dry spots** observed on unripe fruits during **monsoon**, which is hard. Flowers & buds shed. Leaves also affected. Prune affected twigs. Destroy fallen leaves & fruits. **Spray 0.2% Dithane M-45.**
3. **Fruit canker- *Pestlotia psidii.***
4. **Cercospora leaf spot- *Cercospora sawadal.***
5. ***Zinc deficiency:* -** leaves turn **brownish**. The **fruits** become **small & hard**. Spray **0.2% $ZnSO_4$** at 15 days' interval or apply 220 g./ tree

28

Jack Fruit

- Jack fruit/National fruit of Banglades/Poor man's food/Kathal
- B. N.: *Artocarpus heterophyllus*
- Family: Moraceae
- Chromosome No.: 2n = 56 (Autotetraploid)
- Origin: India (Western Ghats)
- Inflorescence: Spike covered by two spathes
- Fruit type: Sorosis
- Edible part: pericarp and fleshy perianth

Introduction

Jack fruit is indigenous to India and is commonly grown in Sri Lanka, Malaysia, Brazil, Jamaica, Myanmar, Singapore, and India. In India, it grows in a wild and semi-wild state in Kerala, Assam, W.B, Tripura, Bihar, U.P. and Western Ghats. Jackfruit is a common and popular fruit crop in the low land tropics of the eastern and southern parts of India. It is the largest edible fruit (20–40 kg). In Sri Lanka, a fruit of about 120 kg was harvested and created the world record. The fruit is also called "poor man's fruit" in the eastern and southern parts. In the Tamil literature, Jack has been given the important second position of the significant 'three fruits', viz., 'Mukkani'. It is mainly distributed in the tropical humid belt. In India, Assam, Bihar, Kerala and Tamil Nadu are the main jackfruit growing states. In Tamil Nadu, the lower Palani hills of Dindigul Anna district and Panruti of South Arcot Vallalar district are known for the best quality jack fruit. India is considered to be the native home of jacks.

Composition and Use

The fleshy carpel (which is botanically the perianth) is the edible portion. Hundred-gram edible portion contains 19.8g carbohydrate mainly as sugars, 1.9g protein, 0.1g fat, 1.1 g fibre, 20 mg calium, 41 mg phosphorus 0.56 mg iron, 175 mg carotene (Vit. A), 0.03 mg thiamine, 0.13 mg riboflavin, 0.4 mg niacin and 7 mg Vitamin C. Natural protein found in Jackfruit-Lectin (used in cancer treatment). One hundred grams of jack carpels supplies 88 kcal of energy. Recently, it has been reported that jack fruit could be very useful in the treatment of the dreaded disease of humans, AIDS. An extract of jack fruit was seen to have inhibited the growth of HIV infection in vitro. The power of this substance, called jacaline, was discovered by Jean Favero. Jacaline is inactive on lymphocytes which have already been infected but has proved its might by protecting the healthy ones. Apart from its use as a table fruit, jack is a popular fruit with housewives for making pickles, for dehydration into jack leather or thin round papad. Canned jack fruit, syrup, jam, jelly and candy have also been attempted.

Climatic and Soil

The fruit tree comes up well in the humid and warm climate of the hill slopes and the arid, warmer plains of south India. It is grown up to 1500 m above sea level and is sensitive to frost and drought. Jack grows well and gives food yield in the warm, humid climate of the hill slopes and the hot, humid climate of the plains. A temperature ranges of 22–35°C will be ideal. Though it tolerates a variety of soils, a deep, rich alluvial or open-textured loamy soil with slightly acidic conditions (pH 6.0–6.5) and perfect drainage is ideal.

Species and Varieties

- ***A. hetrophyllus*** (Bread fruit) previously known as ***A. integnifolia***. related jack fruit species are
- ***A. altilis*** - bread fruit- small sized round fruit, found in T.N, Assam, Maharashtra & Kerala.
- ***A. lakoocha***- Monkey Jack- Small edible fruits- it wild in Assam, W.B, U.P. & Western ghats.
- ***A. hirsute*- Aini /Anjili** / is a semi-wild edible species, native to India. is a tall evergreen tree?
- **Champeden-** edible fruit with strong odour like durian.

Varieties

- Cultivated jack types are classified into two groups (1) firm flesh (2) soft flesh.
- Singapore (or) Ceylon jack It was introduced in Tamil Nadu from Sri Lanka.
- Gulabi' (Rose scented), 'Champa', Hazari, Rudrakshi (small roundish fruits) are also available.

- GKVK-1 and Swarna with superior quality fruits were released from Karnataka and
- Gumless jack, Muttan Varica from Kerala also found to perform better.
- Hybrid jack: It is a cross between Singapore jack x Veliappala, at Fruit Research Station, Kallar
- Burliar – 1: (T Nagar selection) It was developed at Fruit Research Station, Burliar.
- PPI – 1: (Pechiparai – 1) by clonal selection from Mulagummoodu local (2 Crops/year)
- *Rudrakshi* and ***A. hirsute*** are also used as rootstocks.
- Muttam Varikka and Hazar: Bearing large no. of fruit.
- **Champa:** Flavour like champak.
- **Singapore or Ceylon Jack:** Introduced from Ceylon- Off season variety.
- **PLR-1-Jack (Palur):** Spiny surface: Off season: Suitable for HDP, PLI-2.
- **Rudrakshi:** Pumelo sized fruits.
- **Suitable for table purpose:** NJT-1,2,3,4.
- **Suitable for culinary purpose:** NJC-1,2,3,4.
- **Uttar Pradesh types:** Rasdar, Khajwa and Sugandh.
- **Exotic varieties:** Golden Nuggest, Black Gold, Lemon Gold.

Propagation

In India, most of the plantations are still raised by seeds. Soft wood grafting (cleft method) on 1 ½ months old seedling with a scion of 3–4 months old was also found to be successful. To have a regular orchard like mango, the spacing of 10x10 m is followed.

Training

When Jacks are trained to a single stem, early side branches should be removed then and there so that a uniform smooth trunk develops for a height of 1.5–2 m, and then scaffold branches should be permitted to arise.

Manuring

For 1-3-year-old trees, 200:60g NPK/plant/year. For 4-6-year-old trees, 400:240:120g NPK/plant/year. 7th year onwards 600:300:240g NPK/plant/ Year.

Irrigation

Though jack is cultivated under rainfed conditions, it is very sensitive to drought. Hence, irrigation should be done depending on the type of soil, season, etc. so that there should not be any moisture stress, especially during flowering and fruit set.

Intercultivation

The male inflorescence (catkins) is seen in the current season growth, while female catkins are produced as cauliflorus (on trunk and main scaffold). Sometimes there will be more production of female flowers as well as male flowers, but there will be no fruit set. This is mainly due to a lack of proper pollination. The male catkins show a protrusion of anthers on the surface of the spikes. The stigma becomes visible by 8 AM on the 4th day after it comes from the sheath. Every day between 8.30 AM and 9.30 AM, the male catkins showing the pollen have to be collected and rubbed on the sticky surface of the female spikes. Such hand pollination will help with proper fruitset and yield.

Flowering

Flowering starts in December and continues up to March. It is a monoecious plant with a spike type of inflorescence. Male and female inflorescence are present in the same plant. The flowers are cross-pollinated by honey beers. A multiple fruit, known as sorosis, develops following pollination and fertilization. The pericarp around the seed and the fleshy perianth are edible. The fruits take about 90-110 days after the appearance of the spike, mature in June- July and are available almost throughout the year in Assam & South India.

Harvesting

It is a climacteric fruit; harvesting is done by cutting the fruit along with the stalk after developing a sweet and good flavour. Change in fruit colour from green to light yellow, harvesting is done by cutting off the stalk carrying the fruits. Typically, the fruits are available from March to June. In high elevations, harvest extends up to September. Even in the plains, certain genotypes bear an off-season crop during October–December.

Yield and Storage

A 15-year-old tree yields about 250 fruits. The individual fruits may weigh from 1 kg to about 20 kg. Fruits are stored for 2-3 months at 5°C with 85-90% relative humidity and 1 week at room temperature.

Pests and Diseases

1. **Bud weevil (*Ochyromera artocarpi*):** It bores into the tender buds, shoots and fruits.
2. **Management:** Destroy fallen fruits and buds, collect and kill groups, adults and then spray endosulfan (0.035%).
3. **Fruit rot: *(Rhizopus artocarpi)*:** It causes premature fall of young fruits due to rotting and may result in heavy loss in yield under very humid conditions.
4. **Management:** Spraying Diathane M. 45 (0.2% or Bavistin (0.05%) or Fytolan (0.2%) at 15 days' interval during fruit growth.

Pests and Diseases

29

Jamun

- Jamun/Indian black berry/Black plum/Java plum/Rose apple/Pitangu cherry/Diabetic fighter
- **B.N.:** *Syzygium cumini (*Syn. *Eugenia jambolana*)
- Family: Myrtaceae
- Chromosome No.: 2n = 40
- Origin: India
- Fruit type: Drupe
- Edible part: Mesocarp and epicarp

Introduction

It is a tall, handsome evergreen tree of tropical and subtropical regions and has its origin probably in India or the East Indies. It is one of the hardiest fruits and best suited for wastelands. It is drought tolerant and, at the same time, can tolerate water stagnation and marshlands, where other fruit crops cannot be grown successfully. It is more popular as an avenue tree and for wind breaks. The wood is used as timber in buildings and railway sleepers. It is often found growing as a wild and semi-wild in tropical and subtropical parts of India viz., Punjab, Haryana, Uttar Pradesh, Maharashtra, Rajasthan, Gujarat, Madhya Pradesh, Bihar, Chhattisgarh, Jharkhand, Karnataka, Kerala, Tamil Nadu and Andhra Pradesh.

Composition and Uses

One hundred gram of fruit contains 19.7 g carbohydrate, 0.7 g protein, 1.0 g iron, 0.02 g calcium, 0.01 g phosphorus, 0.1 g fat and 0.9 g fibre. Besides being taken as a dessert fruit, it is also used to make beverages, squash, jam, jelly and wine. Fruit syrup is used in curing diarrhea. Jamun is a non-climacteric fruit, used as an avenue or as a wind break and a hardy fruit crop. Fruits are a good source of iron used as an effective medicine against diabetes, heart and liver trouble. Main responsible for astringency**: tannin.** The purple colour of the fruit is due to **anthocyanin.** responsible for the flavour of fruits: **Dihydrocarvyl acetate, geranyl butyrate and terpenyl valerate.**

Climate and Soil

It is adapted to tropical and subtropical conditions, requiring a dry climate during flowering and fruiting. In the initial stages of establishment, there should not be any drought. When the tree has grown sufficiently, it can tolerate drought as well as flooding. The growing ranges of the Himalayas go up to an altitude of about **1300 MSL.** Early rains resulting in better soil moisture will help in growth and development and ripening of fruits. Well-drained deep loam would be ideal, though it can grow on a wide range of soil conditions.

Species

- ***S. Jambos (Rose apple or safed Jamun):*** The tree is **ornamental;** the fruits are light yellow – white in colour. The seeds are **polyembryonic**.
- ***S. fruitecosum*** – The trees are suitable for wind break, fruits are edible and small in size.
- ***S. Javanica*** (***Water apple)***: This species is found in South India and West Bengal.
- ***S. densiflora*** – It is used as a root stock for ***S. cumini.*** It is resistant to attack of termites.
- ***S. uniflora*** (***Surinam cherry*** *or* ***pitanga cherry***): It is small tree bears small sized frits are bright red colour with aromatic flavour.
- ***S. zeylanica*** – Small tree, bears edible fruits in Western ghats of India
- ***Syzygium aqueum*** -Watery Rose Apple
- ***Syzygium malacense***- Malay Rose Apple/Malayan Apple

Cultivars: Goma Priyanka, Thar Kranti**,** Paras (large size fruit)**,** Ra Jamun, NJ-6 (Seedless)**,** SPKV – 15, SPKV- 4, SPKV -14, SPKV-13, J-37, J-42 (Seedless) and KJS – 20.

Propagation

It is successful for commercial raising of plants by the budding method. It is propagated by seeds. For true-to-type progenies, vegetative propagation is recommended. Seeds have no dormancy and fresh seeds can be grown.

Planting: Pits of 1 m^3 size are dug at a distance of 10 x 10 m. for grafts and 12 x 12 m.

Pruning: The plants should be trained initially in such a way that the branches develop from a point 100 cm above the ground level. But pruning is not required.

Manuring and Irrigation

The exact doses and type of manures and fertilizers for jamun is yet to be worked out. Generallly jamun trees ae not manured however, an annual dose of about 20 kg of FYM during the pre-bearing period and 50-80 kg per tree to bearing trees is considered beneficial (Bose *et al.*, 2000). When the soil is fertile, manuring and irrigation are withheld to encourage fruit bud production. Only during the initial 2–3 years do the plants require regular irrigation. In later years, irrigation is required mainly during fruit growth and ripening.

Flowering and Fruiting

Flowering time is a March to April and fruits ripen during June-July, hermaphrodite flowers alight yellow in colour. By natural fruit drop, only 12–15% of flowers reach maturity. There is a heavy drop of flowers and fruits (50-60%) within 3-4 weeks of flowering. A large number of fruits drop off at a very young stage during the 5-8 weeks after blooming. The flower and fruit drop can be reduced with the help of two sprays of GA_3 @ 60 ppm at full bloom and a second spray at 15 days after the initial setting of fruit. Jamun is a highly cross pollinated through honey bees, houseflies and wind.

Harvest and Yield

The seedling trees start bearing after 10 years, while the vegetatively propagated progenies come to bearing in 5–6 years. There are 3 phases of fruit growth and development, *i.e.,* the I phase from 15 to 52 days after fruit set, having slow growth of fruit. The II phase, from 52 to 58 days after fruit set, has fast growth, and the III phase, the last phase, from 58 to 60 days after fruit set, has slow growth. The maturity of fruits changes their colour from green to deep red or violet. The fully grown seedling jamun tree yields **80-100 kg** of fruit on average. They cannot be stored for more than 3–4 days under ordinary conditions. The jamun fruits are highly perishable and can hardly stand 5 days. They should be immediately sent to market. However, if necessary, they can be stored for 3 weeks at 9ºC and 85–90% RH.

Seed Powder

Evenly matured disease free and sound jamun fruits have to select. The pulp and seed of jamun fruit was separated by pulper. Then the seed washed in water and dried in tray dryer at 60°C for 48 hours still complete drying and ground the seed in pulveriser to fine powder of average particle size 0.58 mm.

Plant Protection

1. **Leaf eating caterpillar** – ***Carea subtilis*** Caterpillar attacks the leaves and tree become defoliated.
2. **Fruit fly – *Dialeurodes eugenia*** Affected fruits get wormy appearance on the surface
3. **Anthracnose** – ***Glomerella cingulata*** Affected leaves show small scattered spots,

30

Karonda

- Karonda/Christ thorn
- B N: *Carissa carandas*
- Family: Apocynaceae
- Chromosome No.: 2n = 2x = 22
- Origin: India
- Inflorescence: Terminal corymbose cyme
- Fruit type: Berry
- Edible part: Mesocarp and epicarp

Introduction

Karonda is an evergreen shrub or short stature tree. It grows naturally in the Himalayas and Western Ghats at elevations of 300 to 1800 meters. It is found grown in wild in India, Malaysia, South Africa. It is a well suitedto arid climate and grown well at higher temperature. It is commonly uses for making hedge for orchards. It is suitable for the arid tropics and subtropics and is grown for its attractive, edible fruits. It also grows successfully on marginal and wastelands. It thrives well as a rainfed crop; the plant hardly needs any care and gives yield with the minimum management. The plants can be trained as a strong hedge. Karonda is suitable for growing throughout the sub-tropical and tropical climatic zones of India. Major natural areas of occurrence of karonda have been observed in the states of Maharashtra, Bihar, West Bengal, Chhattisgarh, Orissa, Gujarat, Madhya Pradesh, Rajasthan and in the Western Ghats. In Maharashtra, the major area under this crop is scattered in sub-mountain areas like Kolhapur, Ratnagiri, and Pune district.

Uses and Composition

The fruits are eaten as a dessert when ripe. Mature fruits contain a high amount of pectin and are used for the preparation of different products such as jelly, jam, squash, sauce, syrup, etc. The unripe fruits are sour and astringent and can be used for pickles and chutney. The dried fruits may become a substitute for raisins. The fruits can be candied just like cherries. The wine prepared from ripe fruits contains about 14.5 to 15% alcohol and is very much liked by wine fanciers. The fruits are rich in protein (1.1-2.25%), vitamin C (1.6-17.9 mg/100g) and minerals, especially iron (39.1 mg/100g), calcium (21 mg/100 g), and phosphorus (38 mg/100g). The fruits are traditionally used in the treatments of malaria, epilepsy, nerve disorder, relieve of pain and headache, fever, blood purifier, myopatic spams, dog bite, cough. The roots are used as anthelmintic, stomachic and antiscorbutic agents and for the treatment of intestinal worms and scabies.

Flowering

Flowering and fruiting take place twice a year. The main flowering season is during the months of March-April with fruit maturing during August-September which enables plants to make the best use of monsoon rain. The second flowering season is from October to November. The fruits from the first flowering are mature in the months of August-September. Floral structure, which favours self-pollination, indicates that it is mostly self-pollinated. The initial fruit set by self-pollination was 73.06%, which was reduced to 24.24% at last.

Different Species

- ***Carissa carandas*** -A large evergreen shrub with a short stem grows to a height of 3-6 meters
- ***Carissa spinarum-*** This species bears slightly larger juicy and sweet fruits. It is native of India.
- ***Carissa inermis-*** A large climbing shrub or small tree occurring in West Coast
- ***Carissa suavissima*** -A lofty climber quite glabrous. Spines decurved.
- ***Carissa arduina*** - A **dwarf** glabrous shrub with twice-forked spines
- ***Carissa acuminate*** -Resembles closely to ***C. bispinosa***.
- ***Carissa bispinosa*** - An ever green shrub, hardy
- ***Carissa congesta-*** Distributed in Western Ghats. Fruits are used for preparing jam.
- ***Carissa edulis*** -Native of tropical Africa.
- ***Carissa ibispinosa-***The **flowers** are small and fruits are borne in clusters. The fruits resemble with that of *C. grandiflora.*
- ***Carissa grandiflora*** (Syn. *Arduina grandiflora*) English. -Natal plum, came to India from South Africa, fruits are a rich source of vitamin C.
- ***Carissa ovata*** - Fruits are small and are used for jam making.

- ***Carissa macrophylla*** -Native to India found growing in plenty in Western Ghats.
- ***Carissa paucinerria*** **African species:** Natal plum (***Carissa grandiflora*****).**
- Nakal cherry is a produced prepared from karonda in West Bengal
- ***Carrica ovata***-Jam preparation
- ***Carrica edulis***- Scented flowers

Types of Varieties

Four types of fruit: 1. Green 2. White with pink blush 3. Green with purple blush 4. Maroon

Varieties: Pant Manohar, Pant Sudarshan, Pant Suvarna, Konkan Bold, CHES-K - 2, CZK-2022, CZK - 2031, CZK-2011, CISH Kr-11, CHES- K- V-6, Maroon, Maru Gaurav, Maru Samridi, Thar Kamal.

Propagation and Planting

Seed Propagation

Karonda plants can be multiplied through seed very easily. Seed propagation is mostly commonly used methods in Karonda. The seed should be collected immediately after harvesting. The seeds sown immediately after extraction give higher germinated. Seedsare shown in trays and these seedlings are transplanted in poly ethylene bags at 3-4 leaf stage. The plants because ready for planting in 8-10 months. The germination in seedless or less seeded varieties is low. The plants produced from seeds have lot of variability with respect to fruit size, colour, taste etc.

Planting

Pits of 60 cm x 60 cm x 60 cm size are dug and full of natural manure and soil, in a 1:2 ratio. The planting distance for fence/hedge need to be 1-1.5 meter, requiring 300-400 plant life for planting alongside the boundary of 1-hectare field. If a one kind planting is to be raised, a planting distance of 2 m x 2 m need to be enough.

Manure and Fertilizers

Karanda plant life grown as defensive hedge are hardly ever manured or fertilized. Manuring, however, is beneficial. Otherwise, its plant life slowly gets exhausted after taking 2 years and begin displaying signs and symptoms of die back. Therefore, 15-20 kg well-rotten farmyard manure or compost/ plant to be implemented earlier than flowering.

Irrigation

Water requirement of Karonda could be very low. Irrigation after planting and manuring is essential. However, if there's no rain for the duration of the improvement of fruit, one irrigation can be given. This will boom the fruit size.

Training and Pruning

Regular plantations of Karonda may be trained on single or double stem. Therefore, extra undesirable shoots or laterals are eliminated from time-to-time to present the plant preferred shape. Suckers springing up from diseased dried twigs must be eliminated.

Weed Management

After irrigation, weeding should be done by which development of plant is done properly. It is necessary to weeding after every irrigation in Karanda.

Harvesting and Yield

Harvesting: July to September. Fruit maturity occurs 100-110 days after fruit set. This is an evergreen hardy shrub, indigenous to India. This plant is highly suitable for thorny hedging to provide protection to orchards and gardens. Fruits are very rich in iron (3.91% on a dry weight basis) and are valued for curing anemia. The vitamin A content of fruits amounts to 1619 IU per 100 g of edible portion. They have 87-90% pulp, 13-14.5% total soluble solids, and a pH of 4-6%. Mature fruits are used for making pickles, chutney, preserves, candy, squash, and jelly. In West Bengal, a product called "Nakal cherry" is prepared out of carondas and exported to Bangladesh. Plants start bearing fruit two years after planting. Fruits are available from August to October. An average yield is about 5–6 kg/bush.

Insect and Disease Control

Leaf eating caterpillars: monocrotophos 2 ml/l.

Fruit fly bactrocera dorsalis: Sanitation.

Twig dieback Diplodia natalensis: Pruning.

Stem canker Dithiorella sp.: Effected branches remove and burn.

Anthracnose: Black brown lesions, use copper-based fungicides copper oxide.

31

Kokum

- Kokum butter tree/Ratamba
- B.N.: *Garcinia indica*
- Family: Guttiferae

Introduction

Kokum, is native of the Western. Ghats in India, and Malaysia. In India it mainly grows in the western parts of Maharashtra, Karnataka, Kerala and Goa. It is found in evergreen and semi-evergreen forests and as a home garden tree. Kokum is an evergreen tree it is seldom planted in the form of an orchard and generally scattered trees are found in gardens and on waste lands.

Composition and Uses

The fresh rind of kokam comprises about 50–60% of the whole fruit. Ripe fruit rind is purple; sweet acid in taste contains acids like malic, tartaric and citric and a fat soluble yellow pigment garcinol. *Amrit kokum* is a refreshing drink made from ripe rind and dried rind known as amsole is used in curries in place of tamarind. Kokam seed yields an edible fat, *Kokam butter* (25-30%), which contains fatty acids like myristic, soap industry, suppositories, cocoa butter substitute, and is used for the sizing of cotton yarm.

Soil and Climate

It grows well within a temperature range of 25°C to 30°C. Thrives best in an area receiving annual rainfall between 2000 to 3000 mm. It is a hardy crop and can be grown well in all types of soils like acidic soils, lateritic soils with poor fertility. Well drained soil is essential. Luxurious growth and high yield potential is observed in warm and humid climate.

Varieties

- Vengurla 43
- Konkan Amrita
- CHES Ko- 5/6

Propagation

This tree can be propagated through **seeds,** but because it is dioecious, the sex of the plant can only be identified during the flowering stage. Hence, vegetative propagation methods like approach grafting and soft wood grafting are regarded as commercial propagation methods.

Planting

Pits of 60 x 60 x 60cm are dug in April-May and spacing - 6 x 6 m, pit should be filled up with good soil and about 10 kg F.Y.M. Planting should done in June - July after onset of monsoon, one-year-old vigorously growing seedling or graft is planted.

Manures and Fertilizers

In the first year, each plant should be applied with 2 kg F.Y.M., 50 g N, 25 g P_2O_5 and 25 g, K_2O. This dose is increased every year and from tenth year onwards each plant may be given 20 kg F.Y.M., 500 g N, 250 g P_2O_5 and 250 g K_2O. Manures and fertilizers are applied in a circular trench prepared under the canopy of the plant during the month of August, after the heavy rain are over.

Pollination and Fruit Set

Maximum dehiscence of anthers takes place between 6 a.m. to 8 a.m. The dehiscence of anthers takes place 15 to 20 minutes prior to anthesis of flowers. Wind appears to be an important agent of pollination in kokum. Kokum being predominantly dioecous plant. Cross pollination of the flowers on female trees is essential. Very high percentage (68-78%) of females flowers of kokum set fruits and required at least 10 per cent of male trees in a garden.

Harvesting and Yield

In kokum, flowering starts from October-November and continues up to February. Fruits are ready for harvest during the month of March to June. When the fruits turn from green to reddish in colour, they are plucked carefully by hand. Properly cared well grown kokum tree from 15 years onwards yields about 30 to 50 kg fruits per year.

32

Litchi

- Litchi/Queen of subtropical fruit/Fruit of high commerce
- B N: *Litchi chinensis*
- Family*:* Sapidaceae
- Chromosome No.: 2n = 2x = 30
- Origin: South China
- Inflorescence: Panicle
- Fruit type: Single seeded nut
- Edible part: Fleshy aril

Introduction

Litchi is a popular subtropical evergreen fruit; the white, translucent, flavoured aril is liked very much and is used for table purposes in India. The fruits are available during May-June in northern India. It is a native fruit of the warm subtropical zone of China and has excellent quality juice. Each fruit is a one-seeded nut. The pericarp is the skin, which has a papillate projection like a strawberry. Litchi is commonly consumed as a table fruit, but in China it is very popular in dried (litchi nut) or canned form, which is very popular among Chinese foods. A highly flavoured squash is also prepared from the fruits. The Chinese use the leaves for making poultices; the seeds are used for treating skin disorders. The fruits can be canned, made into squash, pickles, and wine, or dried to form litchi nuts. The tree is a medium sized evergreen with more branches. The fruits are borne in bunches. Flower bark and root decoction are used for gargling throat infections.

Uses and Composition

The Litchi fruit **is a single-seeded nut** and its edible portion is aril**.** The pericarp is the skin, which has papillate projections like strawberries. The fruits are a rich source of sugar (6-20 %), protein (0.7 %), fat (0.3%), minerals like calcium (10 mg), phosphorus (35 mg) and vitamins like vitamin C (64 mg), vitamin B_1 (0.2 mg), B_2 (0.03 mg) and B_3 (0.1 mg) per 100g. The fruits can be canned, made into squash, pickles, and wine, or dried to form litchi nuts. China, Taiwan, Thailand, South Africa and Madagascar are the main litchi-growing countries.

- China is first and India is second producer of litchi in the world.
- Bihar is the leading state in litchi production in india.

Origin and Distribution

Litchi is a native of southern China (Oman). It is cultivated in China, Taiwan, Thailand, India, Myanmar, West Indies, Australia, South Africa, Hawaii, Mauritius, and Hong Kong, South Africa, Australia, and Madagascar are the main litchi-growing countries. India is the largest producer of litchi in the world after China and the important states growing litchi in India are Bihar, West Bengal, Uttar Pradesh, Punjab, Haryana, Assam, Tripura, Orissa, Tamil Nadu and Karanataka.

Climate and Soil

It has successfully grown up to an elevation of 1000 m above mean sea level. This should be followed by a fairly long hot (16-22°C) climate with light rainfall during flowering and then by a high temperature during fruitset (18-24°C) with medium humidity. During the new flush, the temperature should be 28-30°C with high RH and heavy rainfall. During harvest, the temperature should be around 24–28°C with bright sunlight and high relative humidity. Litchi is a fruit crop of the **moist subtropics**. It is a native fruit of the warm subtropics zone of China, having juice of excellent quality. A warm subtropical climate is ideal for the growth and flowering of litchi. The winter should be cool as well as dry and devoid of frost (minimum temperature of 15°C). Wet spring, dry summer, and light winter are desirable conditions for fruiting in litchi. The maximum temperature during the flowering period varies from 21°C to 38°C. Flowering starts from Jan-Feb and fruits ripen in April & May. Temperature has a direct relation to flowering. A night temperature of 15-16°C for 2 months is essential to induce flowering and for vegetative growth. The ideal temperature is 30°C. An alluvial soil with good drainage and easy access to the water table is the best for the growth of litchi. It can be grown on a wide range of soils, i.e., acidic to alkaline soils. Acidic soils produce good quality fruits. The pH range is 5.5-7.0 with adequate soil depth.

Species: The family Sapindaceae and subfamily Nepheleae have about 125 genera and more than 1000 species.

The Genus Litchi has two Species

- ***Litchi philippinesis***- It is a wild type mostly used as root stock
- ***Litchi chinensis***- It is a commercial important species.
- ***Euphoria longana*****-** Langan/Anshpal-Small fruits, inferior quality.
- ***Nephelium lappaceum*****-** Rambutan- hairy fruits.

Varieties

- **Bihar:** Sahi, China, Purbi, Rose Scented, Kasba.
- **Uttar Pradesh:** Dehradun, Saharanpur.
- **Punjab & Haryana**: Saharanpur, Dehra Dun, Calcutta, Mazaffarpur, seedless late, rose scented.
- **West Bengal**, Bombay, China, Bedana, Elachi, Haak Yip, Tai so, Wai chee, Brewster, Gee Kee.
- **Haak Yip:** It is the commercial cultivar of Thailand, Taiwan and China.
- **Taiso:** It is the commercial cultivar of Australia, Queensland and South Africa (Where it is known as Mauritius).
- **Waichee:** It is cultivated in China and Australia.
- **Rose scented:** The aril has a delicate rosy flavour.
- **Muzaffarpur:** It is grown in India.
- **Bombai:** It is an important commercial cultiver of West Bengal.
- **Dehradun**: It is more susceptible to fruit cracking than Calcuttia
- **Early seedless** (Early Bedana) table and processing variety
- **Swarna Roopa:** (Seedless) –Highly resistant to fruit cracking 1st variety developed in India, CHES, Ranchi and commercially grown in Chhotanagpur
- **Early varieties-** Muzaffarpur (large red), Saharanpur, Shahi.purbi
- **Mid seacon varieties-** Dehradun, Rose Scented, Mclean
- **Late varieties-** Calcuttia, China, Bombai, Elaichi
- **Sabour Madhu** - Purbi x Bedana and **Sabour Priya**: Purbi x Bedana
- Regular bearer varieties: Shahi, Rose Scented and Dehradun
- Bunch bearing clone and suitable for canning: **Shah**i
- Table purpose variety: Purbi, China, Elachi, Calcuttia, Bomabai, Shahi (Pride of Bihar)

Propagation

Commercially propagated by **air layering (500 ppm IBA),** Air layering done during the spring and early summer season is commercially used. To avoid the high variability and long juvenile period, seed propagation is not recommended. *Litchi chinensis* and *Litchi philippinesis* (Used as root stock). commercially propagated by air layering, gootee, or marcottage in China (July to September) from 1-year old shoots.

Wind Break

The growth of litchi plants is adversely affected due to hot wind in summer and cold waves in winter. It is advisable to plant suitable wind breaks around the boundary. A windbreak should be established around the litchi orchard using **eucalyptus,** casuarinas, and **jamun,** preferably before one year of the establishment of the litchi orchard.

Planting

Planting is done in the rainy season or even just after the monsoon. Commercial planting system: Square system. Litchi trees are planted by following the square system at 8x8m or 10x10m apart in $1m^3$ pits. Each pit was provided with 30–40 kg of FYM + 2 kg of bone meal + 300 g of MOP. 1 basket of soil from a litchi orchard containing mycorrhizal fungi. A crop that is heavily cross-pollinated Pollinators: honeybees and flies.

Irrigation

The plant is highly sensitive to moisture stress; supplementary irrigation is provided in the hot season and mulching is also practiced.

Manure and Fertilizers

Age of tree	N (g/pl)	P (g/pl)	K (g/pl)	30-40 kg FYM/Tree / Year
Ith Year	75	25	75	
2nd Year	100	25	100	
3rd year	150	50	150	
4th Year	250	25	250	
5th Year	600	25	600	

Training and pruning

A modified central leader system of training is followed and pruning is not done in litchi, except removing dried, diseased and criss crossed branches.

Flowering

The grafted or layered litchi tree comes to bearing in 3–4 years, while the seedlings take 8–12 years to flower. Flowers are terminal and appear in the current season wood. Due to water stress, fruit set is very low and flower drop is also common. Fruit drop occurs 4 weeks after fruit set, which may be due to failure of fertilization, embryo abortion, high temperature, and low humidity. The inflorescence of litchi is a compound raceme emerging from both terminal and axillary buds. Litchi flowers are staminate, hermaphrodite, and pseudohermaphrodite (functionally male). To control fruit drop, apply NAA at 20-30 ppm, GA3 at 20-50 ppm, or 2,4-D at 20 ppm before flower opening. Apart from this, girdling and centering also increase flowering. Flowering: The sequential opening of the flowers in litchi is male (phase I), followed by hermaphrodites (phase II) and pseudohermaphrodites (phase III), introduced in the 17th century to India. The red colour of the skin of fruits is due to anthocyanin. The major organic acid present in the fruits is malic acid, at 80%. It requires high RH and high lime content in the soil. Recalcitrant seed, Viability of seeds: 4–5 days. panicle with branches, Flowers are petalless. Seedlessness is due to stimulative parthenocarpy.

Harvesting

Litchi is a non-climacteric fruit, and harvesting is to be done at a fully mature stage. It takes about 55–60 days from flowering to harvest. The fruits are harvested during May and June. The maturity indices are flatness of tubercles, smoothness of epicarp, and colour development (Green-Pink). The whole bunch is harvested manually.

Yield

The new flush arising in spring, ends in inflorescence. Fruits are harvested after full maturity and ripening on the tree itself. Fruits are harvested as bunches, precooled, and then stored at 0°C to 1°C. About 80-150 kg of fruit/tree in Indian conditions. The fruits cannot be stored for more than 2-3 days at room temperature. It can be stored for 5 weeks at a temperature of 1.6-7.2°C in perforated polythene bags. Fruit colour is improved by dipping it in 250 ppm ethrel.

Physiological Disorder

- **Fruit cracking:** Major problem in the world. The whole fluctuations in diurnal temperature, heavy irrigation/rain after prolonged dry spell, hailstorms during fruit development, Boron deficiency causes injury to the fruit skin.

- **To avoid fruit cracking:** The field to be irrigated during fruit growth and in early summer. Spraying with 2, 4, D (10 ppm), GA_3 (20 ppm) and Boron (0.4 per cent) reduces fruit cracking. Much severe on 'Bombai' 'China' and 'Kasba' variety
- **Chicken tongue** is the physiological disorder of litchi
- Sun burning is a physiological disorder

Pests and Diseases

- **Mites-Eriophyes litchi** Symptoms: Curling of leaves. Leaf galls, caused by *eriophid* mite *Aceria* litchi is a common problem
- **Catterpillar:** *Inderbella tetroanis*
- **The bark feeder** - *Indarbela tertronis*
- **Storage rot-** *Geotrichum candidum*
- **Eriophyid mite** *Aceria litchi:* Major pest of litchi.
- **Leaf rollers** can be controlled by an insecticide like Nuvan which has fumigant action also.
- **Red rust** by alge (*Cephaleuros virescens* can be controlled by spraying lime – sulphur 3 times.
- **Leaf galls**, caused by eriophid mite Aceria litchi is a common prob

33

Longan

- Longan/Eye of dragon/Little brother of litchi
- B.N.: *Dimocarpus longan* with synonyms as *Euphoria longana* and *Nephelium longana*
- Family: Sapindacea
- Chromosopme No.: 2n = 22
- Origin: South China

Introduction

Longan is a close relative of lychee, having similar growth and fruiting habits but being somewhat hardier than lychee. The fruits are smaller, smooth-surfaced with a yellow to brown colour. It originated in the subtropical region of China or in the area between Myanmar and India. Fruits contain 25.2% carbohydrate, 1% protein, 0.5% fat, vitamin A (28 IU/100G) and vitamin C. Longan contributes to the greatest fruit export of Thailand. Longan/Lungan, also known as "Dragon's Eye" or "Little Brother of Lychee", is an important sub-tropical fruit crop.

Climate and Soil Requirement

Longan requires a warm subtropical climate with high rainfall. The winter should be short and frost-free. The summer should be long, hot, humid, and wet. The best temperature for flowering and fruit set is 20 to 25°C. Temperatures above 4°C will cause damage to fruit, resulting in fruit drop. A deep, well-drained, fertile soil with a pH of 5.5 to 6.0 with low salinity will be ideal for its growth.

Cultivars

- **Fu yan:** (Lucky Eye) It is the major cultivar of fujian province of China. It is the best variety for canning.
- **Wu long line:** (Black Dragon peak)
- **Wu yuan:** (Black Round)
- **Daw:** It is the popular cultivar of Thailand.

- **Store Gorge Lungan**
- **Tsao 40 (Early Rice)**
- **She Pi (Snake skin)**
- **Hua Kioh (Flower skin)**
- **Chompoo, Kohala, Degalman, Ponyai, Kay Sweeney, Diamond River, Biewkieware and Fukien**

Propagation and Planting

Propagation and planting Longan cultivars have to be propagated by air layering or marcotting the most recently flushed growth, with maximum maturity at about 45-60 cm from the tip. A planting distance of 6 m x 6 m can be allowed in an HDP system. Plants should be planted in pits the size of 1 m x 1 m x 1 m.

Training and Pruning

Early growth is to be encouraged so as to get a good tree size and canopy within 3–4 years. The trees can be trained to a height of 2 m and 3-4 vigorous young growths with wide crothes are allowed to form the main frame work. Each lateral can be allowed to have two secondaries. The trees should be trained to get light penetration through the canopy.

Irrigation and Manuring

Irrigation is given to encourage bud burst. It should be withheld until after flower emergence. Again, irrigation is to be continued from panicle emergence, flowering, fruit set, harvest, and post-harvest flushing. Again, when the tree undergoes a rest period, irrigation should be stopped. Just like irrigation, the fetilizer application should also be done 3–4 times during the different growth periods in a year. For a five-year-old tree, 625 g N, 150 g P_2O_5, and 800 g K have to be applied and increased by 20% every year.

Interculture and Plant Protections

Alternate bearing is a common phenomenon. In an "on" year, 40% of the flower spikes and 30% of young fruits can be thinned. In an "off" year, a portion of the shoots produced is spring should be thinned to reduce the competition between shoots and fruits.

1. **Fruit fly (*Bactrocera dorsalis*)** is a serious pest, which can be controlled by any safe insecticide like malathion. (Refer Mango section for details).

Harvest and Yield

Longan fruits are harvested after full maturity and ripening. A fully matured tree yields 100 – 120 kg /year. The fruits can be stored for 4-6 weeks at 5°C.

34

Loquat

- Loquat/Japanese medlar/Japanese plum
- B.N.: *Eriobotrya japonica*
- Family: Rosaceace
- Chromosome No. 2n = 32
- Origin: Southern China
- Inflorescence: Panicle (Compound racemose)
- Fruit type: Pome
- Edible part: Fleshy thalamus

Introduction

Evergreen and subtropical fruit crop, used as an ornamental purpose. Japan is the largest producer, followed by China in the world. Loquat is a strictly **sub-tropical** evergreen **pome fruit.** Pome means consisting of 5 carpels united and covered by an edible portion of fruit. Loquat is a delicious summer fruit. Depending on the cultivar, it contains 60 to 70% pulp and 15-20% seed. The pulp is composed of 88% water, 0.6% proteins, 9.6% carbohydrates, 0.5% minerals, and 1.0% fibre. Fruit is a rich source of vitamin-A: 500–2300 IU/100g, potassium, and manganese. Seeds and peels contain amygdalin, which causes toxic symptoms in children when converted to HCN. Some preparations like jam, jelly, and squash are made from the pulp. It is consumed as a fresh fruit. The colour of epicarp usually ranges between yellow and red. The fruit tastes sweet and sour with good flavour.

Origin and Distribution

Loquat is native to China. Many good loquat cultivars also originated in China and Japan. It has become a commercial fruit in Australia, the U.S.A., Hawaii, France and India. **As far as the area under loquat is concerned, U.P.** occupies the first position. Loquat is a highly preferred fruit for kitchen gardens in Punjab. It is also being grown on a small scale in the orchards of Amritsar, Gurdaspur, Hoshiarpur and Patiala districts.

Climate and Soil

Loquat has adapted well to the harsh climate of North India. It is performing well in tropical climate and lower hills, where temperature rarely falls below 0°C. It cannot tolerate severe frost in winter. It is tolerant of drought and heat. It likes a warm and dry climate at fruit ripening time. Frost is a limiting factor for loquat cultivation. Fruits are susceptible to sunburn or a purple spot. Loquat grows well on fertile clay loam soils. It can be grown in loam as well as on sandy loam soils. Waterlogged conditions are harmful to the plants.

Cultivars

- **Improved Golden Yellow:** The fruit colour is similar to the golden yellow cultivar, but the shape is pyriform.
- **Pale Yellow:** This cultivar originated in California.
- **Improved Pale Yellow:** This cultivar originated from pale yellow and have well sized fruits of pyriform shape.
- **Early varieties:** Golden Yellow, Improved Golden Yellow, Large Round, Thames Pride, Pale Yellow-Self incompatible
- **Mid-season varieties:** Fire Ball, Improved Pale Yellow, Large Agra, Mammoth, Matchless, Safeda and Large Round
- **Late season varieties:** California Advance, Tanaka
- **Popular varieties:** Golden Yellow, Improved Golden Yellow, Pale yellow, Pale yellow, Large Round, Thames Pride, Fire Ball, Improved Pale Yellow, Safeda, Mammoth, Matchless, Advance, Tanaka, Ahdar, Akko-13, Asfar
- **Partially fertile varieties:** Advance and Tanaka
- California Advance is the best polloinizer for Improved Golden Yellow.
- **Self-incompatible varieties:** Golden Yellow, Pale yellow, Tanaka, California Advance, Improved Golden Yellow and Agra Large.
- **Partially self-incompatible** – Large Round, Fir Ball, Thames Pride, California Pride and Tanaka.
- **Seedlessness in loquat:** triploid
- **Ist triplopid variety:** Kibou
- To induce seedlessness in loquat by GA_3 @ 100-200 ppm

Propagation

Seeds are sown in the nursery beds after their immediate extraction from fruits. Seeds germinate soon and become seedlings graftable within a year. Don't let the seeds dry after extraction. The seedlings are **inarched** during July-August. Loquat is commonly propagated through **inarching**. Air layering can be done, but success is limited. Air layering is a successful vegetative method.

Planting

It can be planted in February-March and August-September. The planting distance varies from 6–7 metres apart, depending upon the soil fertility level. One-metre-deep and one-metre-diameter pits should be prepared at least one month prior to actual plantation. Lindane 30 g or two litres of chloropyriphos solution @ 10ml/L may be applied to each pit to check for white ants.

Irrigation

Loquat is a hardy plant and can tolerate drought for some time. Irrigation should be applied at longer intervals of 20–25 days during non-fruiting time and this interval may be reduced to 15–20 days during fruit development. No irrigation should be applied during the rainy season and winter.

Interculture

Keep the plant basins clear of weeds throughout the year by hoeing. Rabi crops should not be grown in loquat orchards since fruit is harvested in March-April. In Kharif, pulses like moong, mash, peas and crops like Toria can be grown.

Manures and Fertilizers

As usual, full farmyard manure, super phosphate and muriate of potash should be applied in September. Half the urea should be applied in October and the rest half in January. The manures and fertilizers should be mixed well and uniformly in the tree basins.

Training and Pruning

Loquat is also trained in a modified leader system of training. However, plants are also left to grow on their own and automatically develop a sound scaffold system. The tree trunk should be kept without any scaffolding up to 50 cm of trunk height. The fruits are harvested as whole panicles and small portions of shoots. This much pruning is sufficient to get new growth. If some branches are to be removed, they should be removed in June.

Flowering and Fruiting

Loquat flowers from September to January under subtropical conditions. The panicles producing flowers early in the season also produce well-sized fruits in comparison to those appearing late. Inflorescence is a panicle (compound racemose). Loquat cultivars are self-incompatible or partially self-fruitful. The type of incompatibility is gametophytic. Two to three cultivars planted together give a high yield. Flowers are produced on current season shoots. Cross pollinated crop and mode of pollination: Bees.

Fruit Maturity and Harvesting

Plants start bearing fruit after three years of age. Time to maturity: 70 days after fruit set. Fully matured and well-coloured fruits with a TSS of above 9% should be harvested. All the fruits in a cluster almost ripen uniformly. California Advance, Golden Yellow, and Themes Pride should be harvested at 11% TSS. While harvesting, a small portion of the shoot should also be harvested with a sharp secateur. Separate the clusters into A and B grades.

Yield

An average **yield** of 15 to 20 kg/tree can be obtained. non-climacteric fruit. Application of Paclobutazol (500 ppm) around the base of the trunk is recommended for increasing fruit size. The poor quality fruits may be sent to be local/small market. It is available on the market during mid-march-may.

Packing

The graded fruits are properly packed in boxes. The boxes may be wooden or cardboard. A-grade fruits should be well-sized and unbruised, while B-grade fruits should be small and good. The rest of the fruits may be packed separately. The paper cuttings can be placed at the bottom of each box to provide a cushion between different layers. Wooden boxes of a size of 14 kg should be used for sending the fruit to nearby markets.

35

Mahua

- Mahua/Indian Butter Tree/State tree of Gujrat
- B. N.: *Madhuca latifolia.*
- Family: Sapotaceae
- Chromosome No.: 2n = 24, X = 12
- Native = India
- Inflorescence: Fasicle
- Fruit type: Berry
- Edible part: Mesocarp

Introduction

Mahua is an indigenous tree of high economic value. A characteristic tree of the dry region. It is found in northern and central India. It is commonly grown in eastern Uttar Pradesh, Maharashtra, Bihar, Orissa, Andhra Pradesh and Gujarat. The tree has been well known to rural folk for ages in India. Its flowers, fruits, and oil obtained from seeds are eaten in various ways. The corolla, commonly called mahua flowers, is a rich source of sugar, containing an appreciable amount of vitamins and minerals. Fruit, a berry, is eaten raw or cooked. Pulp is a good source of sugar, while dry husk makes a good source of absolute alcohol. Seed is a good source of oil. Linoleic acid, a major unsaturated fatty acid found in mahua oil, is used to lower cholesterol levels in blood serum, making it useful for heart patients. Madhuca is derived from the **Sanskrit word madhu**.

Use and Composition

Fresh flowers and dried ones contain 21 to 25% and 55 to 58% of sugar, respectively. Vitamin C (30–60 mg/100 g), vitamin A (528–890 IU), and mineral matter (0.8–1.5%) are also present in fresh flowers. Fruits contain 55-65% husk, 10-15 sugar, 1.8-2.4% minerals, 51-72 mg vitamin C, and 586-890 IU vitamin A per 100 g.

Climate and Soil

Mahua prefers tropical and subtropical climate, 1200 m from mean sea level. The optimum pH of soil for phalsa is 6.1 to 6.5. It is often grown in marginal lands. Fruits can be obtained when grown in well-drained loamy soil.

Different Species and Varieties

- **Species** –*Madhuca longifolia, M. latofolia, M. butyracea, M. neriifolia, M. bourdillonii* and *Bassia malabarica*
- **Varieties –Thar Madhu** from CIAH, Bikaner.
- **Propagation:** Plant propagation is generally by seeds. Seeds are sown in the months of June–July and seeds are polyembryony in nature.
- **Planting:** Pit size: 1 x 1 x 1 m and plant spacing required: 10 x 10 m. Planting time monsoon season.
- **Manures and fertilizers:**10 Kg FYM, 100 g N, 50 g P and 75 g K/plant/ year are recommended.

Flowering

Leaf fall occurs from February to April. Fully ripe flower with a shedding corolla. Mahua is a medium-sized to large deciduous tree, usually with a large rounded crown, found up to an altitude of 1,200 metres and of 12 to 15 metres. Fruits are green at maturity and turn pinkish yellow when ripe. Botanically, a fruit is a fleshy berry. The seed contains two kernels, indicating the recalcitrant nature of the seed. Trees flower when the tree is leafless during March-April. pollinated by wind (anemophil).

Seed and Fruit Setting and Yield

The matured fruits fall on the ground in May and July in North India and August and September in South India.

Flower 100-150 kg /tree/year.

36

Mango

- Mango/ King of Tropical Fruits/National fruit of India/Pride fruit of India/ Hisdustan fruit of India/Bathroom fruit/Symbol of love/Ambassador of Indian fruits
- B.N.: *Mangifera indica*
- Family: -Anacardaceae
- Chromosome No.: 2n= 40
- Origin: -South East Asia
- Inflorescence: Panicle
- Fruit type: Drupe
- Edible part: Mesocarp

Introduction

It is the **national fruit** of India. No other fruit can compete with mango in respect to area, production, nutritive value, and popularity. For this reason, the mango is known as the king of fruits. In India, mango cultivation is 4000-6000 years old. Growing states are U.P., followed by A.P. and to some extent in all the rest of the states. It is also grown in India, China, Thailand, Pakistan, Bangladesh, Burma, Malaya, Philippines, Sri Lanka, Egypt, Africa, Florida, Hawaii, Brazil, Mexico, etc. Unripe fruits are used for culinary purposes, such as pickles, powder, chutney, and sarbat. Ripe fruits are used for squash, jam, toffee, juice, slices, and canning of pulp and slices.

Climate and Soil

Mango is a **tropical and sub-tropical** crop that can grow up to an altitude of about 1500 meters. Temperatures range from 24-27°C during fruit development and maturity, and it can withstand temperatures as high as 48°C. Higher temperatures during fruit development hasten the maturity and improve the fruit size and quality. **High humidity** as well as **rainfall** during **flowering** is not suitable, as it **increases pest and**

disease problems. Bright, sunny days and low humidity during flowering are ideal. The two important considerations in mango cultivation are a frost-free **and dry climate at flowering** and sufficient heat during ripening (maturity) periods. Soils from alluvial to laterite types are best. 2.0 to 2.5 m. deep. Well drained and well fertile soils with good water holding capacity are the best. Black cotton soil, as well as saline and alkaline soils, is not good for plant and pH range 5.5 to 7.5 is desirable.

Varieties

- **Pant Chandra**: selection from Dashehari
- **Pant Sinduri**: selection from Dashehari
- **Ambika-** (Amrapali x Janardhan Pasand) Late ripening variety
- **Pusa Lalima**: (Dashehari x Sansation)
- **Pusa Peetembar :**(Amarapali x Lal Sundari) Yellow colour variety
- **Pusa Pratibha:** (Amarapali x Sensation)
- **Arunika** (Amarpali x Vanraj)
- **Pusa Arunima** (Amarapali x Sensation)
- **Pusa Shreshtha** (Amarapali x Sensation) Regular bearer.
- **Mallika** (Neelum x Dashehari) Regular Bearer- 1971, IARI,
- **Amrapal**i (Dashehari x Neelum) Regular bearer, Desert variety, 1978, IARI.
- **Ratna** (Neelum x Alphonso) Regular bearer, 1981 Vengurla.
- **Sindhu** (Ratna x Alphonso) Seedless, Regional Fruit Research Station, Vengurla.
- **Arka Aruna** (Banganapalli x Alphonso) Regular bearer – Suitable for HDP.
- **Arka Puneet** (Alphonso x Banganapalli) Regular Bearer-Suitable for **canning.**
- **Arka Neelkiran** (Alphonso x Neelum) Regular bearer – Suitable for export.
- **Arka Anmol** (Alphonso x JanardhanPasand) Regular Bearer-Suitable for export.
- **Arka Neelanchal Kesari.**
- **Manjira** (Rumani x Neelum), Au Rumani (Rumani x Mulgoa)- 1988.
- **PKM-2** (Neelum x Mulgoa).
- PKM-1 (Chinnaswarnarekha x Neelum).

- **Swarna Jehangir** (Chinnaswarnarekha x Jehangir).
- **Neeludin** (Neelum x Himayuddin).
- Neelgoa (Neelum x Yerra Mulgoa) Regular bearer.
- **Neeleshan** (Neelum x Baneshan) Suitable for canning.
- **KMH-1** (Cherukrasam xKhader).
- **Sai Sugandha** (Totapuri x Kesar) Suitable for pulp.
- **Prabha Sankar** (Bombay x Kalapady) Regular bearer and late season variety.
- **Sunder Langra** (Sardar Pasand x Langra) Free from malformation.
- **Neelphonso** (Neelum x Alphonso).
- **Neelshwari** (Neelum x Dashehari) are evolved at Fruit.
- **Arka Udaya**–Amrapali x Arka Anmol.
- **Akshay.**
- **Pusa Manohar**-(Hybrid 8-11) Amarpali x Lal Sundari, Regular bearing.
- **Pusa Deepshikha**-(Hybrid)- Amarpali x Sensation, Regular bearing.
- **Pusa Surya**- Selection from Eldon, introduce from Brazil.and Haden from Mulgoa.
- **Neelum**-(Kasaladdu) Regul. bearer, late season mango.
- **Totapuri** (Kallamai or Kilimooku) Suitable for processing, grown in south India, regular bearer.
- **Bombay Green**-It is one of the earliest varieties of north India, Susceptible to malformation, pollinizer variety and highest vitamin C.
- **Fazli** -This is indigenous to Bihar and West Bengal, late-maturing (August) mango.
- **Gulab Khas:** It is indigenous to Bihar, (Rosy flavor) Regular and heavy bearer.
- **Himsagar**-Very popular in West Bengal, it is a regular bearing mango.
- **Kesar-** Popular in Saurashtra region of Gujarat, Suitable for processing, known is Queen of mango.
- **Sonpari:** Alphanso x Baneshan.
- **Kishen Bhog**-Indigenous to West Bengal.
- **Mankurad**- It is a mid-season variety, popular in Goa.
- **Pairi** -A native to coastal Maharashtra including Goa, regular bearer, Early maturing.
- **Dashehari-51, 35**: from Dashehari selected at the CISH, Lucknow.

- **Paiyur -1**, has been made from Neelum, in addition to few dwarf polyembryonic selections made in the north-eastern region.
- **Sucking purpose**: Indira Mohini and Indira Raswanti.
- **Pickle purpose:** Indira Achar, Indira Maya, Indira Bhusan, Ram kela.
- **Canning purpose**: Alphonso and Dashehari, Padiri.
- **Neelum** a regular bearing cultivar is a best combiner with most cultivars.
- Manjeera (Rumani x Neelum) are developed at Sangareddy.
- **Ratna, Arka Puneet, Arka Anmol and Arka Neelkiran** are Free from spongy tissue.
- **Niranjan** - off season mango selection made at Parbhani.
- **Rosica-** is the only mutant cultivar of mango.
- **Madhulica-** is the most precocious cultivar of mango and off season,
- **Rumani-** Apple shaped variety.
- **Alphonso** (Khader, Badami, Gundu, Haphus), Suitable for export and susceptible to spongy tissue.
- **Dashehari**- Mid season, Susceptible to mango malformation.
- **Export varieties: Alphonso, Kesar, Gulabkhas, LakhanBhog, Safdar Pasand.**
- **Langra**: (Turpentine flavour), -Most popular in North India, Biennial, more prone to fruit drop than Dashehari.
- **Xavier –Variety have highest TSS -24.8 Brix.**
- **Chausa-** Late maturing variety.
- Dwarfing cultivars: Ambalavi, Kalapady.
- **Konkan Ruchi**, **Konkan Raja, Lily, Austin, Suvarna**.
- **Banganapalli-** Commercial var. in A.P., Early maturing variety.
- Mainly growing var. in Kanyakumari district due to microclimate; Neelum, Rumani, Bangalora.
- **Regular bearer varieties:** Totapuri or Banglaora, Bangapalli, Neelum, Ratna, Sindhu.
- **Most popular varieties in North India:** Dashahari, Chausa.
- **Gomera:** (Gomera -1, 2, 3 Salt tolerance), Nekkare salt tolerant, (13/1 salt tolerant, commercially in Israel).
- **Promising dwarfing genotype**: Kerela Dwarf, Janardan Pasand, Amrapali, Creeping and Nileshwar.
- **Colored mango cultivars:** Tommy Atkins, Zilette, Haden Sensation and Julie are introduced from Miami, Florida (USA).

- **Indian colored varieties:** Janardan Pasand, Suvarnarekha, Fernandin and Rosa, Sindhura, Vanraj, Kesington, Sensation, Gulabi, Arka Puneet, Nazuk Badan, Raspuri.
- **Mulgoa:** is the mother of all Floridan coloured mango, Late season and shy bearing and useful for preserved.
- **Jawahar:** Gulabkhas x Mahmood Bahar -from Bihar.
- **Menaka:** selection from gulabkhas.
- **Subhas:** selection from Zardalu.
- **Chhattisgrh Nandraj**

Other Indian Varieties

Gaddamar, Ottu Mangai, Langra Banarsi, Badshah pasand, Surkha, Husan Nara, Cheruku Rasalu, Chinna Rasalu, Pedda Rasalu, Fajri Kalan, Fernandian, Himayath, Imam Pasand, Jehangir, Kalami, Komanga, Lalbagh, Maldah, Malgis, Mankur (GOA), Mankurad, Moovandan, Nattuma, Rajapuri, SammarBahisht, Suvarnarekha, Vanraj, Zardalu, Alampur Baneshan, Puliyan, Kuttiyattor, Ela Manga, Nannari.

Propagation

- Seed
- Vegetative propagation:
- Inarching/ Approach grafting
- Soft wood wedge grafting
- Veneer grafting and side grafting
- Epicotyl / stone grafting

Inarching or approach grafting

Inarching: This is a complicated and time-consuming method that has a 90-95 % success rate.

Disadvantages: This is a laborious and time-consuming method, costly method and rootstocks have to be carried to the mother plant for grafting. I usually get pot bound. For the actual inarching operation, the stock is brought close to the scion. A thin slice of bark about 6–8 cm long and about 8 mm in thickness at a height of about 20 cm above ground level is removed with a sharp knife from the stock. A similar cut is made in scion. Thus the cambium layer of both stock and scion are exposed. These cuts are brought together and tied firmly with the help of a 200 mm polyethane strip.

Softwood Wedge Grafting

In this technique, grafting is done with a mature, procured scion on the emerging soft coppery red shoot of the rootstock. The leaves on the stock must be retained while grafting to achieve high success. The technique is effective in **dry, hot weather in the** area. Arid and semi-arid regions in arching grafting are not well established due to disturbance of the tap root system, so *in situ* soft grafting is recommended.

Planting

Square, rectangular and Contour are common methods of planting but square system is the most popular in mango the planting distance will vary with the cultivar and the location ranging between 10-12 m and high Density Planting of cv. Amrapali could be planted at a spacing of 2.5 m x 2.5 m with a density of 1600 plants per ha. Dashehari at 3.0 x2.5 m.

Preparation of Land

Usual ploughing, harrowing and leveling as well as preparation of irrigation channels. **Square** system of planting is most popular for mango cultivation.

Preparation of Pits and Planting

Planting distance will vary according to vigor of variety and type of soil. Generally, it ranges from 10 to 12 meters. Pits of 1×1×1 meters are dug during summer (two months before monsoon). Kept Open under the sun. Pits filled with mixture of 50 kg. FYM and soil. Mix 100 g. termite killing dust. Planting is done during the rainy season when the soil in the pits has already settled. The planting should preferably be done during cloudy weather and in the evening. The plants should be irrigated immediately after planting.

Training and Pruning

Normally mango trees require very less or no pruning however, the training of the plants in the initial stages is very essential to give them proper shape. Especially when the graft has branched too low, the process of training becomes very important. At least 75 cm of the main stem should be kept free from branching and the first main branch may be allowed after that. The main branches should be spaced in such a way that they grow in different directions and are at least 20-25 cm apart.

Manuring and Fertilization

Manuring mango plant start right from the inception of planting in the field. Fertilizer application during the first year of planting may be given as 10 kg

FYM 300 gm urea + 300 g SSP + 100 g MP per plant. Above dose should be increased. Accordingly, a 10-year-old tree should nutrients contained in 100 kg FYM + 3 kg urea + 5 kg SSP + 1 kg MP and should be followed for subsequent years. Fertilizers may be applied in two split doses, one half immediately after the harvesting of fruits in June-july. Half in October, followed by irrigation if there are no rains.

Irrigation

Irrigation during the period of first four to five years of the early growth. Stages is necessary mainly to encourage the rapid and vigours growth of the tree and stronger root system. 1st year – young trees should be irrigated after every 10-15 days in winter and 5-8 days in summer.4. 4th – 5th year- 14-15 days in winter and 10-15 days in summer. Bearing trees- irrigation at regular interval (10-15 days).

Inter-Cropping

Grow some intercrops in the orchard during initial years (nonbearing period) to get extra income. Uncultivated land will lead to weeds. Onion, tomato, radish, carrot, cabbage etc. and fruit crops like papaya, phalsa, banana etc. can be grown as intercrop.

Cover-Cropping

Grow sunhemp, dhaincha etc. to protect orchard soil from erosion and also to enrich the soil.

Flowering

- Nov - Dec. in A. P. and states of S. India.
- Feb - March in Northern India.
- Jan. - Feb. in Eastern parts of India.
- Dec. - Jan. in Gujarat

Generally period of flowering of two weeks. Low temp. extends it, and higher temp. shorten it. Two types of flowers in each panicle, (**Male and Perfect**). Number of flowers/panicle-1000-6000. Only **0.1%** fruits reach to maturity. Highly cross pollinated crop. Pollination through insects & wind.

Harvesting

Most satisfactory method available at the moment is the one which makes use of a bamboo hand tool, called mango picker locally known as Zela. Consists of long bamboo pole fitted with a cutting shear at the distal end, and under which a fruit collecting net is tried.

Yield

Seedling tree after 8-10 yrs. Grafted tree after 4-5 yrs. Av. yield **80-100 kg. & up to 200 kg/tree.**

Marketing

Freshly harvested matured fruits are packed in 10 kg volume card-board box. & send for market. Mango is climacteric fruit, so stored properly for ripening. Mature fruits ripen after 5-6 days. Ethylene (3000 ppm) can accelerate ripening.

Biennial/Alternate Bearing

Biennial, alternate, intermittent or irregular bearing generally signifies the tendency of mango trees to bear a heavy crop in one year (on year) and very little or no crop in the succeeding one (off year). e.g. Alphonso, Kesar, Deshehari, Langra are biennial bearers. Banganpalli, Neelum, Sindhu, Ratna Niranjan are regular bearers. Biennial bearing had been attribute to the causes like genetically, physiological, environmental nutritional.

Factors Affecting of Biennial Bearing

A. Internal Factors

- Varietal Character: Alphanso, Pairi, Langra etc.
- C: N ratio: During 'On' year most of CHO are utilized for development of fruits due to which mango flowers alternate year.
- Sex ratio: Low percentage of perfect flowers may turn to 'off' year.
- Age of plant: Up to 10 years generally bears regularly and it sets in alternate bearing after 10 years.
- Age of shoots: Shoot which is 8 to 9 months old in Jan only flower.

B. External Factors

- Adverse climate factors: High temperature, frost and rains strong wind at flowering result 'off year'.
- Pests and disease: Powdery mildew, hoppers result in 'off' year.

Nutrition and management also is in external factor.

Control of Biennial Bearing

- Proper maintenance of orchard.
- Deblossoming in 'On year'.
- Smudging & chemical regulation. Smudging (smoke) is age old practice in Philippines. In India not followed. Spraying of 200 ppm Etheral during 'off year' induce flowering. Auxins, KNO_3 also beneficial.

- Pruning.
- Planting of regular varieties like Bengalora, Totapuri, Neelum, Romania, Redsmall, Rumani, Mallika, Amrapalli.

Pests

- **Mango hopper**

  Nymph & adult suck sap from tender fruit, Panicle etc. seen throughout year. But damage during flowering & in rest period hibernate under the dried bark of stem. They secrete honey-dew on which sooty mold develops. Dusting 0.15 % carbaryl. or 0.05 % phosphomidion or 0.15 % melathion or 0.04 % endrin or nuvacron, once at panicle stage & next at fruit-set stage.

- **Mealy Bug**

  Nymph emerge from soil in **Dec-Jan**. & climb. Suck juice from tender **shoot, panicle** etc. Affected part dry. Controlled by (1) Mixing 10% BHC in soil around trunk. (2) Stick 30 cm wide Greece band at 30 cm above on the trunk in Dec-Jan. (3) Spray monochrotophas 0.04% or melathion 0.08%.

- **Mango Stem Borer**

  It **tunnels trunk & branches**. Controlled by (1) Any wound on branch/ trunk should not be kept open. (2) Clean the tunnel with hard wire & pour petrol or kerosene or endosulfan& seal the hole with mud.

- **Fruit fly: Fly lays eggs under the rind of fruit, before fruit matures**.

  The affected fruit begins to rot & drop down. Controlled by (1) Keep the orchard clean. (2) Collect & destroy damaged fruits. (3) Use Methyl euginol trap.

- **Termites:**

  Destroy galleries of termites from trunk. Mix 10 % BHC at planting in soil mixture.

Diseases

1. **Powdery mildew** : Fungal disease. Favoured by high humidity, cloudy weather and low night temp. During panicle development & fruit-setting period Greyish white powdery bloom on flower buds & fruits. Which results failure of crop? Controlled by (1) Spray wettablesulphar 0.2 %, Bavistin 0.1 %, Benlate 0.1%. First spray at pre-bloom stage & two sprays at full bloom & fruit set stages.

2. **Anthracnose** : Fungal disease More in humid & high rainfall regions. Panicles, leaves, tender shoots etc. are affected with black necrotic areas & die back. Controlled by pruning affected dead twigs. Spray Bordeaux mixture (3:3:50) or Bavistin 0.1 % or Benlate 0.1 % in Feb, April & September.
3. **Bacterial canker or bacterial spot:** Affected fruit crack & gum oose out. Spray Bordeaux mixture (3:3:50) or streptocycline.

Physiological Disorders

Mango malformation: production of thick vegetative shoots and transformation of floral parts into a compact mass of sterile flowers. Two types, vegetative and floral Vegetative malformation resembles a "bunchy top", which may dry and die in due course. Floral malformation results in enlargement of flowers with new flowers being produced even after fruit set but with a lower percentage of hermaphrodite flowers. Malformed panicles may be two types: loose and compact. The severity of malformation is greater in the north than in the south. It may result in a loss of about 50–60% of the crop. Kishen bhog, collecter, Langra, Neelum are tolerant (seedling trees are found to be tolerant). Virus, fungus, mites, nutrients, C/N ratio, carbohydrates, nucleic acids, amino acids, proteins, phenolic compounds, enzymatic activity in the plant, phytohormones, and the occurrence of malformation-like substances are all supposed to be the probable causes for malformation.

Control Measures

- Application of plant growth regulators and phenolic compounds (NAA, Ethrel, GA, Paclobutrozol, etc.
- Deblossoming: at bud brust stage-ethrel
- Use of antagonists and antimalformins: Glutahione, Ascorbic acid, Silver nitrate
- Application of nutrients: High NPK added with FeSO4, Cobalt sulphate
- Pruning of malformed parts.
- Application of pesticides: Parathion, Kelthane, Kerathane.
- Covering panicles with polythene film to raise the temperature around the panicle. Inspite of this, malformation is still a puzzling problem.
- It is therefore concluded that malformation can be kept under check by maintaining
- Orchards cleanly using disease free planting materials only.

- Regularly inspecting the orchard
- Regularly removing all malformation parts and
- Spraying of insecticides and after each pruning.

Fruits Drop

The intensity of fruit drop amounts to about 99% at various stages of growth, especially during the initial few weeks. It also varies from variety to variety. The fruit drop is more or less a continuous process and can be classified into three phases: **i.** Pin head drop, **ii.** Post setting drop, and **iii.** May month drop. The drops are insignificant compared to the May month's drop, which affects the final yield significantly and needs more attention. Embryo abortion, climatic factors, disturbed water relations, lack of nutrition, disease and pests, and hormonal imbalance are the major factors that lead to fruit drop. Regular irrigation during the fruit development period The spray of NAA (25 ppm) or 2,4-D (30 ppm) at the pea stage of fruit is beneficial.

Black Tip

The distal end of the fruit becomes black and hard due to a physiological disorder. Pollution of the atmosphere with smoke, carbon monoxide, carbon dioxide, sulphur dioxide, acetylene causes this disorder. Spraying borax 0.6% from fruit set at 10-15 day's intervals controls this (Punjab, U.P., Bihar, W.B.).

Clustering (Jhumka)

Clustering of fruits without growth at the tip of the panicle caused by adverse weather (low temperature) during Feb-March. Most of the fruits drop + shrivelled and aborted embryos.

Spongy Tissue

Fruit appears normal but contains yellowish, sour spongy tissue inside. High temperature, convective heat, and exposure to sunlight after harvest are supposed to be the causes. Remedy lies in sod culture, mulching in the orchard and harvesting fruits at $3/4^{th}$ mature stage only.

Soft Nose

A physiological disorder caused by Ca deficiency causes flesh breakdown near the fruit's apex before ripening.

37

Mangosteen

- Mangosteen/Queen of tropical fruits/Finest fruit of the world/Fruit of god's/Mystery fruit/Ice fruit /Energy tablet/National fruit of Thailand
- B N: *Garcinia mangostana*
- Family: Clusiaceae (Guttiferae)
- Chromosome No.: 2n = 2x = 24
- Origin: Indonesia or South East Asia
- Inflorescence: Solitary
- Fruit type: Berry
- Edible part: Pericarp and fleshy perianth

Introduction

The mangosteen is one of the finest fruits of the tropics. It is a sweet, soft, and delicious fruit with an exciting flavour. They are formed from the nucellar tissue in the 'parthenocarpy' fruits. broad-leaved evergreen tree. A natural staple food for man.

Composition

The fruit is a rich source of Vitamin-C (66 mg), A, B_{12}, minerals (Ca, P, Fe), carbohydrates (19 g), protein (0.5g), citric acid, fibre, fat (0.6g), phytin (0.68 percent). The delieate aril flavour is due to hexyl acetate and cis-hex-3-enyl-acetate, and 6 sesquiterpenes have also been identified.

Uses

Mangosteens are usually eaten fresh as dessert. It serves as a topping for ice-creams, canning, acid fruits for preserving, jam. The seeds are sometimes eaten after boiling or roasting. The twigs are used as chewstics in Ghana. The fruits contain 7–14% catachin, tannin and rosin and are used for tanning leather in China. It also yields black dye. The rind and bark are used in traditional medicine. The wood is dark red, coarse and strong and can be used in carpentry work. The bark extract and rind powder called **"amibiasine"** are used to treat amoebic dysentery and diarrhea.

Origin and Distribution

Mainly cultivated in Malaysia, Indonesia, Thailand, Srilanka, Myanmar, India, Vietnam, Philippines, China, California etc. In India, its cultivation is mostly confined to the lower slopes of the Niligiris. Limited cultivation is also noticed in Kerala, Karnataka and West Bengal.

Climate and Soil

The tree grows on a wide range of soils. Deep clay loam or silt loam with good drainage is suitable. The fruit tree requires a tropical climate with high humidity, high temperature, and abundant rainfall. A shaded environment is required during the early growth stage. It cannot tolerate temperature below 5°C or above 38°C. It grows up to an elevation of 1500 m above mean sea level.

Species and Varieties

The important species of *Garcinia.*

1. ***G. mangostana*** is only cultivated plant and may be allopolyploid hybrid, between *G hombroniana* (2n = 48) x *G. malaccensis* (2n = 42).
2. ***G. cowa-*** available in the tropical forests of Assam, Bengal, Orissa and Andaman, Fruits are edible.
3. ***G. schomburgkiana-*** fruits are used as a preserve.
4. ***G. hanburyii-*** Gum resins used for colouring.
5. ***G. xanthochymus*** fruits are edible young shoots & mature fruits as vegetables.
6. ***G. dulcis-*** Grown as a fruit tree in Southern part of Thailand.
7. ***G. indica*** -Kokum found in tropical forests of India, gives acidic sweet fruits, mainly used as a souring agent for vegetable curry and dal.

Variety: **Jolo**-from Philipines

Propagation

Mangosteen is mainly propagated by apomictic seeds (non zygotic). Such seeds produce plants true to type, hence seed propagation is commonly practised since the seed loose viability very quickly. also propagated by grafting, budding or air layering.

Planting

Planting dimensions are 7 x 7 m and 10 x 10 m, depending on location and soil fertility. The young trees should be provided with shade to protect them from the direct scorching sun.

Manures and Fertilizers

About 20–25 kg of FYM/tree/year is to be applied. The young trees receive 70 g N, 6 g P & 50 g K/plant/year. For 15 years onward, about 2.7 kg of complete fertilizers (10:10:19)/year/tree is required.

Irrigation

The trees can stand water logging but not drought. Stress followed by irrigation is required to induce flowering.

Pruning

The regular pyramidal crown and slow overall growth limit pruning. Dead branches, base suckers, water sprouts, and small inner branches should be pruned, to open the canopy.

Removal of Male Plants

Dioeciousness is a major problem in mangosteen cultivation. At present, no method is available to detect the sex of the plants in the seedling stage. About 7-8 years after planting, the sex of the seedling plant is known. The male plants serve as pollinizers and, hence, it would be adequate to keep 10 percent of male plants well scattered in the orchard.

Flowering and Fruit Set

Flower initiation is noticed at the tip of the swellings. From bud stage to anthesis, it takes 25 days. Fruit development takes 100-120 days from anthesis. It may take 180 days in a cooler area. The trees tend to bear fruit in alternate years. The trees bear fruit from May to July at low elevation and from September to October at high elevation. Under Indian conditions, flowering occurs twice a year, i.e., during the monsoon, July-October, and during the summer, April-June. Fruit set is not a problem, however, when 10-15% of flowers are set.

Harvesting

Fruits are picked at a mature stage; soft and dark red/purple patches on the skin of the fruit. Fruits are harvested with the attached peduncle. The plants yield up to 200–2000 fruits/tree, or 20–25 kg/tree (4-5t/ha).

Storage

The presence of yellow dried latex oozed from the latex vessels on the fruit skin is known as **gamboge.** It may be scrapped off. The thick fruit wall hardens as the fruit ripens and the fruits are dipped in fungicides, and fruits can be stored at 8–10°C for up to 8 weeks.

Pests and Diseases

1. **Mite's attack-** the fruit surface and make it unattractive.
2. **Caterpillars** and **grasshoppers** cause leaf damage.
3. **Cankers on stems**/ branches, Zingarella garcineae the affected part should be cut and burn.

38

Mulberry

- Common name: Shahtoot
- N.: Morus sps.
- Family: Moraceae
- Chromosome no.: 2n = 308
- Origin: Japan
- Fruit type; Sorosis

Introduction

Mulberry is one of the important minor fruits found to grow all over India from tropical to temperate regions. It is mainly cultivated in India for rearing silk worm. Silk worms feed on the leaves of the mulberry plant. Its sour-sweet taste of fruits is becoming very popular in India and abroad. It is cultivated in South eastern and Southern parts of Asia, whole Africa, Southern-Europe, Australia and some parts of North and South America. It is mainly found to grow in the sub-Himalayan tracts. It is grown in the states of Kashmir, Karnataka, Punjab, Uttra Pradesh and some parts of West Bengal. In Indian its cultivation was introduced from Persia. In India it is widely cultivated in about 40,000 hectares of area. Karnataka has the largest area of about 28,000 hectares and Kashmir occupies the second position.

Composition and Uses

Mulberry is a nutritious fruit. It is reported that fresh fruit contains about 8 g sugar and 1 g acid per 100 g of edible pulp. The fruts are consumed frsh. Fruit juice is a cool and refreshing drink during summer and it is very effective for quenching thirst. Fresh fruits are used to process for making jam, jelly, wines and beverages. Dried fruits are ground to make powder which is an excellent poultry feed.

Soil and Climate

It can be grown in wide range of soil. It is also grown well in poor soils even in roocks. However, clay-loam or sandy-loam soil is ideal for mulberry cultivation.

Varieties: Thar harit and Thar lohit.

Propagation

Mulberry is usually propagated by seeds. But vegetative method of propagations is always preferred to get seedlings of true to type. Generally cutting, shield, ring or flute budding and inarching are also practised in mulberry. Hardwood cuttings of 20-30 cm in length are taken from matured stems. The cuttings should have about 4-5 active buds. Cuttings are planted in well-prepared nursery beds in rows 10-15 cm apart. Cuttings are planted in slanting position and 2/3 portions should be embedded in the medium in the month of December-January. The cuttings become ready within 3-4 months and transplanted in the field when the monsoon sets in. Optimum time of budding is March and early April and buds are initiated by May-June.

Planting

Pits of 45 x 45 x 45 cm size should be dug out in the well ploughed and levelled land 15-20 days before planting. It is then filled with soil and F.Y.M. Pits are made 6-7 m apart in the orchard. Pits are watered atleast 10-15 days before planting. Best time of planting is July.

Manures and Fertilizers

Before planting 10-15 kg F.Y.M. or well rotten cow dung manure is applied per pit. However, 30-40 kg F.Y.M. should be incorporated per plant before commencement of flowering and fruiting. Application of 500 g suphala (15:15:15) per tree at the start of rainy season helps to increase fruit bearing.

Irrigation

Though mulberry is a drought resistant tree, it also responds well to irrigation which results in better yield. Under arid conditions it would be desirerable to collect the run-off water near the root zone.

Pruning

Pruning encourages better fruit production if done properly. The stem is kept clear of all the lateral shoots upto a height of about one metre. Best time of pruning is when the trees shed thier all the leaves. Root pruning is also advocated after shoot pruning to encourage new growth and flowering.

Fruiting

Mulberry tree flowers in February-March and the fruits ripen in May-June.

Harvesting and Yield

Matured fruits are harvested after judgement of colour change on the fruits. Harvesting is over in 3-4 pickings. A full grown tree can yield about 10-15 kg of fruits.

Post-harvest Handling

Fruits of mulberry are highly perishable in nature. Fruits can not be stored for a long time. After harvesting, fruits should be immediately graded and sent to the market. It can also be used for preparation of various products like jam and jelly. Fruits should be packed in the shallow baskets lined with newspapers or mulberry leaves.

Pests and Diseases

Pesta:

Leaf-eating Caterpillar

Larvae feed on the young leaves and tender twigs voraciously.

Control

Spray Navacron @ 1.5 ml/itre of water to the affected plants to control the pest. Also collect egg clusters and destroy them.

Diseases

Dieback

Affected shoots dry up from the tip downwards. During raing season, a film of silky thread of mycelium develops on the branches and later fungus develops a pink growth.

Control

To control the disease, remove all the affected plant parts and destroy them. Aply Bordeaux paste on the cut surface.

39

Olive

- Olive/Emblem of United Nations
- B. N.: *Olea europaea*
- Family: Oleaceae
- Chromosome No.: 2n = 46
- Fruit type: Drupe
- Edible part: Epicarp and mesocarp

Introduction

Olives are found growing wild in the Himalayan region comprising Kullu, Shimla, Solan, and Sirmour districts of Himachal Pradesh. It is extensively used for the extraction of oil from its fruits. Its oil is edible, possessing a valuable therapeutic character. About 92% of its production is used for oil extraction. The main producers of olive oil are Spain, Tunisia, Italy, Greece, Turkey, Syria, Algeria, Morocco, Egypt, and the adjoining gulf countries. Olive cultivation in India has been limited to the north Sub-Himalayan region, which includes Jammu and Kashmir, Himachal Pradesh, and Uttar Pradesh.

Climate and Soil

Normally, olives are grown in warm, tropical to subtropical climates. It can be grown up to 800–1400 m above mean sea level. Temperature is the most important factor influencing its cultivation. Olive cultivars require 7–35°C, but 15–20°C is ideal. The occurrence of prolonged hot and dry summers coupled with acute water stress causes post-anthesis flower and fruit drop. Adequate chilling at 7-20°C is required during the winter to break the rest period and promote fruitfulness. Olive trees are not irrigated, but they require 950 mm of water during their entire growing season. A well-distributed rainfall is conducive to the growth of trees. Fruit drop is caused by a lack of adequate soil moisture caused by a prolonged drought during the summer. Inadequate or delayed rain during the winter season delays the emergence of the new flush, causing a substantial reduction in flower bud differentiation. The occurrence of autumn frost is more harmful since it affects fruit maturation and oil accumulation.

Varieties

Its varieties can be divided into two categories.

Oil type: Carolea, Coratina, Pendolino, Frontio, Ascolanterena, Aglandeace, Leccino, Biancolilla, Moraiolo, Canino, Metoides, Koroneiki.

Pickle type: Ascolano, Mission, Grosseune, Picholin, Cornicobra, Coratina, Manzanilla, Etenea, Toda Ibleu, Nocellara.

Propagation

Olives have for thousands of years been propagated vegetatively, probably starting with rooted suckers or "ovuli" and later by grafting and cutting. Propagation by seeds is not advisable due to the heterogeneity of progeny and an extremely long juvenile phase. Olive propagation by both sexual and asexual methods is relatively easy. It can be propagated by seed, cutting, layering, budding, grafting, sucker and root cuttings.

Planting

For better establishment, olives are planted from July-August. Planting during the rainy season helps the plants tolerate transplant shock quickly and reduces post-transplant mortality. Planting could also be undertaken during February–March if irrigation facilities are available.

Plant Spacing

A planting distance of 8 m x 5 m or 6 m x 5 m, corresponding to a planting density of 250 or 333 trees/ha, can also be used.

Training and Pruning

At the time of planting, a support system of about 10 feet should be erected. Remove weak laterals. Eliminate all other undesirable scaffold branches. Young non-bearing trees can be pruned any time of the year if the weather is clear and free from intermittent rain. In young plants, only corrective pruning should be done as and when necessary. Pruning is not recommended during the first 2-3 years.

Manures and Fertilizers

Under Indian conditions, trials to determine the optimum doses of N, 0.80 kg of P_2O_5 and 0.40 kg of K_2O per tree were applied. One full compliment of phosphatic and potassic fertilizers and one third of nitrogenous fertilizers has to be applied as a basal dose during January. The remaining 2/3 has to be applied in two split doses during February–March and July–August. Regular use of organic manure is advocated for maintaining the fertility status of soil at a fairly high level and for improving soil structure. It has been said that an autumn application of organic matter will help to keep the soil warm during the winter months and also reduce the harmful effects of drought conditions.

Irriagtion

Baldini and Pisani (1962) reported that, in general, three irrigations need to be applied at the critical period viz., (i) before flowering for reducing premature pistil abortion; (ii) at the time of fruit set to avoid excessive fruit drop; (iii) at the time of maturity to reduce the pre – harvest fruit drop. Watering has to be avoided during flowering or after the onset of fruit ripening.

Interculture

Intercropping with pulses or potatoes, cereals, or green manure crops could be practiced during the rainy season. If planting is done with a wider spacing, fruit crops like figs and pomegranates can be grown as fillers.

Harvesting and Yield

Olive fruits mature four months after blooming. The fruits attain maximum size by the middle of September, when the colour changes from dark green to a yellowish-light green. The colour changes from pale green to violet at the apical end. The olive is a **non-climacteric fruit**. Olives are harvested by different methods, viz., by natural dropping of ripe fruits, picking by hand, beating with wooden poles and mechanical harvesting using mechanical shakers. The yield varies widely from 1 to 5 tonnes per acre.

Pests and Diseases

1. **Olive canker: (*Pseudomonas sayastanoi*),** Chlorosis leading to necrosis and premature fall of the leaf. Tumours are found on all tupes of buds and branches, occasionally on root.

   Contro: (i) Avoid use of infected plants, grafts and cuttings and not to wound the trees. (ii) Application of Bordeaux mixture during winter. (iii) Planting of species and cvs. Like Ascolano and mission.

2. **Peacock eye: (*Cycloconnium oleaginum*),** Fungus produces lesions on the upper surface of the leaves and causes yellowing followed by early leaf fall. Severe attack occurs in abnormal spring and autumn.

   Control: Use of Bordeaux mixture is very effective. Modern synthetic organic products such as phthalic acid amides and benzymidazoles are also useful.

3. **Anthracnose: (*Gleosporium olivarum*),** Spryinng a mixture of copper and zinc oxymidazole is recommended.

40

Papaya

- Papaya/Melon tree/Wholesome fruit/Wonder fruit of tropics and subtropics/ Backyard fruit/Breakfast fruit/Continuous flowering and fruiting crop/ Pawpaw
- B.N: *Carica papaya*
- Family: Caricaceae
- Chromosome No.: 2n = 2X = 18
- Origin: Tropical America
- Inflorescence: Male flowers in panicles and female in cymes
- Fruir type: Large hollow Berry
- Edible part: Mesocarp

Introduction

Papaya has long been known as the wonder fruit of the tropics and is one of the five major fruit crops of the tropical world. Papaya fruits throughout the year are a remunerative fruit crop which is grown in almost every part of the country. Next to bananas, it has the highest yield and income per unit. It is generally grown as a fruit plant in kitchen gardens, home gardens, or nutrition gardens and later as a commercial fruit. Papaya is usually dioecious, but hermaphrodite and gynodioecious types are also Recognized. It is the most remunerative crop (highest net return/unit area) of any other fruit crop. India is the 4th largest producer in the world. Karnataka has the highest production and productivity in India. A valuable proteolytic enzyme is called papain, which helps in the digestion of protein-rich foods. Transgenic PRSV resistant varieties: Sunup and Rainbow were created from coat protein genes grown in Hawaii in 1992.

Origin and Distribution

The papaya is a native of tropical America and was introduced to India in the 18th century. It is now grown in almost all tropical and subtropical countries in the world. In India, it was introduced by the Portuguese in 1611. India is the second largest producer of papaya in the world after India. It is a fruit native to tropical America and is commercially grown in many parts of the world, viz., USA, Fiji, Jamica, Australia, Hawaii, Taiwan, Puerto Rico, Peru, Florida, Texas, California, South Africa, Pakistan, Bangladesh and India. In India, it is largely grown in Gujarat, Madhya Pradesh, West Bengal, Karnataka, Bihar, Assam, Maharashtra, and Andhra Pradesh. It is also planted as a filler plant in a fruit orchard. The milky latex, which is obtained from immature fruits, is known as papain. It is also being used for the extraction of oil from the liver of tuna fish.

Composition and Uses

Papaya is a very delicious, wholesome, refreshing fruit with a unique food value. The ripe fruits are used for dessert purposes only; preparation of jam, jelly, soft drinks, ice cream, flavoring, crystallized fruits, canned in syrup, fruit salad. The unripe fruits are consumed as vegetables and also as a contraceptive. The seeds also have medicinal properties. The yellow pigment in papaya is **caricaxanthin**. Papaya is a wholesome fruit and is rich in vitamin A (2000 IU/100g). Papaya yields a valuable proteolytic enzyme, papain, which has several and varied uses in medicine and industry. The other uses of papain are in the treatment of ulcers, diphtheria, pre-shrinking of wool, manufacture of chewing gum, degumming natural silk and rayon, in cosmetics, dental paste preparation etc. The milky latex, which is obtained from immature fruits, is known as papain. It also contains Vit. C, riboflavin, niacin, and is a good source of Ca, P, Fe, fat, and carbohydrates.

Uses of Papain

As a digestive aid and for the treatment of ulcers and diphtheria. In the industry of pre-shrinking of wool and degumming of natural silk and rayon, tenderization of meat and it is used for tanning of leathers. used as a clarifying agent in the brewing industry, etc. used in the preparation of drugs for treating intestinal cancer, tape worms, round worms, and kidney disorders. Papain is used to correct certain digestive ailments. The carpaine alkaloid present in green parts and in seeds is used as a heart stimulant, as a diuretic, and for the treatment of amoebic dysentery. It is also being used for the extraction of oil from the liver of tuna fish. Apart from papain, raw fruits can also be used for the extraction of pectin. Pectin is generally extracted using the alcohol precipitation method. It is used in the food industry as a flavouring extract and an emulsifying agent.

Tutti Fruity

Papaya bits (immature) are first soaked in brine, boiled in sugar syrup, and immersed in sugar syrup for 48–60 hours with permitted colours.

Climate and Soil

Papaya is tropical in its climatic requirements. It is susceptible to frost. It requires a humid climate and can be cultivated up to an elevation of 1000-1200 m above MSL. It thrives well at temperature between 38-44°C, but the optimum range is 22–36°C. A day temperature of 35°C and a 25°C night temperature are most suitable. It cannot tolerate very hot summers or frost. This limits cultivation in Northern India. It cannot tolerate very hot summers or frost. A dry, warm climate tends to increase the sweetness of the fruits. In strong wind-prone areas, wind breaks have to be provided to save the trees from wind damage. Tamil Nadu is an ideal home for growing papaya because of the mild temperatures and freedom from mosaic and leaf curl virus diseases. sensitive to frost, strong winds and highly susceptible to waterlogging or stagnation. It does well in varied soil types, the best performance is observed on loams of uniform texture up to 1.8 m in depth. Even two to three cm of water stagnation around the tree for a few hours is likely to damage them due to the collar-rot disease occurrence. These features help all the year round cultivation of papaya grow well under well-drained soil with a pH range of 6-7.

Species

The genus Carica contains 48 species of which, only 3-4 species are important.

Badillo (1993) divided the genus *Carica* into two sections, *Carica* and *Vasconcella*

- *Cariaca papaya* – It is a cultivated commercial species and polygamous in nature.
- *C. monoica* - Monoecious and is suited for mild climate, found in Amazon basin.
- *C. candamarcensis* – This species is known as mountain papaya is suited to cold climate.
- *C. gracilis* - Ornamental
- *C. pubescens* - Mountain papaya
- *Carica dodecaphylla* - Jaracatia
- *Carica goudotiana*
- *Carica stipulata* - Chamburo

- *Carica quercifolia*- Oak Leaf Papaya
- *C. stipulate*.
- *Carica quercifolia* is the hardiest of all the species.
- *C.cauliflora, C. microcarpa, C. monoica* and *C. candamarcensis* are cross compatible but all are incompatible with *C. papaya.*
- *C. cauliflora* and *C. candamarcensis* are virus resistant.
- Mountain papaya- *Vasconella candamarcensis (Previously known as Carica candamarcensis)*
- Frost resistant species- *V. pentagona*

Varieties

1. **Honey dew:** It is dioecious with less per centage of male plants and breeds true to the type from seed. This variety is popular all over India, also known as **Madhu Bindu**.
2. **Sunrise Solo:** The variety is named 'solo' because one man can easily consume one fruit. This is gynodioecious variety from Hawaii with small sized fruits, pyriform and yellowish orange pulp
3. **Coorg honey dew:** selection from **Honey Dew** at IIHR, Chethalli. This variety produces no male plants. The plants are either **hermophorodite** or female. Fruits borne on female plants are almost seedless and are of excellent quality.
4. **Washington:** Fruits are medium to large, round to ovate, sweet, pulp orange in colour with good flavour.
5. **Ranchi-** All 3 sex forms- Male, female and hermaphrodite and first varity of India.
6. **Surya**: IIHR-39- Sunrise solo x Pink flesh sweet, Cold tolerance.
7. **Punjab Sweet** –Sunrise solo x Pink flesh sweet, Gynodioecious

Varieties Released from Tnau

1. **Co-1** It is a selection sibmating from variety **Ranchi**, dwarf type.
2. **Co-2:** It is a dioecious, semi dwarf and papain extraction. It yields 5-6g. of papain per fruit.
3. **Co-3:** A hybrid between Co-2 x Sunrise solo. It is gynodioecious in nature.
4. **Co-4:** A hybrid derivative from Co-1 x Washington, it is dioecious.

5. **Co-5:** Selection from **Washington** and the plants are exclusively selected for papain extraction and yield about 14–15 g. of papain/fruit. Papain extraction (1500–1600 kg/ha). Highest papain variety.
6. **Co-6:** Selection from **Pusa Majesty**, Dual Purpose (both table and Latex-Papain), it yields about 80-100 fruits and is also a high papain (7.5-8 g/fruit) yielder, i.e., 890 kg of dry papain/ha.
7. **Co-7:** It is also a gynodioecious developed through multiple crosses between CP-75 (Pusa delicious x Co-3) and coorg Honey Dew, yielding 160-180 tons/ha.

Varietis are Released IARI

1. **Pusa delicious** - It is a gynodioecious plant with medium-sized (1.5kg) fruits and a high yield; the male has been genetically eliminated by sibmating with a hermaphrodite.
2. **Pusa majesty:** A gynodioecious, medium-sized fruit with good keeping quality. plants resistant to viral diseases. The fruit can stand transport over a long distance.
3. **Pusa giant-** Dioecious variety with big sized fruits (2-3.5 kg), plants are very vigorous and resistant to wind damage. The fruits are suitable for canning industries. suitable for tooty-fruity and candy
4. **Pusa dwarf:** The plants are dwarf and bears fruits at 38cm from the ground, they are dioecious and medium yielder, medium sized fruits, suitable for HDP and home gardens.
5. **Pusa nanha:** A dioecious dwarf mutant that is ideal for use in the kitchen garden, nutrition garden, and pot cultivation. It yields about 60–65 tonnes of fruit per ha. It is developed by Gamma radiation and is suitable for high density planting.

Others Varieties

1. Pant papaya-1, 2 and 3.
2. IIHR-57- Waimanello x PPS.
3. Pink Flesh Sweet, Good dessert variety.TSS-12-14Brix.
4. Pant Papaya-1, Dioecious.
5. Taiwan-Blood Red colour.
6. Hortus Gold cultivar developed by Dr. J.D. Hofmeyer.
7. Gynodioecious Varieties: Sunrise Solo, Wilder, Walmanalo, Thailand, Taiwan, Washington.

8. Arka Prabhath- Gynodioecious hybrid.
9. Sunrise Solo type of papaya produces no male plants.
10. Popular exotic variety in India: Red Lady, Zinda.
11. High carotene content: **Sunrise Solo.**
12. CO_2, 5,6, Pusa majesty cultivars suitable for papain production.

Seed Production

Papaya is a highly cross-pollinated crop. Seeds taken from a fruit will rarely breed true to type. If a variety is to be maintained pure, controlled pollination between selected female and male progenies of the same parent, i.e., crossing of sister and brother, called sib mating, has to be done. This consists of the collection of pollen from the male parent and applying it to the previously bagged female flower. Seeds from such sib-mated fruits should be used for further multiplication. Failure to observe this precaution leads to the deterioration of the variety resulting in the progeny being a mixture of all kinds of types within a few years.

Propagation

Papaya is usually propagated by seeds. The seeds should be collected from healthy ripe fruits, essentially from plants free from pests and diseases. Commercially propagated by seeds, **250-300g/ha (Gynodioecious)** and **400-500g/ha (dioecious**). Germination takes place in 15 to 20 days. Induction of femaleness in papaya is enhanced by spraying of 50 ppm of $GA_{3.}$ Planting during February shows more male plants, while planting in March and April produces an equal number of staminate and pistillate plants. One g seeds contain 20 seeds, papaya seed viability: 3 months, for seed germination:30°C. 6 weeks before transplanting, sow seeds in poly bags (22 x 15 cm with 150 gauge). The media should be disinfected by using a 2 per cent formalin solution. 4 seeds per polybag should be sown if it is a dioecious variety, or 2 seeds if it is a hermaphrodite variety. The seeds should be rubbed with ash to remove the mucilaginous coating (Aril/Sarcotesta) to facilitate better germination. The fruits are cut open and the seeds are carefully extracted in trays. They are washed and dried in the sun or shade and are stored in bottles. Fresh seeds may be mixed with fine cold wood-ash which absorbs the slimy coating on them and helps to keep the seeds separate on drying. About 500 g of seed is required for one hectare. Seedlings can be raised in the raised nursery beds or in polythene bags, however the seedlings from the latter one are good. Two seeds in the gynodioecious type or 5 to 6 seeds in the dioecious type should be sown per poly bag. However, seeds can be stored for 45 days in an airtight container if stored at 10°C. Soak seeds in a 100 ppm GA_3 solution before sowing to increase germination percentage.

Planting

Generall spacing 1.8 m x 1.8 m but for papain production: 1.6 mx1.6 m. and High density planting (HDP) spacing is require **1.25 m x 1.25 m (6000 plants/ha)**, suitable variety for HDP is Pusa Nanha. Planting is done during the monsoon season. Due to sex variation, more than one seedling is required per pit. About 6-7 weeks old, seedlings will be ready for transplanting. Due to sex variations, about 40 to 60 per cent of the plants may turn out to be male in the case of dioecious varieties. Therefore, in such a case, 2 to 3 seedlings per hole at 30 cm apart in the pit should be planted, so that when they reach the flowering phase, the unproductive male trees can be removed to keep the population ratio of one male tree for every 15 to 20 female trees. The best time for planting papaya is the beginning of the South-West monsoon in most parts of India. In south India, June to October and January to March are suitable for planting, as the other months are either too hot or too rainy.

Manures and Fertilizers

As three stages, *i.e.*, flowering. fruit development and harvesting, concurrently occur in papaya plants, regular fertilizer application, i.e., 10 kg of FYM/plant as basal, is essential. Besides, 50g each of N, P and K per plant at a bimonthly interval is recommended by TNAU. At the Indian Institute of Horticulture Research, Bangalore. A dose of 250 g each of N, P_2O_5 and 500 g K_2O per plant per year in six split applications is recommended to get a higher yield.

Irrigation

Regular irrigation helps fruit development and induces the tree to bear larger-sized fruits. Water stagnation should be avoided. In most parts of India, papaya is irrigated once in 8 or 10 days.

Intercultivation

Different short-duration vegetables can be profitably grown for about six months from planting. For control of weeds, weedicides like fluchloralin or butachlor at 2.0 kg/ha have to be applied. Post emergent herbicides: Glyphosate has also been found effective. It is not possible to identify the sex at the early stage until they put forth flowers, which may take 4-5 months from planting. At this stage, male trees should be removed, maintaining one male tree for every 10 female trees for proper pollination and fruit set. In each pit, only vigourously. The growing female/hermaphrodite tree should be retained and the other plants removed.

Thinning of Male Plant

Remove male plants by maintaining 1 male plant for every 10 plants after flowering. Stamen carpelloidy is related to floral structural variation (catfaced fruit). Carpellody formation: below 20°C, low elevation, cool winter. Male plants bear fruits in the summer season, which is called carpellody.

- Male: Female ratio in gynodioecious variety: 2:1.
- Male: Female ratio in dioecious variety: 1:1.
- Best nutrient analysis: Petiole 4th leaf.

Sex Expresion

- Classification of sex forms in papaya was proposed by Storey (1958)-8 sex form
- Sibmating is practiced -Sibmating means crossing of female and male progenies of same parent
- Papaya sex is controlled by single gene with 3 alleles: M_1, M_2, m
- Sex reversing male: M_1RR_m or M_1Rr_m
- But sex-reversing males are heterogygous (Rr) for a pair of alleles while pure male is homozygous(rr) recessive.
- Sex reversal in papaya is governed by a single gene (R)
- Red skin colour of papaya is governed by inherent heterozygosity and recessive gene
- Maleness is controlled by satellite chromosomes
- Pulp colour is governed by recessive gene
- The satellite chromosome controls maleness in papaya
- High temperature leads to female sterility:35°C
- Metaxenia is found in papaya

Papaya is a Continuous Flowering and Fruiting Crop Sex Distribution in Papaya

Papaya is a polygamous species, many forms of inflorescence have been reported by **Frankel** and **Galun, 1977**.

- Polygamous plant, Predominant 3 sex Forms-Male, Female and Hermaphrodite

- Sex expression Many sex forms such as dioecious, hermaphrodite, gynodioecious etc., have been reported in papaya.
- There, are no distinct or definite methods to ascertain the sex of the plants at the early stage itself.
- In papaya change in the sex expression is influenced by environmental factors
- Like, ow temperature favoring production of perfect flowers on the male tree.
- It was observed that fertile hermaphrodite types had some pistillate flowers which showed male tendency in summer and female tendency in winters.
- Seedless fruits are developed from the pure female flowers which are not cross pollinated After flowering the fruits mature within 5 months after flowering;
- Besides, many factors have been reported to influence the sex expression.
- Planting during February shows more male plants while planting in March/ April produces an equal number of staminate and pistillate plants.
- Growth regulators: such as GA (50 ppm), ethrel (200 ppm) SADH (250 ppm) and phosphon - D (2500 ppm) increase the femaleness in dioecious types
- Under adverse climatic condition flowers & fruit drop is very common, to avoid this problem spraying of planofix @ 1ml/liter.

Flowering and Fruit Set

Papaya is usually dioecious but hermophrodite type and gynodioecious types are also recognized. In dioecious type, both male and female plants are separate. The male flowers are found on long pendulous panicle. Female flowers are solitary and are much larger than male. The ovary usually large in female flowers. In the case of hermophrodite flowers, two kinds are often observed viz. one with long corolla type and 10 stamens and another type with a short corolla and 5 functional stamens. Fruit is a large hollow berry elongated or globular in shape. In gynodioecious type, the female and bisexual flowers are borne one on the same plant. The fruit develops from female flowers are globular in shape while fruits that develop from bisexual flowers are elongated in shape. Selfing regular and prolific bearing hermaphrodite and/or crossing (sibmating) the female (mm) with hermaphrodite (M_2m) may follow breeding for gynodiecious lines. Type of placentation: Parietal placentation, highly cross pollinated crop; pollination by wind.

1. **Staminate or male flowers:** Spleder spoon-shaped buds with long, narrow tubular flowers on long peduncles from leaf axil petals fused at the base, stamens but no ovary; they cannot develop into fruits.

2. **Pistilate/female flowers:** Large yellow flowers, borne singly or in groups of three in the leaf axils, close to the trunk. These flowers have five large twisted and fleshy petals surrounding an ovary, which swells and develops into a papaya fruit.
3. **Hermophrodite flowers:** These are bisexual flowers, having both female and male organs. They are creamy white in colour and have large sized flowers with no peduncle. The bisexual flowers are also borne in the same manner as the female flowers, i.e., closer to the trunk in leaf axils.

Papain Extraction and Collection

Papain is a proteolytic enzyme extracted from immature papaya milky latex. About 400 tons of papain are produced throughout the world. The USA and the UK are the major consumers of papain. The enzyme found in dried latex is papaya protein (papain). Non-metallic instruments should preferably be used in tapping and collecting, as the juice acts upon metals and gets discoloured. An ivory blade, a sharp edge, or a piece of bamboo splinter may be used. The latex should be collected in porcelain, glass, or earthen containers. After about 2 to 4 hours, the latex is scraped out of the tray and dried in the sun. Always collect the papain in glass vessels or Aluminum trays. The immature fruits of 90-100 days old are used for the extraction of milky latex by giving 5-6 cuts of about 2 mm depth. The quality and grade of papain is determined by colour and enzyme activity (Tyrosine unit). Oblong fruits relatively yield more than round ones. There will be a 4-fold increase in papain yield by using ethrel at 200 ppm. A hectare of land could yield 450g per plant and 250-375 kg of papain. The varieties CO-5 and CO-2 are preferred for papain extraction. Tapping should be undertaken early in the morning so that drying in the sun can be done before mid-day. This makes the material sufficiently dry by the evening. When thoroughly dried, the latex becomes crisp and flaky. It may be then ground into a powder, preferably. The dried papain is powdered and sieved in a 10 mesh sieve. The cream-coloured powder should be placed in air-tight bottles or poly bags. Papain can also be dried artificially at a temperature of 50 to 55°C, which will attain a better colour and quality. Potassium metabisulphite (KMS) at 0.5% may be added to it for better colour and to keep quality. K is required for the TSS content of latex and the enzyme activity. Osmotic dehydration is done for the concentration of fruit pieces. Latex is obtained from ½ to ¾ th mature fruits. Papain contains 72.2% protein.

Harvesting Indices

Papaya comes to bearing 10 months after planting. Light green with tinge of yellow at apical end. Economic life is only 3–4 years. Fruits are harvested when the fruit turns a slight yellow in colour. Manual harvesting is commonly practiced. Wash the fruits with water or fungicide, i.e., Benlate at 0.05 percent, and pack them in boxes with cushioning materials. After harvesting in about 5-6 days' fruit ripens depending upon room temperature.

Yield

The average yield is 30–80 tons/ha. After harvesting, the fruits are graded on the basis of their weight, size, and colour. Fruits are highly perishable in nature. The cropping is practically continuous during the life of the tree. In the plains of North India, fruits continue to mature through the spring and summer, but in the cooler places in the hills, they only 3 to 4 months from February to May.They can be stored for a period of 1-3 weeks at a temperature of 10–13°C and 85–95% relative humidity. The first crop of fruits becomes available in 12–14 months from the time of planting. Fruits could be stored for more than 4-5 weeks at °C. Also, individual fruits can be wrapped in newspaper for ripening. After harvesting, the fruits are graded on the basis of their weight, size, and colour.

Pests

Few insects like scales, mealy bugs, aphids and thrips have been reported infesting the fruit tree. Scale insects and mealy bugs on stem and leaves. Mite incidence may be occasionally noted especially during summer.

Root knot *(Meloidogyne incognita)* and Reniform, *Rotyenchulus reniformis* are serious problem.

Papaya mealy bug *(Paracoccus marginatus)* outbreak occur in Tamil Nadu during 2009

Diseases

1. **Caller rot and wilt:** *pythium aphanidermatum* and phytopthora palmivora.
2. **Anthracnose:** (*colletotrichum gloesporiodes*): The initial symptoms are small round water soaked area on fruits which later develop into sunken or concentric lesions. The disease also affects the petioles of lower leaves leading to their shedding.
3. **Stem or foot rot** caused by *Phytophthora aphanidermatum* and *Rhizoctonia solani* affects young and old trees.
4. **Powdery Mildew:** (*Oidium caricae*)- The fungus is found mostly growing on the upper surface of the leaves with drawing nutrients from the cells of the leaf surface. Under severe attack the top portions of the seedling may die.

5. **Foot rot/collar rot/stem rot:** *(Pyhium aphanidermatum)*: Soil borne- More severe in rainy season
6. **Damping off:** *(Rhizoctonia, Pythium, Phytophthora) is* most serious or common disease in nursery

Viral diseases

1. **Ring spot**. - The papaya ring spot virus is common in North India, Karnataka and Andhra Pradesh and after it has become a major threat to papaya production in several tracts.
2. **Papaya leaf curl-** is transmitted by whitefly *(Bemisia tabaci):* Major problem in North India.
3. Papaya is affected by several virus diseases like papaya mosaic, ring spot, leaf curl, distortion virus and leaf crinkle viruses. Sap and vectors from the major source of diseases spread.
4. **Papaya Ringspot virus (PRSV)** is transmitted by aphids: Major problem in India Papaya ringspot virus is 1st recorded in Western India (1958).

Nematodes

1. Root knot (*Meloidogyne Sp*) - Affects the root systems and cause yield reductions.
2. Recommended bio-control agent: *Acerophagus papaya* (Parasitoid)

41

Passion Fruit

- B N: *Passiflora edulis*
- Family: Passifloraceae
- Chromosome No.: 2n = 2x = 18
- Origin: Brazil
- Inflorescence: Solitary
- Fruit type: Berry
- Edible part: Mesocarp

Introduction

Commercial in South Africa, Srilanka, Kenya, Cameroon, Hawai, Zimbabwe and Australia. In India, it is growing wild in Nilgeria, Wynad, Kodaiknal, Coorg and Malabar. It is being cultivated in H.P. Nagaland and Mizoram. **Brazil** is the world's largest producer. A rich source of vitamin A, it also contains fair amounts of sodium, magnesium, sulphur and chlorides. It gives a very good squash and is used in jam, jelly, nectar and ice cream making. Dried flowers contain an alkaloid called 'Passiflorin', which is used for relieving pain and inducing sleepness.

Climate and Soil

It can be grown in infertile soils too. Deep, well-drained, fertile loam soils are best for good yields. Optimum pH is 5.5 to 7.5. It cannot tolerate water logging over long periods of time. Passion fruit cannot tolerate frost. Can be grown up to 2000 m. The purple one does well at higher altitudes and the yellow one at lower elevations. Optimum temperature range is 20- 30°C. Below 15°C growth and flowering are restricted and above 32°C vegetative growth increases at the cost of fruiting.

Species

About 60 species of Passiflora produce edible fruits. *P. edulis* is most popular.

- **Purple passion fruit- *Passiflora edulis,*** More productive, higher elevation, susceptible to diseases.
- **Golden/yellow passion fruit-***Passiflora edulis var flavicarpa,* Resistant to Fusarium wilt, Phytophthora blight and resistance to nematodes and Alternaria leaf spot.
- **Blue passion fruit-** *Passiflora caerulea,* Resistant to Fusarium wilt, Phytophthora stem rot, tolerant to viruses.
- **Banana passion fruit-***Passiflora mollissima.*
- **Water lemon/Golden or Bell apple-***Passiflora laurifolia.*
- ***P. incarnata, P. coerulea*:** Cold, Disease tolerant stocks and for breeding.

Cultivars

- **Var Noels special-** Golden, early bearing, self incompatible.
- **Kaveri** (Purple x Yellow)- resistance to nematodes, passion fruit woodiness virus (PMV) and Fusarium wilt and high yielding, relesed from IIHR.
- **Australian Varieties**: Purple gold, Lacey, E-23 Black beauty.

Propagation

Propagation: seeds, cuttings and grafting on resistant stocks.

P. incarnata, P. coerulea: Cold, Disease tolerant stocks and use for breeding.

Flowering

Flowers are borne singly in the axils of leaves at the terminal region of the **current season's growth**. Flowers are protandrous in nature. It produces hen's egg-sized fruits. Passion fruit is a perennial woody vine, or long-day plant. Self incompatibility is reported in yellow passion fruit. Type of pollination: Cross (Pollinators: Bees).

Planting

Spacing is 2 m x 3 m wider spacing between plants, which improves yield. The monsoon season is best for planting. Passion fruit plants can be retained for 3-6 years only.

Manure and Fertilizer

100 g N, 50 g P_2O_5, 100 g K. per vine/year in two splits for purple variety. 110 g N, 60 g P_2O_5, 100 g K per vine/year in 2 split doses for Kaveri (hybrid).

Irrigation

Irrigation should be regular keeping the soil at field capacity to a depth of about 60 cm. Water stress during growth and development reduces quality and yield to a greater extent.

Interculture

Weeding should be regular. Oxyfluorfen 1.92 lit/ha, can be effectively used for controlling the weeds. Maintaining be hives helps in improving fruit set specially in golden-passion fruit as it is self incompatible.

Training and Pruning

Passion fruits are trained on a well built **kniffin system trellis (2 arms).** Two leaders are trained on either side of the vines from the main stem. Laterals on these leaders are allowed to hang freely. These are pruned systematically to encourage new fruit-bearing shoots. Pruning is done after harvesting in April and November - December.

Harvesting and Yield

Passion fruit takes 60-90 days from fruit set to maturity. Plants produce fruit from the 10th month, but good yields will be given from 16-19 months. The main seasons of fruiting are **Aug - Dec** and **March-May.** Passion fruit does not ripen well after harvest. The yield will be 8-10 kg/vine for the Kaveri variety. On an average, 12-20 tonnes/ha/year can be obtained.

Grading, Packing and Storage

Passion fruits are graded by size and colour. Normally packed in cartons or polythene bags. can be stored for a few days only at room temperature as they start losing moisture and becoming shrivelled. They can be stored for 4-5 weeks at 5°C and 80-90% relative humidity keeps the quality well for 15-30 days.

42

Phalsa

- Phalsa/Dhamani/Star apple
- B.N: *Grewia subinequalis*
- Family: Tiliaceae
- Origin: India
- Chromosome No.: 2n = 36
- Inflorescence: Cymose
- Fruit type: Multiseeded Berry
- Edible part: Mesocarp and epicarp

Introduction

Phalsa is a shrub or small tree of Indian origin. Subtropical bushy fruit plants. It is a very hardy fruit, and hence can be planted on marginal soils, where the rest of the fruits can not be grown. Its cultivation is more popular in the vicinity of cities and towns. Phalsa fruit is liked for its acidic taste, coloured squash and syrup. A woody perennial, arid zone fruit crop. suitable for intercropping in mango orchards. Phalsa leaf extract has antibacterial activity against *Escherichia coli* and *Staphylococcus aureus*. Phalsa fruit has a short shelf life, suitable only for local marketing. It is cultivated on a commercial scale, mainly in the northern and western states of India.

Area and Distribution

It is minor fruit and is being cultivated on very small scale in each state. However, in Punjab, Haryana and Uttar Pradesh it is cultivated near cities commercially. The popularity of this crop is restricted due to small fruit size, prolonged ripening period, repeated harvesting and the highly perishable nature of fruits. Apart from India, it is cultivated in Pakistan, Nepal, Bangladesh, Laos, Sri Lanka, Thailand, Philippines, Vietnam and on an experimental basis in some provinces of the United States of America.

Use and Composition

The ripe phalsa fruits are purple in colour at maturity and may turn black at ripening on the bushes. Phalsa is considered to be antioxidant in nature due to its coloured fruit and juice. The whole fruit is eaten along with the seeds. The mildly acidic fruits are rich in vitamins A, C and minerals. Fruit contain 55 percent juice with 11-12% TSS and 3% acidity. Ripe fruits are consumed as fresh. Juice and squash are relished in the summer. The shoot can be used to prepare baskets. Citric acid is the major acid in the fruit, with traces of malic acid.

Nutrient Value per 100 g Edible Portion

Moisture 80.8 % Carbohydrate 21.1%, Protein 1.5 %, Fat 0.9 %, Fibre 1.2 %, Calcium (129 mg), Phosphorus (39 mg), Potassium (375 mg), Iron (3.1 mg), Vitamin B3 (0.3 mg), Vitamin C (22 mg), Vitamin A (carotene) 419 IU.

Species

- *Grewia subinaequalis*
- *Grewia asiatica*
- *Grawia elastica var. vastita*
- *Grewia glabra,*
- *Grewia micrococas*
- *Grewia optiva,*
- *Grewia tilifolia*
- *Grewia belosa.*

Climate and Soil

The phalsa is a subtropical fruit plant but can be grown in a wide range of climatic conditions, except high altitude. Tolerate temperature above 45°C. It prefers a hot and dry environment during fruiting. In winter, it goes dormant and sheds its leaves. It resprouts in March. The high temperatures of June help in fruit ripening. It is not fastidious about its soil requirements. It can very easily be grown on poor soils. Loam soils are considered best. It makes good growth under scanty irrigation conditions. The plant grows satisfactorily up to an elevation of 1,000 m. The plant does well in the areas of where there is distinct summer and winter. The plants are deciduous and normally shed leaves on the onset of the winter season and go into dormancy. But in warmer regions, plants do not shed leaves and there is no dormancy. It can grow at temperature ranging from 3°C to 45°C. Plants can tolerate light frost. but requires protection from the very low temperatures. Adequate sunlight and warm or hot temperatures are required for fruit ripening, development of appropriate fruit color, and good eating quality. The optimum pH of soil for phalsa is 6.1 to 6.5. Phalsa is often grown in marginal lands close to city markets to facilitate prompt marketing of fruit. But the higher yield and quality of fruits can be obtained when grown in well-drained loamy soil.

Varieties

Two types: Tall and Dwarf.

Variety: Sharbati and Thar Pragati.

Propagation

Phalsa is commonly propagated through seed. Bold seeds give 90 percent germination during July. Sow seeds on raised beds 2 cm deep. Cover the seeds with mixture of sand + F.Y.M 50: 50 ratios. The seeds are extracted from ripe fruits. Seeds are very small, around 5 mm in diameter. And 0.5 to o.75 gm in weight. Around 1-2 kg of seeds are required for a one-hectare area. The seed remains viable for 90 to 100 days at a normal temperature and 6 months in cold storage. The distance between the rows should be 30 to 45 cm. Seed germinates in 15–20 days. Seedlings of 7 to 8 months are planted in the field.

Planting

Spacing 2 m x 2 m (250 plants/ha), leaf samples were taken from position 4th from the shoot tip, considered the best to evaluate nutrient status. A higher yield can be obtained from high density planting. The pair row planting with 60 cm distance between rows and a 3 m distance between two pair rows has also been reported to give more yield. Therefore, pits of 60 cm x 60 cm x 60 cm should be prepared. Planting of phalsa can be done in July-August or February. In that area, phalsa trees shed their leaves in winter and the plant remains in dormancy.

Irrigation

Phalsa is a drought-tolerant plant and does not require frequent irrigation. But it has been found that irrigation during fruit set and fruit development increases yield and quality of fruits. Irrigation is usually recommended after fertilizer application. Thereafter, for better yield and quality, irrigation should be given at a 15-20-day interval during the fruit growth and development period.

Pruning

Phalsa is usually trained as a bush. It can also be trained on a head system with a single stem. The height of the single stem should be kept at 90 cm to 1.0 m. The plant produces shoot above this height just as in the bush system. In North India, Phalsa is pruned every year from December to January. When it is trained as a bush, the shoots are cut from the ground level by leaving 2–3 cm stubs. From here, many sprouts emerge on the stubs Flowering occurs on these shoots in March-April. Phalsa needs heavy pruning, Severe annual pruning to a height of 0.75 to 1.25 m is recommended for Phalsa. While low-growing dwarf and/or bushy types of phalsa plants, which develop a good blend of sugar and acid in the fruit flesh, are preferred for cultivation. Phalsa bears fruit on the current season's growth and for good yield, there is a need for regular annual pruning to cut the old growth and to enhance the new growth. Annual pruning to a height of about 1 m. encourages new shoots and a higher yield of marketable fruit than does more drastic trimming.

Manures and Fertilizers

Apply 5 kg of farmyard manure to each bush after pruning in January. The bushes may be applied 50 to 100 grams of urea in two parts, *i.e.,* during March and April, depending upon the age. High N doses lead to profuse shoot growth, which is not desirable for good fruiting. When the bushes reach four years of age, increase the dose to 200 gm in a split dose. Apply 100 gm in March and rest 100 g in April at a month's interval.

Weed Control

Phalsa requires two hoeings. One after pruning in January and the second in April-May. If the intensity of weeds increases in the rainy season, then spray Gramoxone (paraquat) @ 6-7ml/L. of water in vacant places in plantations that are trained on head system. However, no herbicide needs to be sprayed on bush trained plants. Shade of the foliage keeps a check on the weed growth.

Flowering and Fruit Development

Phalsa plants start fruiting after 15 to 18 months of planting, but a good yield can be obtained only after three years of planting. The flowering starts in the month of February-March. The inflorescence has 3 to 5 flowered axillary cymes. They are grouped into groups of 2 to 8, which are 16 to 25 cm long. The flowers are small, 1 to 2 cm in diameter, and bright orange-yellow in colour. The fruits are very small and purple to crimson red in colour when ripe. Fruits are born in the current season's growth. Phalsa flowers are hermaphrodite. The use of ethephon at 500 ppm during the colour break stage significantly reduces the number of picks. Phalsa is a highly **self-pollinated crop.** It is an ideal plant for multi-storey cropping.

Harvesting and Yield

For uniform ripening of phalsa fruits, the apices of shoots may be pinned. Fruits start ripening in the first week of June and continue for a month. Fruits should be harvested twice a week. Fruits are packed in small baskets of size 2kg or in packs. Phalsa fruit is perishable in nature, hence should be transported to the market soon after harvesting. Phalsa fruits ripe 40 – 55 days after flowering. The fruits are harvested in April and May when they turn crimson red from green. Harvesting is done manually as they are very small and all fruits do not mature at the same time. Mature fruits can be stored for 48 hours, while ripe fruits can be stored for 24 hours. The average fruit yield is 5 to 10 kg per plant.

Plant Protection

1. **Leaf spot-** Fungus *Cercospora grewiae.*
2. **Rust-** *Dasturella grewiae.*
3. **Leaf-cutting caterpillar-** *Euproctis fraterna*
4. **Bark eating caterpillar-** *Inderbela tetrani*
5. **Mango mealy bug-** *(Drosicha mangiferae*
6. **Iron chlorosis** of phalsa is main problem in calcareous soil

43

Pineapple

- Pineapple/Heaven fruit/Friendship fruit/Symbol of friendship and hospitality/Dessert fruit/Golden queen
- *B. N.: Ananas comosus*
- Family: Bromeliaceae
- Chromosome No.: 2n = 2X = 50, 75, and 100
- Origin: Paraguay (Brazil)
- Inflorescence: Panicle, Compact spike (100-200 flowers)-Self sterile
- Fruit type: Sorosis
- Edible part: Bracts and Perianth

It was introduced in india in 1548

Introduction

It can be grown as a mixed crop in most of the orchards, as a hedge plant or as a pure crop. monocotyledonous, monocarpic, xerophytic, CAM, herbaceous perennial herb. It is drought-tolerant and suitable for marginal lands. **Pineapple does not contain starch**. Pineapple is a Spanish word, botanically it is called **syncarpous** (Sorosis) fruit. Brazilian Pineapple, ***Ananas microstachys*** is considered to ancestor **of cultivated *(A. comosuss)***. The pineapple is a tropical fruit and is known as the "Golden Queen" all over the globe. It can be grown as a mixed crop in most of the orchards, as a hedge plant or as a pure crop. Fruits are relished as dessert in the form of slices, either fresh or canned, juice, squash, jam and mixed jam. Candy is prepared from fruit cores.

Composition and Uses

Pineapple has been described as an excellent source of vitamin C and also a good source of vitamins A and B. Pineapple is a rich source of Vitamin-A: 50 IU/100g and Vitamin C: 50 mg/100g. Sugar 13 per cent, Acidity-0.6-1.0 per cent mineral matter-0.05 per cent, Fiber, 0.3 per cent, Ca, Fe, P. The fruit contains a protein digestive enzyme, **Bromelin.** The most common processed product of pineapple is canned. Fruits are relished as dessert in the form of slices, either fresh or canned, preparation of juice, squash and jam, or mixed jam. Candy is prepared from fruit cores. Fruits are used in the preparation of alcohol, vinegar, calcium citrate, citric acid, gum, and pineapple flavors. The extraction of strong white leaf fibres is used for solid 'pina' cloth and cordage (rope)preparation in the Philippines and Taiwan. The pineapple plants and the fruit residues after extraction of juice are used as cattle feed.

Origin and Distribution

It is believed to have originated in North Brazil, from where it spread to other tropical parts of the world. The major pineapple-growing countries are: Cote d'voire, Swaziland, Fiji, Mexico, Malaysia, Kenya, Kenya, Taiwan, Australia, Sri Lanka, Singapore and India. Its cultivation is confined to the high rainfall and humid coastal regions of peninsular India and to the hilly areas of the north-eastern region of the country. The important states are Assam, Manipur, Tripura, Karnataka, West Bengal, Maharastra, Tamil Nadu, Meghalaya, Andhra Pradesh and Orissa.

Climate and Soil

The optimum temperature is from 22–32°C with an optimum rainfall range of 100–150 cm. A humid crop is a drought-resistant crop. Pineapple leaves are silvery white in colour due to the presence of trichomes. North East India produces the world's best quality pineapple due to its high TSS and less fibre. Pineapple flowers open in the early morning 6-9 a.m., Mode of pollination: Humming birds only in South America other parts by parthenocarpically. An edible portion develops from the peduncle. Pineapple can be grown successfully from the seacoast to an altitude of about 1500 m above sea level. The higher temperature is beneficial for fruit development. The low temperature is harmful and does not improve the colour of fruits. A mild tropical climate, as found in the humid hill slopes, is best suited. It can be grown in plains under shade. Elevation from 500 m to 700 m is ideal. The higher temperature is beneficial for fruit development. The low temperature is harmful and does not improve the colour of fruits. The fruits grow well in sandy loam soil. Fruit size is larger on heavier soil, but the flavour of the fruit is better when grown on lighter soil. It prefers soils that are acidic (pH 4.5-5.5). While high Mg and Mn content are injurious to the crop, fruit size is larger on heavier soil, but the flavour of the fruit is better when grown on lighter soil.

Genera

Smith (1979) established two genera

1. **Ananas:** Ananas bears conspicuous coma of foliaceous bracts and plant produce slips but not stolons and fruits remains seedless.
2. **Pseudananas:** Pseudananas at maturity bears a minute inconspicuous coma of bracts and plants produce elongated stolons and no slips, where as

Ananas has 8 Species

Ananas comosus: The cultivated species with large fruits and a short thick spike.

1. ***A. monstrosus:*** Crownless- There is no leaf formation on fruit.
2. ***A. bracteoutus*****-**Resistant to wilt, heart rot and root rot, bright red flower bracts, fruit large.
3. ***A. Fruitzmuelleri:*** It bears pale green, developed bracts.
4. ***A. ananassoides:*** With small fruits and a long fairly thin spike.
5. ***A. nanus:*** A dwarf species.
6. ***A. parguazensis:*** with curved leaves.
7. ***A. lucidus:*** It has smooth (Spineless) good quality fibre can be obtained.
8. ***Pseudoananas sagenarius***is tetraploid 2n = 4x = 100.

Varieties Classified in to 5 Group

1. Spanish group
2. Queen Group
3. Cayenne group
4. Abacaxi group
5. Maipure Group

1. **Spanish group:** Fruits are of 1-2 kg in wt, globose in shape, fibrous, sub acid, spiny, resistant to mealy bugs. Ex: Mauritius Red Spanish, Singapore Spanish, Green Selangor, Castilla, P.R.1- 67
2. **Queen group:** Conical fruits, spiny, deep eyes, low in fiber 0-5-1.1kg, sweet with yellow flesh. **Ex:** Queen, Mac gregor, Natal, Ripley and Alexandria.
3. **Cayenne group:** Cylindrical fruits with mild acidic, yellow (Spiny tip) 2-3 kg. flesh sweet, uitable for sweet canning, low fibre. Eg: cayenne, Baron, Rothschild, Smooth Guatemalan, Kew, Giant Kew and Typhone.

4. **Abacaxi group:** Concal fruits, pale yellow- white flesh, spiny 1.4 kg sweet tender and juicy grown in Brazil. Eg: perola, Abakka, Sugar loaf, Papelon, Venezolana, Amarella.
5. **Maipure group:** Fruits ovoid to cylindrical, sweeter, fibrous 1-2.5kg with yellow flesh, very juicy. Ex: Maipure, Bumuguesa, Rondon, Perolera, monte lirio.

Indigenous Cultivars

- Kew, Giant kew, Queen, Mauritius.
- Jhaldheep (Sweet type)
- Bakhat (Assam)-Sour
- Lakhat (Nagaland)
- Baruipur local (West Bengal).
- Natural tetraploid variety: James Queen or Z Queen (By mutant of Netal Queen)
- Charlotte rothchild
- Jaldhup and Lakhat: both are under queen gourp.
- Jaldhup is alcoholic flavour.
- Natural triploid variety: Carbenzona or Bull Head.
- Smooth Cayenne is triploid $2n = 3x = 75$, artificially obtained in Hawaii and Philippine.
- Smooth Cayenne is a quantitatively short day variety.
- Shy suckering cultivar: Smooth Cayenne and free from slips variety: Hilo.

Commonly Grown Varieties

- **Smooth Cayenne-** Spineless leaves and resistance to gummosis-ideal for canning
- **Kew and Giant Kew** are commercially grown; Queen on hills,
- **Mauritius**-Mid season variety of Queen Group-Red skinned type, mainly grown in Kerala
- Clones are 81 and 2/83 are promising.
- Through hybridization, H-8 H-7, H-3 and H-1 hybrid have been developed for high yield of quality fruits.
- **Queen:** is an earliest variety ripening in June to July, hilly, best desert and produces more slips and suckers.

- **Kew** (Fibreless variety)- Leading commercial variety and highly suitable for canning, Shy suckering habit and late variety.
- **Table purpose:** Golden yellow colour.
- **Sugar Loaf-** Sweetest and flavourved fruit.

Propagation

Pineapple is mainly propagated by vegetative methods, viz., suckers, slips, crowns, and stumps. Among these, suckers, slips, and crowns are the three important parts used for planting. The suckers arise from the axils of the leaves below the ground level and come to flowering early (15–18 months), and the slips formed on the fruit stalk below the fruit are usually preferred for planting (15–18 months) and 20–22 months after planting, respectively, followed by the crowns (24 months). Ideal plant material: Slips (350g) and Sucker (450g), best planting material: Slips (300-450g). The planting materials should be treated with a mercurial fungicide. The planting materials should not be stored for more than 14 days.

Planting

Planting is done during the kharif season (July–September). Before planting, suckers or slips should be sun-cured. Dry leaves and scales at the base should be removed. The basal end of planting materials should be dipped in 0.4 per cent difolton and 0.05 per cent ekalux to avoid fungal and mealy bug infestation. HDP: 22.5 cm x 60 cm x 90cm (63400 plants/ha) is ideal for subtropical and mild humid condition where 25x60x90 is ideal for tropical condition (53,300) plant /ha. The suckers, or slips, are planted in 10-15 cm deep holes. While planting, the growing part should not be buried. In a single row system, we can accommodate about 15,000–20,000 plants/ha. In a double row system (25 x 35 x 90 cm), High Density Planting (HDP) is adopted in almost all pineapple-growing regions of India and can accommodate about 64000 plants/ha. Flat bed planting is suited for the western plains, while furrow planting is practiced in the hills. A single row system is most often used for ratoon crops. Trench planting is a widely practised method in India.

Culture Practices

Earthling up is an important cultural operation, followed in pineapple. Ratooning has been done in Assam for more than 25–30 years. The commonly recommended ratooning system is 4-5 years. A single sucker/plant with a regular package of practises can give a normal yield. Natural flowering period: December to March.

Manures and Fertilizers

FYM is 40–50 t/ha. N 16 g, P_2O_5 4 g and K_2O 12 g/plant in two equal splits at the 6th and 12th months after planting. Apply as a foliar spray 0.5%-1.0% sulphate of zinc and ferrous solutions at a 15-day interval to overcome the deficiencies in the early crop phase. Nutrient or fertilizer analysis in pineapple: D leaf stage (450 angle)

Micronutrient Deficiency

Iron, Zinc and copper deficiency are common in pineapple, the deficiencies can be corrected by spraying $FeSO_4$ (3 per cent), Zn SO_4 (1 per cent) as foliar spray. A copper deficiency can be corrected by drenching $CuSO_4$ (1.5-2 per cent) at 30–50 ml per plant. Aluminium sulfate-the best fertilizer for pineapple

Irrigation and Weed Control

Although pineapple is a rainfed crop, 4-6 irrigations at 20-25 day intervals may be required during the dry season. Combination sprays of Bromocil and Diuron @ 2kg a.i/ha as pre-emergent herbicides can control weeds efficiently.

Flowering and Fruiting

A pineapple plant generally attains the flowering stage 11–12 months after planting, by which time the plant should have produced at least 40 leaves. Irregular flowering behaviour is one of its major drawbacks. NAA (10-15 ppm), β – hydroxyethyl hydrazine, ethylene, chlorohydrins, ethrel etc. induce uniform flowering in pineapple. "Hermaphrodite" is a flower type. stage for flowering induction: 39–42 leaf stage and ready for harvest after about 15–22 months. To have uniform flowering, apply the following when the crop attains 35-40 leaf stage. NAA 10 ppm + 2% urea (20 g in 1 lit of water) @ 50 ml / plant poured into crown or 2% urea + 0.04% Sodium carbonate + 20 ppm Ethephon (ethrel) @ 50 ml/ plant poured into the crown. To increase the size of the fruit, 200-300 ppm NAA should be sprayed after fruit formation.

Harvesting

At maturity with a slight colour change at the base of the fruit. The lowest eyelets have an orange-yellow colour and are flattened. Pre-harvest application of Ethrel at 500 ppm induces uniform ripening and colour development, but fruits will be acidic and lack flavour The bracts become loose and brown. The fruits should be harvested along with a 5-7 cm stalk at full maturity because they are non-climacteric fruits. Fruits can be harvested from 18 to 24 months. A slight colour change at the base of the fruit indicates maturity. Increase the fruit size: NAA @ 300ppm. Fruits take a long time to ripen: 4 -5 months.

Yield

Yield 75 – 105 t/ha, a plant crop and two ratoon crops are normally taken and in the Mauritius variety, up to five crops can be taken. Pineapple exhibits vegetative parthenocarpy and gametophytic self-incompatibility. Canning and distant markets fully matured fruits

Storage

At 10-13 C, the fruits stored for more than 20 days at 10-13°C. Do not store at less than 8°C since it results in browning of the pulp.

1. **Soft rot, Storage rot and Fruit rot-** ceratostomella paradoxa -This disease is prevalent in the lower pulney hills of India causing sever loss in region with high rainfall and low temperature.
2. **Heart rot or Stump rot (*Phytophthora parasitica*):** Emits foul smell – Predominant in alkaline Soil-Severe in dry and wet regions
3. **Fasciation and multiple crowns**: Multiple crowns may occur due to genetical factorincayenne group(kew) as well as due to soil and environmental reasons. It is also due to excess Nitrogen.
4. **Pineapple wilt virus** – transmitted by mealy bugs etc., also causes damage.

Physiological disorders

1. **Sunscald:** The cells under the fruit skin that are exposed to sunrays get damaged; care should be taken to control lodging and the fruit should be covered with dry straw or its own leaves during April-May.
2. **Fasciation and multiple crowns:** Multiple crowns may occur due to genetical factors as well as soil and environmental reasons. It is also due to excess nitrogen. Fruits get flattened and fascinated. In fertile virgin soil of warm areas, more abnormal fruits occur as compared to less fertile soil.
3. **Black heart:** It is also known as endogenous brown spot or internal browning. It is characterised by the formation of brown spots at the base of fruitlets and further black discoloration of the centre core. Low temperature or exogenous application of GA_3 can induce this disorder. Fruits exposed to a high temperature (40°C) for 24 hours reduce black hearts in cold-stored pineapple.

44

Pomegranate

- Pomegranate/Anar/National fruit of Iran/Apple of carthage
- *B. N: Punica granatum*
- Family: Punicaceae
- Chromosome No.: 2n = 2x = 16, 18
- Origin: South West Asia-Iran (Persia)
- Inflorescence: Hypanthodium
- Fruit type: Balausta (Modified Berry)
- Edible part: Juicy seed coat (Aril)

Introduction

Winter hardy and highly drought tolerant. This is an excellent choice crop for the arid and semi-arid conditions of India. Maharashtra is the leading state in India for pomegranate cultivation: Maharashtra. **NRC of Pomegranate, Kesaon, Solapur, Maharashtra: Established in 2005.** Because of the presence of anthocyanin, the aril and skin have a red color. Dried seeds with pulp are known as 'Anardana': It is widely used as a condiment, prepared from a sour type of wild pomegranate is called as Daru. Commonly found in the Himalayan region, has a long productive cycle and is resistant to bacterial blight. Double flower type ornamental plant in it modification of stamens into petals makes them look like rose flowers. Nana or dwarf type is called Bonsai pomegranate, exclusively used for ornamental purposes.

Area and Distribution

The pomegranate is a favourite table fruit in tropical and subtropical countries. Pomegranate is one of the ancient and important table fruits in tropical countries. It is grown on a large scale in the Mediterranean countries such as Iran, Spain, Morocco, Egypt, etc. It is grown all over India and commercially in Maharashtra (more than 60 per cent), followed by Gujarat, Rajasthan, Uttar Pradesh, and Andhra Pradesh.

Use and Composition

Anardana: An acidulant product used in souring and culinary preparations (condiment). Pomegranates are a rich source of carbohydrates (14.5%), protein (1.6%), calcium (10 mg/100 g), phosphorus (70 mg/100 g), iron (0.3 mg/100 g) and vitamin C (65 mg/100 g). Indian local product Anar rub (TSS: 70-75°B): used as a sauce, prepared from the juice. The fruit juice of pomegranates is valued medicinal properties for leprosy patients. The pomegranate's taste is due to citric acid. The bark and rind of the fruit are commonly used in treating dysentery and diarrhea. Tannin is obtained from fruit rind, leaves, stems, and root bark. One pomegranate provides approximately 40% of an adult's daily Vit-C requirement. It is also rich in riboflavin, protein, fat, sugar, pectin, calcium, and iron content.

Climate and Soil

Despite being adapted to a wide range of climatic conditions, pomegranate yields the best quality fruits in areas of cool winter and hot, dry summer. It can be successfully grown right from sea level right up to 1850 metres. It is fairly tolerant of low temperature, of course, with differences among varieties. For proper fruit development, maturity, and sweetness, a temperature of 35–38°C is needed. Under humid conditions, the quality gets affected. At higher elevations and in areas of low temperature during winter, the tree behaves as a deciduous one. It is both evergreen (subtropical and tropical) and deciduous (temperate regions). The fruits are borne terminally on short spurs arising from mature shoots. The tree is hardy and bushy, with a tendency to develop multi-stems. It can be grown in limy, alkaline and saline soils and also thrives well in shallow rocky gravel soils. Deep, heavy, loamy, well-drained soil with a pH range of 5.5-7.5 produced the best fruit yield and quality.

Species

1. ***Punica protopunica*** *is* resistant to diseases: **Used as a rootstock**
2. **(*Punica granatum var. nana*)** Species most suitable for pot plant production dwarf pomegranate.

Varieties

1. **Paper shell:** The flesh is pinkish and the seeds are soft.
2. **Jyothi-** (Bassein seedless x Dholka) Red arils
3. **Mridula-** (Ganesh x Gul-e –Shah Red), Blood red and soft seeded
4. **Jalore seedless** – It is a soft seeded variety developed and recommended by CAZRI.

Hard Seeded Types

1. **Kandhari:** Fruits are large with deep red rind, aril deep blood red or deep pink with sweet, slightly acidic juice. Seeds are very hard.
2. **Musker Red:** Medium sized fruits with medium thick red rinds. Aril is fleshy with moderately sweet juice; seeds are medium hard.
3. **Alandi or Vadki:** It possesses medium sized red fruits, aril fleshy, blood red or deep pink with sweet acidic juice. Seeds are very hard.
4. **Kabul:** Large fruits, dark red with yellow parches, aril dark red fleshy seeds hard with slightly bitter juice.

Soft seeded types

1. **Dholka:** This cultivar has large fruits with It is the commercial variety of Gujarat.
2. **Spanish Ruby:** Flesh rose colour and seeds soft.
3. **Ganesh:** It is a selection from **Alandi** developed by **Dr. G.S. Cheema** at Ganeshkhind Fruit Research Station, Pune in 1936.

Special varieties

1. **Ruby:** Ganesh x Kabul x Yercard, a multiple cross hybrid developed at IIHR, Ruby derived certain fruit quality attributes from Ganesh, Russian variety 'Gulsha Rose Pink.
2. **Amlidana:** It is an F1 hybrid (Ganesh x Nana), its fruits provide more acidic (16.18%) 'anardana' an acidulant.
3. **Yercaud-1 (YCD-1):** This was found suitable to mid elevation of Shevroys hills.
4. **CO-1:** It is a selection, with purple aril and soft seeds.
5. **Bhagwa**: (Ganesh x Gule e- Shah Red) open pollinated variety, semi-hard seeded variety
6. **Super Bhagwa:** selection from Bhagwa.
7. **Amlidana:** It is a cross between **Ganesh x Nana,** dwarf, suitable for HDP, (Most suitable for anardana purpose)
8. **Bedana Bosec** and **Khog** cultivars of pomegranate **are tolerant** to fruit cracking.
9. **Ganesh, Basseein Seedless** and **Paper Shell** are soft seeded cultivars of pomegranate.
10. **Shirin Anar-** Resistant to bacterial leaf spot
11. Solapur Lal – Table purpose

12. Solapur Anardana-Processing purpose
13. **Phule Arakta** (Ganesh x Gul- e-Shah Red), Dark red aril
14. Musket in Maharashtra, **Bassein Seedless** in Karnataka; **Dholka** in Gujarat and **Jalore Seedless** in Rajasthan.
15. **G-132, G-107** and **G-137** are high quality strains selected from Ganesh.
16. **P-13, P-16, P-23** and **P-26** are suitable strains selected from Musket.

Propagation

Presently, pomegranates are successfully propagated by hardwood stem cutting and air-layering.

Stem Cutting

Cuttings are collected from high-yielding plants that are one-year-old & fully matured shoots. About 25–40 cm long cuttings should be planted by removing leaves and treating the bottom end of the cuttings with rooting hormone (IBA-2000 ppm) and inserting them into the soil. Plants will be ready in 55–60 days.

Air Layering

Pomegranates may also be propagated by air layering. It should be done during the rainy season; the bottom of the uppercut can be treated with IBA 10,000 ppm for better rooting. It can also be propagated by air layering or gootee, as well as ground layering.

Planting

Land is prepared thoroughly during the onset of the monsoon. Pits of 60-75 cm^3 at a spacing of 5 x 5m, planting should be done during the monsoon season. High density planting in pomegranates can accommodate 1000 trees/ha (5 x 2 m), giving 2.5 times higher yield compared to the 5 x 5 m standard (400 plants/ha). Planting of 1-2-year-old rooted cuttings in the centre of pits and providing support with staking. So, at the time of planting, inoculating the roots with 50 g of phosphorbacterium and 150 g of Vesicular Arbuscular Mycorrhiza will aid in root growth and establishment.

Irrigation

A newly planted orchard requires frequent and regular irrigation. During flowering and fruiting, the orchard should be irrigated regularly to avoid cracking of fruits and for better development of fruits. Drip irrigation using a pitcher pot or tube will keep the soil moisture constant without much fluctuation. This helps the pomegranate to get better establishment in the early stages as well as regular bearing in the fruiting phase.

Intercultivation

During the pre-bearing stage, legumes, vegetables and cereals can be taken during the pre-bearing stage. Papaya can be used as a filler plant. Basins should be maintained free from weeds.

Manures and Fertilizers

Besides, every year 20 kg of FYM should be applied/tree. The manures and fertilizers are mixed and applied in a round basin 1 metre away from the trunk. A week before application of fertilizer, the soil around the root zone is slightly removed and 50 g of phosphobacterium and 150 g of Vesicular Arbuscular mycorrhiza have to be applied near the feeder roots. Besides farmyard manure, each tree is supplied with 10 kg of pressmud by spreading it on the top layer of soil. In sandy soils, press mud serves as a mulch and prevents moisture loss during the summer season. Besides this, press mud also supplies some of the nutrients. Press mud applied plants have come to bear in sandy soil conditions. application of 375 kg N, 1875 kg P_2O_5 and 187.5 kg K_2O.

Training and Pruning

Training of pomegranate plants is important to allow a certain number of shoots/ stems per plant. It may be trained as a multi-stemmed or single-stemmed tree.

1. **Multi-stemmed tree**

   This method is preferred in Maharashtra, where in 3-4 stems are retained on a hill and the remaining shoots are removed. But yield has not been found to be affected. This will give a frame to the plant.

2. **Single stemmed tree**

   Train the plants, remove all the side shoots upto 2-3 feet and a single stem is left. This operation begins at the time of planting. The main stem is pinched at a height of 1 m, which results in the formation of branches. Only well distributed 4-5 branches on all sides are allowed to grow. Pruning of water shoots, weak crotches, dead twigs, and old spurs is done regularly. After 10 years, old main stems should be removed by cutting back to make them more productive.

Flowering

To obtain a higher fruit yield during a particular period, the pomegranate plants are given a resting period. Pomegranates bear a spring crop on the previous year's growth and a summer crop on the current season's growth. Hyphanthodium is a type of inflorescence. Type of pollination: often cross-pollination.

- Mode of pollination: Insects-Honey bees.
- Both self and cross pollination is observed.
- Predominant pollination: Cross pollination.
- Heterostyly is common in pomegranate.
- Flowers has **3 types: Male, Hermaphrodite, Intermediate.**
- Hermaphrodite flowers usually homostlyous or pin eyed and male flowers are thrum eyed
- Fruits are borne terminally on short spurs from mature shoots.
- Seedlessness in pomegranate: Lack of **lignifications of testa**

Bahar Treatment (crop regulation)

1. **Ambe bahar** – Flowering can be induced in **February March** – This is taken in the areas where, enough water is available during hot weather.
2. **Mrig-bahar-** Flowering can be induced in **June-July**, coinciding with the outbreak of monsoon, this treatment is taken in the areas where, water is scarce during the hot weather.
3. **Hastha Bahar** - Flowering can be induced in **September October**, where the trees have to be subjected to stress during August – September. This is rather uncertain because of the monsoon rains that occur during this period. The tree starts bearing fruits from 3-4th year and continues for about 25 to 30 years. Economic yield is generally obtained from 6th or 7th year onwards.

Bahar treatment, or flower regulation, is followed. The best quality fruits are produced in cool winters and hot, dry summers. Three distinct flowering seasons: Subtropical, Central and Western India have three distinct flowering seasons. It is done by withholding water for about 60 days in advance of the normal flowering, roots are exposed and it is known as bahar treatment. Fruits of the 'Mrig' bahar are more prone to cracking.

Harvesting and Yield

5-7 months after blossoming, pomegranate fruits become ready for harvesting. Adding anti-transparents such as 10% kaolin, 10-5m phenyl mercuric acetate, and 1% liquid paraffin is beneficial for increasing its productivity. After 120-130 days, the fruits are ready for harvest. Mature fruits become slightly yellowish and further pink to red. On tapping, the fruits give a metallic sound and when pressed they give a "Crunch" sound and flattened during maturity. The fully grown up tree of about 10 years old produces 80-120 fruits (16-20

Kg). The fruits can b stored for 15 to 20 days under ambient temperature. At 0°C and 80 % RH, they can be stored even for 2 months. Hardening peels are due to prolonged drought conditions. 12 A grades are preferred in northern and southern India.

Pests

1. **Anar butterfly or fruit borer** (*Virachola isocrates):* Serious pest all over India
2. **Bark eating caterpillar** – *Inderbela tetraonis*
3. **Stem borer:** *Aleurodes species.*
4. **Mealybug:** *Ferrisia virgata*
5. **Whitefly:** *Siphonimus phyllyreae.*
6. **Thrips:** *Rhipiphorathrips creutatus*,
7. **Aphids:** *Aphis puniae*

Diseases

1. Black spot / fruit rot: (*collectotrichum gloeosporoides*)
2. Fruit rot:
3. Bacterial leaf spot: (*Xanthomonas campestris pv. punicae*)
4. Bacterial blight (*Xanthomonas oxonopodes pv. Punicae)* is serious problem in Maharashtra
5. Fruit rot – *Phomopsis sp*
6. Leaf spot – *Colletotrichum gloeosporiodies*

Physiological Disorders

Fruit Cracking

It is common in pomegranates and also serious. This disorder is due to the prevalence of high temperature and moisture stress followed by rain, which causes fruit cracking during fruit development. In the young fruits, it could be due to boron deficiency. While in Mrigbahar (July), grown fruits might be due to a sudden fluctuation in diurnal temperature. Also, prolonged dry spells cause hardening of the peel. If this is followed by heavy rainfall or irrigation, then the pulp grows fast and results in cracking of the pulp. The cracked fruits are also liable to be invaded by certain fungi and bacteria. The fruits lose their market value and become unfit for human consumption. Cracking tolerance is higher in cultivars such as "Bedana Bosek," "Jalore seedless," and "khog."

Control Measures

1. Planting of varieties tolerant to fruit cracking.
2. Early harvesting of fruits immediately after maturity
3. Maintain optimum soil moisture during Mrig bahar fruiting
4. Spray calcium hydroxide on foliage after fruit set. If boron deficiency - spray borax at 0.1 per cent to pl/20g/pl/year.
5. Spray GA3 at 250ppm in June.
6. Plant windbreak around the pomegranate plantation.

Internal break down/Blackening of arils: Disintegration of arils in matured fruits is serious malady

45

Rambutan

- Rambutan/ Hairy litchi/High temperature loving plant
- B. N: *Nephelium lappaceum*
- Family: Sapindaceac
- Chromosome No.: 2n = 22
- Origin: Malaya Peninsula
- Inflorescence: Panicles
- Fruit type: Nut
- Edible part: Aril

Introdution

Rambutan is called "hairy litchi" as the fruits have red-soft spines on the surface. Fruits are produced in bunches and possess long, thick, soft hairs on their surface and are hence called rambutans after the word 'rambut', meaning strands of hair in Malay. The external colour of the fruits is attractive in red and yellow. The white is translucent and sub-arid. Its appearance differs from the litchi by having red and soft spines (hairs) covering the whole surface of the fruit. It has since spread to Southeast Asia, Central America, and Africa. The Rambutan-growing countries are Thailand, Myanmar, Sri Lanka, and India.

Composition and Uses

The fruit contains 14.5 CHO, 0.9 g protein, 0.1 g fat, 31 mg/100 g vitamin C, and TSS-14.9–21.2% in 100 g. Rambutan is mainly served as fresh fruit or as a dessert mixed with other fruits, or used for canning or made into fruit syrup. The oil seed is used for soap making. The root has medicinal uses while the tannin it contains is used for dyes. The whole tree, with its beautiful leaves, flowers and colourful fruits also serves as an environmental tree in landscape. Every tree has many branches, and every green tree reaches up to 15 m or more in height. Fruits are one-seeded; 10-12 fruits are borne in clusters. The soft curved spines (hairs) covering all the pericarp may be red or yellow in colour. Inside the fruit, a big seed is surrounded by pearl-while aril/flush) which is fixing, insipid as compared with the sweeter litchi.

Climate and Soil

A warm, humid tropical climate would be ideal. Mean minimum temperature -5°C. The minimum temperature should not be below 13-15°C. It comes well under humid condition. It requires 2000–2800 mm of annual rainfall. A rich sandy or clay loam with high organic matter. And good drainage will be ideal for the growth and flowering.

Cultivars

- Rohug-Rian – Thailand cv. Oval bright red fruits spines are red which green tip.
- Seematjan- Fruits deep red with soft curved spines.
- BR.1 (Boting Rambutan No.7) - New cv. Developed in China.
- Rongrien
- Chompu (Thailand)
- Bingjai
- Lebak Bulus
- Rapiah (Indonesia)
- Azimal
- Kelip
- Singapura (Malaysia).
- Arka Coorg
- Arka Coorg Arun

Propagation

Seed: Both monoecious and dioecious types are seeds in rambutan. and production results in wide variation among seedlings. Dioecious tree, about 1/3rd of the total number of progeny become male plants. Used only for raising seedling rootstock for grafting seeds are sown immediately when taken out from fleshy fruits because they loose viability within a few days, if left to rot.

Vegetative Propagation

Patch or forkert budding on 1-2-year-old seeding rootstock is recommended.

Planting

Spacing given for planting is 10 m x 7 m. Planting of different cultivars is advised to encourage pollination. Temporary shading of young plants is necessary. A modified leader system of training is practiced. Light, unusual pruning of crossing branches, dead and diseased twigs and water shoots is suggested.

Manure and Fertilizers

Age of tree	N (g/plant)	P (g/plant)	K (g/plant)	
I year	75	25	75	30-40Kg FYM/Plant
2 year	100	25	100	
3 year	150	50	150	
4 year	250	75	250	
5 year	600	200	600	

46

Sapota

- Chikku/Sapodilla bully/Sapodilla plum
- B. N.: *Achras sapota*
- Family: Sapotaceae
- Chromosome No.: 2n = 2x = 26
- Origin: Mexico (Central America)
- Inflorescence: Cymose Solitary
- Fruit type: Berry
- Edible portion: Mesocarp

Introduction

Sapota is a delicious fruit introduced from tropical America. It is also known as sapodilla or chiku in India. Sapota fruits are recommended to adult patients ailing from tuberculosis and children from primary complex. This offers high economic returns even under marginal land and low input management.

Importance and Uses

Sapota is a good source of sugar, protein, fat, fiber and minerals (Ca, P, Fe). Sapota is a delicious dessert fruit. The latex from stems and immature fruits is used in the preparation of chewing gum. Fruits can be dried and made into nutritious powder, which can be used in milk shakes and sweets. In some countries (Indonesia), young leafy shoots are used in salads or as vegetables. Sapota is supposed to be medicinal as well as seeds as diuretic, bark as tonic, antipyretic, febrifuge and in curing biliousness and febrile attacks. Chickle, which is made from the milky latex of bark and immature fruits, is used to make chewing gum and statues. There is a lot of vitamin A (410 IU/100 g) and potassium (344 mg/100 g). Sapota fruit is a good source of digestible sugar, 12–18%.

Origin and Distribution

India is considered to be the largest producer of sapota in the world. However, sapota is considered to be a minor crop in India. Sapota is commercially grown in Karnataka, Gujarat T.N., AP, WB, Maharashtra, and Gujarat is the leading state in production and Tamil Nadu is the leading of productivity, followed by Andhra Pradesh. Cultivation was taken up for the first time in Maharashtra in 1898 in the village of Gholwad. Central America cultivated in West Indies, Philippines, Malaysia, Indonesia, Srilanka and India.

Soil and Climate

Sapota being a tropical fruit crop can be grown from sea level up to 1200 M. Optimum temperature range is 11 to 30°C. High temperature 41°C causes drying of stigmatic surface. Dry and strong winds also damage Sapota. Sapota a drought-tolerant plant. It prefers warm and moist weather and grows in both dry and humid areas. A coastal climate is the best suited. Areas with an annual rainfall of 1250–2500 mm are highly suitable. Sapota, being a hardy tree, can be grown on a wide range of soils. The soil should be well drained without any hard pans. Deep sand loams, alluvial soils, or medium black soils are best. Sapota grows well on calcareous soils (pH 6–8).

Types of Tree

Sapota is grouped into 4 types based on nature of branches and colour of foliage as follows:

1. **Trees with erect growing habit:** Branches appearing in whorls, leafy dark green, broad and oval, fruits large and superior.
2. **Trees with drooping habit:** Branches in whorls, leaves light green narrow and elliptical, fruit small, inferior.
3. **Trees with spreading habit:** Branches irregular, leaves dark green, broad and oval, fruits medium to large, superior.
4. **Trees with spreading habit:** but with inferior quality fruits

Cultivars

- **Kalipatti**- Commercially grown in Maharashtra, Gujarath, North Karnataka. Has spreading branches,
- **Chhatri**: Similar to kalipatti with droping branches in whorls.
- **Dhola Diwani:** Whitish oval fruits with superior quality
- **Long:** Has narrow and small leaves, fruits long poor bearer.

- **Bhuri or Bhuripatti:** Medium bearer, fruits large superior.
- **Jingar**: Medium sized tree, small leaves, fruits in clusters.
- **Vanjeet:** Slow growing, knots on stems shy bearer, superior.
- **Pala**: Popular in A.P. and TN fruits small, oval, heavy bearer in clusters.
- **Kirthibarthi:** Popular in A.P. Fruits small to medium with 4-5 ridges, oval, superior.
- **Dwarapudi:** Popular in A.P. fruits round like cricket ball, superior.
- **Cricket ball:** Popular in TN, Karnatataka, Maharastra, W.B A.P. Fruits large, round, superior, shy bearer.
- **Oval:** Fruits small to medium, oval, inferior shy bearer.
- **Vanivalasa:** AP, fruits oval, medium sized, medium quality.
- **Calcatta Round:** WB, Karnataka, AP fruits large, medium quality.
- **Jonnavalasa-1:** AP fruits medium, ovate, superior and no ridges
- **Jonnavalasa-2:** AP, fruits medium, ovate, depression at stalk end, whitish flakes on skin eight ridges superior.
- **Baramasi:** WB, Bihar, UP fruits medium, round medium quality.
- **Pot Sapota:** Fruits small, oval, superior, bears in pots itself
- **Gavarayya:** AP, TN fruits small with varied shoulders 8-10 ridges, superior.
- **Thagarampudi:** TN fruits medium sized, round or oval, superior, good for export.
- **Ayyangare:** TN fruits medium to large, round or obovate, rose scented, superior.

Hybrid and Special Varieties

- **CO1-Cricket Ball** x **Oval.**
- CO-2 is a clonal selection from **Baramasi.**
- CO-3 is a hybrid between Cricket Ball x Vavilavalasa, Suitable for HDF
- PKM-1 is a clonal selection from Guthi. It is dwarf, bearing throughout the year.
- PKM-2 is Guthi x Kirth Barthi.
- PKM-3 Kalipatti x Cricket Ball.
- DHS1- Kalipatti x Cricket Ball.

- DHS2-Kalipatti x Cricket Ball.
- PKM-4, Clonal selection from PKM-1, Spindle shaped fruits, Suitable for dry flakes production.
- PKM-5, Clonal selection from Virudhunagar, Off-season bearer.
- CO1, DHS1, DHS2, PKM-3 is resistant to leaf spot of sapota.
- Calcuttia special Round is most susceptible of leaf spot.
- Murabba- Popular in MH.
- Kirthi Bharti- Popular in AP, Good transport value.
- **Exotic varieties:** Alano, Hassia, Makok and Molix.

Propagation

Inarching and softwood grafting is best technique (Ideal time: July to August). Number of seeds/ fruit: 0-12 (Most common 3 to 5 seeds). Commercially propagated by grafting on Rayan or khirni /Pala (*Manilkara hexandra*) Stocks.

Other Rootstock used are

Sapota seedlings

- Adams apple *(M. kauki).*
- Mahua *(M. latifolia).*
- Mee Tree *(Bassia longifolia).*
- Star apple *(Chrysophyllum cainito)*
- Miracular fruit *(Sideroxylon dulcifieum).*

Planting

Spacing is 8-10, 1m cube pits are dug Pits are filled with a FYM and soil mixture containing 1.2 kg of bone meal. High density planting with 5 m x 5 m spacing improves yield. It is better to have a wind break around the Sapota plot. The monsoon season is the best.

Manures and Fertilizers

A bearing tree of 11 years is given 400 g N, 260 g P_2O_5 and 450 g K_2O per year in addition to 40 kg FYM and graded doses are applied from the beginning according to the growth of the plant. Application of more "N" is reported to increase the yield. Manuring should be done in 2 spilt doses, coinciding with the monsoons.

Irrigation

Sapota can tolerate drought to some extent, but irrigation helps in improving the yield. Regular irrigation should be given from planting till the plant is established well. Later irrigations may be scheduled according to need and soil and weather conditions. Insufficient irrigation results in the dropping of flowers. Drip irrigation can be more useful.

Interculture

Intercrops like bananas, papaya, and leguminous vegetables can be grown profitably during the pre-bearing period. Weeding should be regular. Spraying SADH (Succinic acid-2, 2-Dimethylhydrazide) 100 ppm gives good fruit set and (Planofix) NAA 300 ppm gives fruit retention.

Training and Pruning

It is better to remove the lowermost branches on the trunk up to 60-90 cm as they will be touching the ground and are mostly unproductive. Stock sports should be removed from time to time. No regular pruning is needed for Sapota.

Training System

The Central Leader System

Flowering and Fruiting

Fruit setting is a problem; natural fruit set is 10–12%. Fruit drop is due to seed incompatibility. For improvement of fruit set: spraying of NAA @ 100-300 ppm during flowering. Flowers of sapota are protogynous. Sapota fruit follows a double sigmoid growth curve. Fruits are borne in the axils of the leaves of the current season's growth. Types of pollination: allogamous-aneomophilous (wind pollinated). Spraying SADH 100 ppm gives good fruit set and (Planofix) NAA 300 ppm gives fruit retension. Spraying occurs twice before flowering and once during the pea stage.

Maturity Indecis

Milky latex on scratching will be reduced and will show a yellow streak rather than a green streak. harvested by hand picking. Fruits at full maturity develop a dull orange or potato colour. Sapota takes 7-10 ½ months from anthesis to maturity of fruits. Brown scaly material gets reduced. The dried stigma at the tip of the fruit drops easily. It develops into a dull orange or potato colour.

Harvesting

Fruits should be harvested with the stalk intact. Peak harvesting periods are Jan-Feb and May-June in Maharashtra; March-May and Sep-October in Karnataka and A.P. Sapota normally produces fruit throughout the year. It is a climacteric fruit.

Yield

30-year-old tree yields 2500-3000 fruits per plant per year. Temperature between 12-14°C fruits ripen slowly and keep well for about 5 weeks. Ripe fruits can be stored at 2-3°C and 85-90 % R.H. for 6 weeks and firm fruits for 8 weeks at 3 to 5°C and 85-95 % R.H. Fruits are dipped in GA_3 300 ppm + Bavistin 1000 ppm solution at the pre-packing stage to extend shelf life and avoid storage loss. The fruits are ripening by keeping the fruits in an airtight chamber with 5000 ppm Ethrel + 10 g NaOH pellets. Fruits ripen after harvest in about 4–13 days, depending upon the cultivar. Ripening can be hastened by treating with ethrel (250-750 ppm).

Pests

1. Stem borer: *(Isocrata tetraonis)*
2. Mealy bug: *(Phenacoccus icerjoides)*
3. Scale insects: *(Pulvinaria psidil).*
4. Fruit borer: *(Virachola isocrates).*
5. Chiku moth: *Nephoperyse eugraphela* causes 40-60% yield losses. is most destructing pest
6. Fruit fly: *Daucus correctus*
7. *Anarsia achrasella* are serious pests of sapota.

47

Tamarind

- Tamarind /Indian date/Imli/Girl fruit
- B. N.: *Tamarindus indica*
- Family: Fabaceae
- Chromosome No.: 2n = 24
- Origin: Tropical Africa
- Inflorescence: Racemose
- Fruit type: Lomentum or Pod
- Edible part: Mesocarp

Introduction

It is the most popular tree found cultivated throughout India as stray plantations or avenues. Regular orcharding is wanted but common in waste land. The wood is used in cart wheels and rice pounders. Oil mills, etc. The bark is used in tanning. The tender leaves and flowers are eaten as a vegetable. The pulp of fruit is used for culinary purposes. The seeds are rich in pectin. India is the major producer and consumer of this fruit in the world. The seeds are rich in pectin. India is the major producer and consumer of this fruit in the world. Immature fruits are used for chutney, and ripe fruit pulp is an important ingredient in south Indian vegetable dishes to induce sourness. Tamarind is indigenous to tropical Africa and some parts of South India.

Composition

Fruit is a pod that contains 55 % pulp, 20.6 percent water, 3.1% protein, 0.4 % fat, 70.8 percent carbohydrates, 3% fiber, and 10 % tartaric acid. Seeds have 63% starch, 16 % protein, minerals (2.9 g) per 100 g of edible portion. Pectin (4.4 %), tartaric acid (6-10 %), minerals like calcium (170 mg), phosphorus (110 mg), magnesium (40 mg), iron, copper and vitamins like beta-carotene (60 mg), thamine (44 mg), niacin and riboflavin are other components in pulp.

Uses

Climate and Soil

Semi arid tropical and subtropical region but can be grown in monsoon region with god drainage. It prefers a warm climate. self-sown in forests and wastelands. A compost of fibrous loam and sand suits him very much.

Cultivars

Baily recognized two types of tamarind: East Indian type having long pods with 6 to 12 seeds. The West Indian type having shorter pods containing 1 to 4 seeds. Paulos (1975) recognized a tamarind type known as «Valakatchi» that bears long and rectangular pods.

On the basis of pulp colour: the yellow or brown pulp type turns dark brown on storage. The reddish pulp type is locally known as 'Raktichinch'. A high-yielding red type (Yogeshwari) has been released by MAU, Parbhani.

On the basis of taste, the cultivated types could be broadly classified into two groups, viz., sweet and sour types.

1. **Sweet type:** Makham, Waan, Secthong, Manila sweet, Pratisthan'.
2. **Sour type:** No. 263, Urigam and cumbum

- PKM-1 (Periyakulum-1) - Suitable for high density.
- DTS – 1 and DTS – 2 released by Dharwad Agricultural University
- ***Valakatchi*** is a type that bears long and rectangular pods.
- ***Yogeshwari*** **and** ***Raktichinch*** are two Indian types having red pulp.
- Goma Prateek variety.
- Ajantha Rudhira.

Propagation

Propagation seed is common and 2.0–2.5 kg of seeds are needed to raise seedlings for 1 hectare. softwood grafting for arid regions, in situ soft wood grafting on one-year-old seedlings.

Planting

Spacing 12 x 12 m, 10 x 10 m, and 5 x 5 m at the latest. Pit size: 1 x 1 x 1 m and planting is done at the beginning of the monsoon. No pruning is required.

Fruiting and Yeild

Flowering takes place in May to June and the fruits are available from February to March. Inflorescences appear on the current season's growth. About 230-250 days are required from fruit set to maturity. Productivity has increased and continued to be productive for more than 60 years.

Harvesting

When harvested, the fruits are almost semi-dried and do not spoil due to high acidity. A fully developed tree can give Ripe pods are harvested by shaking tree branches, and a fully grown tree produces 200 kg of fruit per annum. An alternate bearing tendency is noticed in a few trees. After harvest, the fruits are allowed to dry and their hard shells are removed. Tamarind pulp is also exported to Europe and America for preparing chutney and meat sauces. Under ordinary conditions, the pulp remains good for about a year provided it is kept in dry conditions. The tree is affected by a number of rots, such as saprot, brownish saprot, and white rot.

Pests

Scales, Mealy bug white grubs, caterpillars that eat leaves, Citrus aphids, black and white flies, Thrips, Cow bugs, bud and flower pests and fruit borers.

48

Wood Apple

- Kavet/Cuard apple/Curd fruit
- B.N.: (*Feronia limonia L.*)
- Elephant apple (*F. elephantum)*
- Family: Rutaceae
- Chromosome No. 2n: 18
- Origin: India and Sri Lanka
- Inflorescence: Panicles
- Fruit type: Hard shelled seeded berry (Amphisarica)
- Edible part: Aril

Introduction

The fruit is a hard-shelled, many-seeded berry with its pinkish brown, aromatic sour and sweet pulp being the edible portion of the seeds embedded in it. The tree is one of the hardy trees of arid and semi-arid regions.

Uses and Composition

The pulp contains 18.1% carbohydrate, 7.1% protein, 3.7% fat, 5.0% fiber and 1.9% mineral matter. The pulp is a rich source of calcium (130 mg/100 g), phosphorus (110 mg/100 g) and iron (0.48 mg/100 g). The vitamins supplied by one hundred gram of pulp are carotene 61 μg, riboflavin 0.17 mg, niacin 0.8 mg, thiamine 0.04 mg and vitamin C 3 mg. The ripe fruit pulp makes excellent chutney and it is also consumed a fresh along with sugar. It is used as an adjunct in jelly preparation along with the pulp of guava. Fruit contain 3 to 5% pectin.

Climate and Soil

Wood apple can be grown in dry tracts of tropical and sub-tropical regions. And right from sea level up to 1400 m above MSL. It is adapted to a wide range of soil conditions, including degraded soil. It can also tolerate salinity to a certain extent. It is an ideal tree to be exploited for growing in wasteland.

Propagation

Propagation, though seed propagation is done mainly. Budding has been reported to be successful if done during the late summer and early monsoon. Budded plants are dwarf and precocious in bearing.

Cultivars: Two types are reported. (sweet and sour)

Thar gaurav have polyembryony (up to 60%) is reported in this crop,

Planting

Pit lines are drawn across the slope and pits can be dug at a spacing of 8 m x 8 m and each pit has a size of 1 m x 1 m x1 m. Planting should be done at the onset of the monsoon after filling the pit with 20 kg of FYM, sand and top soil.

Interculture

Training is done by the central leader method, allowing well spaced branches in all directions. Intercrops can be taken during rainy seasons for the first 5 years. In the post monsoon season, the basins can be mulched with dry leaves. Every year, 25 kg of FYM is to be applied to each tree at the beginning of the monsoon rains. This will help in increasing fruit size and quality.

Harvest and Yield

Budded plants come to bear 3 - 4 years after planting. But to reach optimum productivity, it will take about 10 years. The crop flowers from February to May and fruits will be available from July to December depending on the flowering month. The yield of a grown-up tree (10 years or more) varies from 200 to 250 fruits per annum. After harvesting, the fruits are graded according to size and marketed.

49

Minor Fruit Crops

KER/KAIR/TEET

***Capparis deciduas*. Edgew.**

(Capparidaceae)

- Ker, a native of India, is a perennial shrub seen in arid and semi-arid
- Plants are very hardy and withstand extreme hot and dry conditions.
- Often considered a desert plant,
- which grows widely in the Thar Desert of Western Rajasthan in India.
- This can be successfully grown in wastelands with sandy and rocky soils.
- Immature fruits are used for pickling and for culinary purposes.
- Immature flower buds and flowers are used as vegetables.
- Plants are valued for their medicinal properties against cardiac trouble and biliousness.
- Bark is used for rheumatism.
- Immature fruits are dried and kept by the people of Rajasthan for use in the off-season for culinary purposes. Fruits contain 54% moisture, 4% protein and 85 mg/100 g of ascorbic acid.

WHITE SAPOTE

***Casimiroa edulis* Leav.**

(Rutaceae)

- White sapote, indigenous to Mexico and Central America belongs to sub-family Toddalioidae.
- It is a hardy, strong, vigorous, medium-sized tree of about 15 cm in height with a spreading habit.
- Small flowers are greenish white in colour and are produced terminally in the axils of mature leaves.
- The apple-like fruits produced in clusters are spherical or oblong in shape, 7 to 10 cm in diameter, and their skin is yellowish green and slightly ribbed.

- Flesh is creamy white or yellowish, custard like, sweet, melting, aromatic and pleasing, with seeds 2 to 5 in number, enclosed in a hard, bony endocarp.
- Along with lemon juice, white sapote can be used for making jelly.
- It makes a delicious flavouring for milkshakes.
- Leaves are used to relieve high blood pressure and also as a remedy for diarrhea.
- Two alkaloids—Casimiroine and Casimiroedine are obtained from seeds.
- It grows from the tropical highlands to the subtropics and comes up to an elevation of 900 to 3000 m.
- Grafting or budding is also possible in this crop.
- Past seasons procured shoots are the best scion material.
- Seedlings/grafts are planted at a spacing of 5-6 m distance.
- There are two distinct types of white sapote:
- *Ever bearing:* Almost continuous flowering – produces green fruits with excellent flavour – varieties – Suebelle and Wilson
- *Seasonal:* Main period of blossom is in spring and crop matures in summer. Prolific yields are given by Golden, Pike and Dade varieties.

CAROB
Ceratonia siliqua
(Fagaceae)

- Carob is a perennial leguminous tree native to Mediterranean basin.
- Also called as 'St. John's bread," it is grown mainly in Spain, Italy, Portugal, Morocco, Cyprus, Greece, Turkey, Algeria, Tunisia, Israel, and Australia.
- Staminate, pistillate and hermaphrodite flowers are produced on different trees in this trioecious species.
- In commercial orchards, the ratio of staminate or hermaphrodite trees to pistillate ones should be 1:8. In commercial plantings, male branches are grafted onto female trees to provide pollen *in situ.*
- Though this is a leguminous crop, plants are unable to fix nitrogen. Important cultivars are Negra, Rojal, Matalafera, Duraio, Gibiliana, Amele, Mullata, Galhosa, Hemere, Santa Fe and Tylliria.
- Carob beans are dark brown edible pods, fleshy and 10 to 30 cm long with 5–15 hard seeds. Pod is very rich in sugar, up to 45% and low in protein.

- Finely ground pods make a sweet, nutritious food and are used as cocoa, in cakes, drinks, puddings, breakfast cereals, and candy.
- Carob powder has more vitamins and minerals and less fat, carbohydrates, and calories.
- Carob gum or locust bean gum, which surrounds endosperm, is a popular commercial stabiliser and thickener in a variety of food industries, including bakery, ice cream, baby foods, jams, sauces, cheese, and jellies.

STAR APPLE
Chrysophyllum sp.
(Sapotaceae)

- Large or medium-sized trees grow in warmer parts of India. There are mainly two cultivated species – *Chrysophyllm cainito* and *C. roxbughii,* of which the former one is valued both for its beauty and its delicious fruits.
- Latter one is valued much for its wood. Fruits can be eaten fresh or after making preserves.
- Fruits contain 111.47% dry matter, 2.34% protein, 1.39% fat, 4.4% total sugars, and traces of fibre and ash. Seeds have 6.6% fatty oil and 0.2% saponin.
- Star apple (Cainito), a native of Tropical America, is commonly grown in the West Indies, Central and South America, Tropical Asia, Africa and in Florida.
- It is a tall, evergreen, very attractive tree, 15 m high, with fine foliage that is dark green and shiny above and golden brown beneath.
- Flowers are brone in axillary clusters on mature twigs. They are small in size and purplish-white in colour. Fruits are smooth, globose, green, purple, or copper-coloured, and can be up to 10 cm in diameter.
- The white and sweet pulp are eaten fresh after removing the skin, which contains an unpleasant tasting latex. The pulp is arranged in eight segments and has hard, brown, glossy seeds embedded in it.
- Even though several types of plants differing in size and quality are known, two races are the most common, one bearing green fruits and the other with purple fruits. These fruits differ in taste and flavour.
- It grows well in tropical climates and requires a deep, well-drained, rich soil. The tree is sensitive to frost.
- Seeds are common propagating materials and propagation by cuttings, layering, budding and inarch grafting is also recommended.

- In a few varieties, two crops per year are obtained. The time of cropping varies from early spring to midsummer.
- Immature fruits are astringent and contain sticky white latex. When allowed to ripen on the tree, the fruit develops a sweet, delicious flesh with a sapotaceous flavour.
- A full grown tree yields about 65-70 kg of fruit per year.

LASORA/GONDA/INDIAN CHERRY
Cordia myxa. L.
(Boraginaceae)

- This is a naturally grown fruit in arid and semi-arid areas. A small or medium-sized hardy tree with a crooked stem.
- It is generally propagated by seeds. Budding is successful on its seedling rootstocks. Fruits are valued for their medicinal properties as diuretics, demulcents, antihelmatics, and expectorants.
- Fruit pulp content ranges from 94 to 97%, total soluble solids from 6–7%, acidity from 0.08 to 1.0%, and vitamin C from 32–48 mg/100 g. Unripe mature green fruits are widely used for making good quality pickles and for culinary purposes.
- Ripe fruits are sometimes eaten fresh, as they are rich in sugar, proteins, calcium and phosphorus.
- Lasora fruits contain a high percentage of gum also, which has to be evaluated for varied uses. An adult Lasora tree yields about 200–250 kg of fruit.
- *Cardia gharaf* is a related species, commonly called "goondi".
- The fruits are smaller and have medicinal values. *Cordia sebestena* is known as' lal lasora 'and is valued for its ornamental value.

TREE TOMATO
Cyphomandra betacea (cav.) Sendt
(Solanum betaceum)
(Solanaceae)

- A tree tomato is an evergreen, subtropical, semi-woody shrub or small tree, 2-4 m high, growing at elevations ranging from 500–2500 m.
- In low-lying tropical areas, growth is poor and it seldom fruits. The plant is propagated through seeds or cuttings and it grows well in well-drained humus soils.
- Growth of plant is rapid, comes to bear within 1.5 - 2 years and continues to yield for 5 or 6 years.

- Even though it fruits year round, main season is from November to March. It produces heavy yields under irrigated lands.
- Fruits are egg-shaped, 4-6 cm long and 3-5 cm in diameter.
- It is preferable to allow fruits to ripen fully on the tree, indicated by reddish colouration.
- Fruits have a musky acid and a tomato-like flavour and can be eaten raw or cooked.
- They can be preserved for several weeks without any loss in edible quality.
- They contain protein (1.5%), fibre (4.2%) and minerals like calcium (0.1%), phosphorus (0.03%), iron (0.7 mg) and vitamin A (540 IU/100g).
- Fruits are also rich in pectin content, so they can be processed into jelly after removing the rind and seeds, as they are bitter and have a disagreeable flavour.

KEI APPLE
Dovyalis caffra **Narb.**
(Flacourtiaceae)

- This is a native of South Africa. adapted to tropical and subtropical conditions.
- Grown commercially in Florida, California, and southern Europe.
- The plant grows as a shrub or small tree.
- Fruits are thin-skinned, oblate, bright yellow to orange with similar coloured flesh, having an acidic flavour.
- Pulp is very juicy and is often eaten fresh after sprinkling sugar and for jellies and preserves. 4-6°C with an RH of 85-90%.

VELVET APPLE
Diospyros discolor **Wild.**
(Ebenaceae)

- The synonym is Mabola persimmon. This is a medium sized evergreen, dioecious tree with a straight trunk, about 30 m high.
- It is native to the Philippines and cultivated in the tropics. In India, trees are grown in Bihar, Assam and the southern parts of the peninsula.
- It has ornamental value because of its attractive foliage and showy fruits.
- The fruit is ellipsoid or spherical and attains the size of an apple.
- It has a brown or reddish rind, covered densely with soft brown hairs.

- Seeds, 4-8 in number, are embedded in aromatic pulp. Types having sweet pulp and less sweet pulp as well as seeded and seedless ones are known.
- A plant is propagated either by seed or by grafting on its own seedlings.
- It is common at low and moderate altitudes, grown as a shade tree and suitable for planting on roadsides.
- Quince like fruits ripen in June-September and are edible.

LOVI-LOVI
Flacourtiaceae sp.
(Flacourtiaceae)

- Lovi-lovi, also known as' tomi-tomi ', is valued for its attractive, colourful berries.
- Tree is native to Malaysia and is cultivated in many parts of South India.
- There are mainly two species, *Flacourtia cataphracta* producing sweet fruits, dioecious in nature with spines on the trunk, and *inermis* spineless bisexual trees with sour fruits.
- Fruits are small, roundish and turn dark red on ripening. They have a tart flavour, are rich in pectin and acids and are suitable for making marmalade, jam, jelly, syrup and preserves.
- Fruits contain calcium, magnesium, potassium and iron. Leaves and young shoots taste like rhubarb, are astringent and stomachic, and are useful against bleeding gums and toothaches.
- This plant prefers a tropical climate, comes up well in a wide range of soils and is moderately drought tolerant.
- It can be propagated by seeds or by vegetative means—air layering (on current season shoots), budding and grafting.
- Softwood grafting using sweet type seedlings as rootstock gave a higher percentage of establishment in both types.
- Trees are not manured, but applying compost or farmyard manure at a rate of 40 kg per plant results in vigorous growth and a high yield.
- Seedlings require 5-7 years for flowering, while vegetative plants bear early.
- Flowers appear as terminal or lateral cymose cluster on past season shoot.
- Sour types flower and fruit almost throughout the year, with two main flowering seasons, June-july and October-November.

- In the sweet type, only one flowering season, from October to November, is noticed. Fruits require about 70-80 days in the sweet type and 120-140 days in the sour type for attaining full maturity. A grown up tree produces about 80-100 kg of fruit.
- Fully ripened fruits are harvested. In the sweet type, browning occurs once the fruit is detached from the stalk. Hence, they are to be harvested along with the twig.
- In the sour type, browning is not a serious problem. Harvested fruits can be stored for 2-3 days at a normal temperature, but their storage life can be extended to 10 days at a low temperature.

MALABAR TAMARIND
Garcinia gummigutta Roxbs.
Clusiaceae (Guttiferae)

- The Malabar tamarind (Kodampuli) is a medium-sized tree that originated in the Western Ghats and is found in the Southern States of India.
- It is a popular tree in Kerala, especially in Central Travancore and grows mostly as a homestead crop. Plants are also grown in the Shola forests of the Nilgiris up to an altitude of 180 m. Fruits are edible but too acidic for raw consumption.
- The dried rind of fruit is used as a condiment for flavouring curries in place of tamarind or lime.
- In Kerala, it is used in fish curry preparation to add special taste and flavour.
- Rind contains sugars and acids, especially (-)-hydroxycitric acid (HCA – 18.9%).
- Derivatives of this acid are potent metabolic regulators of obesity. Rind also has antiseptic properties.
- A decoction of fruit is used for rheumatism, enlargement of the spleen and bowel complaints.
- It is also used in veterinary medicine. The translucent yellow resin of the tree has purgative properties.
- It is soluble in turpentine and gives a yellow varnish. Seeds yield 31% edible fat, resembling kokum butter.
- Wood is grey and used for posts, making matchboxes and splints.
- This plant is propagated through seeds. Under normal conditions, seeds take more than six months for germination, but seed coat removal reduces this time drastically and it germinates within 2-3 weeks.

- Since this tree is androdioecious in nature, sex is revealed only at the flowering stage, for which it takes about 7-8 years.
- The vegetative propagation method (soft wood grafting) is standardized in this crop. Orthotropic shoots should be used as scion material for getting normal growth habits.
- Plagiotropic scion shoots produce dwarf and bushy trees with drooping branches.
- *By adopting top-working techniques, Garcinia* can be converted into a productive female.
- Flowering commences from January to April, with a peak period from February to March.
- The harvesting period was between May and July. The fruit rind is the economic part, which turns yellow at ripening.
- Wide variations in morphological characteristics of fruit were reported, especially in shape (round, oblong, cordate, pear and oval), colour (light yellow to dark orange) and thickness of rind.
- Fruit weight varied from 60 to 200 g and fresh fruit yield from 25 to 100 kg per plant.
- After removing the seeds, the rind is either sun dried or smoke dried. The yellowish brown to black-coloured dried rind is mixed with salt, which has a shelf life of more than one year.

LANGSAT
Lansium domesticum Correa
(Meliaceae)

- Langsat (Duku) is tropical/subtropical fruit plant native to Malaysia.
- It produces fruits that resemble loquat and appear in bunches like grapes.
- They have an aromatic skin which does not adhere to the inner pulp, and the pulp is separated into 3-5 segments.
- fruit is round, oblong, ovoid or ellipsoid in shape with thin pericarp and milky juice, while another *pubescens* has sub-globose fruits with thick pericarp and more latex content.
- It is a fairly tall tree with a fluted trunk, leaves pinnately compound, 25–45 cm long with alternate leaflets, and flowers yellow, solitary and in clusters borne on hairy spikes arising from the trunk or old branches.
- Fruits are 2.5–5 cm in diameter, pubescent; seeds 1–5 are surrounded by a sweet, white, juicy and translucent pulp.
- Fruits are delicious, eaten raw or preserved in syrup or candied. Pulp contains 2.3% fiber, 0.8% protein, minerals such as calcium, phosphorus,

and iron, and vitamins such as carotene, thiamine, riboflavin, and ascorbic acid.

- Dried peel yields an oleoresin, which is useful in diarrhoea and intestinal spasms. Seeds are used as vermifuges and febrifuges.
- Langsat is commonly propagated by seeds. Vegetative methods like grafting (cleft and veneer) and budding are successful.
- Planting can be done by giving a spacing of 8–10 m. They cannot be grown in areas with a prolonged drought or dry season.
- The fruiting season is from April to September and the yield is about 15-20 kg per plant. Storage life is very poor for these fruits.
- Low temperature storage (8-15°C and 85-90% RH) can extend shelf life by two weeks.

TROPICAL CHERRY
Muntingia calabura Linn
(Tiliaceae)

- *calabura* also known as Japanese cherry or Birdseye cherry, is a small to medium-sized tree with spreading branches, native to South America and cultivated in tropical regions as an ornamental plant.
- The leaves are serrated and obliquely lanceolate, and the flowers are small, white, and 2.5 cm long. The fruit is a berry, smooth, yellow or red, globose or ovoid, 1.5 cm in diameter, with many small seeds embedded in a juicy pulp.
- Fruits are sweet and can be eaten raw or made into jams and tarts. Edible pulp with seeds contains total solids (24.6%), protein (1.98%) and reducing sugars (8.1%).
- Decoction of leaves is used as tea, with flowers for incipient colds and headaches.
- Plants can be propagated by semi-hard wood cuttings. It grows rapidly, is a fairly drought-tolerant tree.

GUIANA CHESTNUT
Pachira aquatic. Aubl.
(Bombacaceae)

- It is a native of Mexico and North – South America. Shaving brush tree, money tree and Malabar chestnut are synonyms. The medium to large-sized, spreading tree produces large flowers with hundreds of showing stamens.

- It is a fairly hardy tree, grown in sub-tropical climates such as Southern California, although it prefers a warmer year-round climate.
- Fruit is a large, heavy brown woody capsule which splits open into five sections, exposing large seeds embedded in fleshy pulp.
- Seeds are edible and are often eaten raw or after roasting.
- They are ground to yield flour and used for baking bread. It is usually propagated by seeds or by rooted cuttings.

CACTUS PEAR
Opuntia ficus-indica
(Cactaceae)

- Cactus pear is also known as' Barbary 'Fig', 'Tuna', and Indian Fig.
- A plant is a fleshy bush or small tree, native to Mexico. Italy, Spain, Egypt, Morocco, Israel, S. America, S. Africa, S-W. Asia, USA and Australia are other growing countries.
- Common cultivars are Reyna, Rojo Pelona, Esmeralda, Gialla, Rossa, Bianca, Verdales, Morados, Sanguions, and Blancos.
- It prefers semi arid climate with temperature ranging from 18-26 ^{0}C and grown as a windbreak and live fence. The fruit is pear-shaped with a number of small spines, a soft juicy pulp, and numerous black seeds.
- Fruits are harvested by hands during July to December. Propagation is possible by vegetative means using 'cladodes', or shoots.
- Plants flower in two to three years. Fruits are used as fresh and for the preparation of paste, jam, salads and refreshing drinks.
- Fruits have antidiabetic effects and are also used for industrial products such as alcohol, soap, pigments, pectins and oils.
- Fruit pulp contains 85% water, 38 calories, 0.5 g protein, 0.1 g fat, 11 g carbohydrates, 1.8 g fibre, 60 mg Ca, 34 mg P, 0.8 mg iron, 0.8 mg Na, 161 mg K, 40 IU vitamin A and 30 mg vitamin C.
- This can be cultivated in marginal lands and the development of superior and spineless cultivars is the future demand.

SCREW PINE
Pandanus tectrius Soland
(Pandanaceae)

- Native to Southeast Asia. seen in northern Australia, Hawaii, India, and the Andamans.

- Synonyms include Hala, Tourist pineapple, Kewda and Nicobar breadfruit.
- medium-sized tree. Screw pines form thickets of support roots near the lower part of their trunks.
- The leaves are long and fan out in clusters from the branches. Male and female flowers are borne on separate plants.
- It is mildly hardy and adapted to salt water, seaside conditions, so it does well in poor or highly alkaline soil. It grows well in full sun or partial shade.
- Propagation is by seed. Spadices have a very attractive fragrance and are used for perfuming purposes.
- Produces clustered fruit looking much like a pineapple and is large with 40-80 segments, each of which usually contains one or two seeds.
- Fruits are eaten raw or cooked. The tree is also prized for its fiber, found in its long leaves.

MIRIKA-TENGA
Parmeria polyneura
(Apocynaceae)

- One of the popular indigenous minor fruits of North East India.
- This plant is an evergreen climbing shrub.
- Fruits are very attractive, oblong shaped, red on the outside, blood red inside, juicy with a sub-acid taste and are used extensively for pickling.
- Medicinal values are also attributed to the plant.
- Mature fruits contain 85% moisture, 8.8°B TSS, 3.4% titrable acidity, 4.96% total sugar and 10.48 mg/100 g ascorbic acid in the pulp.

TOMATILLO
***Physalis ixocarpa* Brot.**
(Solanaceae)

- It is a glabrous annual shrub, 1–1.5 m tall, native to Guatemala and Mexico.
- Tomatillo is a hardy plant that comes up well in warmer climates.
- It is rarely attacked by pests and diseases. The fruit is a yellowish or purplish violet berry like a big tomato, firm and mildly acid in flavour.
- They are acidic, sweet and can be eaten raw or as a salad and used in the preparation of jam, pickles and soup.
- Another related species is *minima.* It is an herbaceous pubescent annual, 15-30 cm high. The leaves and fruits are edible.

- Flowers are solitary, and the berries are round, green, many seeded, and completely enclosed in an enlarged membraneous calyx.
- In parts of Africa, fruits are used for preserves. Fruits are used as tonics, diuretic and purgative.

CAPE GOOSEBERRY

***Physalis peruvia* Linn**

(Solanaceae)

- It is a perennial erect-branching shrub, native to tropical America, widely grown in Uttar Pradesh, Punjab and Rajasthan.
- The plant requires a warmer climate and dry weather at the time of fruit development and maturity stages. The shrub comes up well in a wide variety of soils, ranging from sandy loam to laterite.
- It is grown as a pure crop (duration of 3–4 years) or as an intercrop (duration of one year) in orchards.
- Seeds are small and seedlings planted when they are 15-20 cm height during September-October.
- Plants raised from cuttings (taken from one-year-old shoots) flower early and produce more yield, but seedling plants are more vigorous.
- The yield per plant is about 3-5 kg. Small seedy berries are enclosed in a papery husk developed from the calyx.

ABIU

***Pouteria caimito* Radlk.**

(Sapotaceae)

- Formerly known as *Lucuma caimito* or *Achras caimito* and synonyms include Yellow Star apple, Camio, Luma, and Temare. Native to the Amazon headwaters and the Eastern Andes Mountains,
- Seen as a backyard tree in Brazil.
- Tree produces round, oval, soft yellow-coloured fruits with creamy, jelly-like flesh. Fruits are relished for fresh consumption.
- Different varieties like Gray, Z2, Z4 and E4 are available.

MAMEY SAPOTE /MARMALADE PLUM

Pouteria sapota (Calocarpum sapota Jacq.)

(Sapotaceae)

- A very popular fruit in Florida, the Caribbean islands, and Central America.
- Native to Southern Mexico and Northern South America.

- It is now extensively cultivated in Central America, the Caribbean, and south Florida. Many varieties are available and the fruit is highly esteemed.
- The fruit is large with orange flesh, tasting somewhat like a flavorful pumpkin.
- It is almost always eaten fresh out of hand or used in desserts and drinks.
- The large trees flourish in tropical climates.
- Young trees are highly susceptible to frost.
- The best varieties are usually propagated by grafting.
- Grafted trees may bear fruit in 1-4 years. Seedling trees take 8–10 years to fruit.

WATER LEMON
Passiflor laurifolia
(Passifloraceae)

- Synonyms include Golden Apple, Bell Apple and Yellow Granadilla.
- This is a native of tropical America and is an invasive plant in tropical regions that has spread to many parts of the world.
- Vines grow wild and fruits are often available in the markets of a few countries.
- Vine grows quickly and thrives in slightly drier tropical climates. Propagation is by seed and rooted cuttings. Bottom heating seeds at 70-80°F promotes rapid and improved germination. Fruits are medium-sized, ovoid, with deep orange skin and a white yellow, extremely juicy pulp. The taste is mild but attractive.

PILU
***Salvadora oleoides.* Decne**
(Salvadoraceae)

- It is commonly grown in Rajasthan and Gujarat in India.
- Trees have great potential for wasteland development, as they are highly adaptable to salinity and drought conditions.
- Trees are suitable for windbreaks, shelterbelts, and avenue planting.
- Fruits are harvested during May-June.
- Ripe fruits are highly perishable and pulp contains glucose, fructose and sucrose. Seeds yield oil, which is non-edible but used for industrial purposes.

PEPINO/PEPINO MELON
Solanum muricatum
(Solanaceae)

- This is a native of the Andes Mountains and is popular in New Zealand and the USA. A small bush or shrub, similar to the tomato, that fruits in 9–12 months. A frost-free climate is preferred for optimal survival.
- Pepinos withstand a few degrees of freezing.
- Medium-sized fruits have juicy, soft flesh, a mild taste and are valued for fresh consumption.
- They have thin, creamy-colored, sometimes purple-spotted skin, much like tomatoes.
- Popular in New Zealand, the pepino is becoming common in some Unitted States markets.

INDIAN HOG PLUM
Spondias sp.
(Anacardiaceae)

- *Spondias pinnata* and *Spondias mombin* (yellow mombin) are the two species.
- Trees are medium to large in size and widely seen in the humid tropics of South India. Fruits, produced in bunches are pear-shaped, turning yellow when ripe.
- Sweet, sour and bitter types are available. Immature sour fruits give excellent pickles.
- Trees exhibit leaf shedding during December and by February they start flowering. Propagation can be done by seeds and through hardwood cuttings.

MIRACLE FRUIT
Synsepalum dulcificum
(Sapotaceae)

- A relatively tasteless berry with an amazing side effect. After eating one miracle fruit, sour things will instantly taste sweet. Eating even the sourest of lemons, one will taste only sugary sweetness.
- The effect lasts an hour or two. The miracle fruit is a remarkable natural sweetener that is virtually unknown to much of the world.
- This is a native of tropical West Africa. A slow-growing bush or small tree, which needs acidic soil and is intolerant of alkaline conditions.

- Fruit is produced throughout the year and hundreds of berries can be harvested from a single plant. Fruits are red, resembling coffee berries and contain 'Miraculin', which gives a miraculous effect on the mouth.

WATERY ROSE APPLE
Syzygium aqueum (Burm) Alston
(Myrtaceae)

- A tree up to 8 m in height, with leaves opposite, sub-sessile and oblong-lanceolate.
- Flwoers are produced terminally or as axillary cymes, 2-3 cm long, white in colour.
- Fruits flattened at both ends, white, rose or red in colour, faintly fragrant, juicy, spongy, sweet-acidic, watery, seeds 2–6 in number. There are two types, one which bears pale-rose coloured fruits and the other with fruits of white colour.
- It prefers a tropical climate and is propagated by seeds.
- Vegetative propagation methods like layering and budding are also successful.
- Compatible rootstocks reported are Kopo (*densiflorum*) and Semarang roseapple (*S. samarangense*).
- The flowering season is February-March and the fruits become mature during May-June. Ripe fruits are eaten fresh or can be made into syrup and beverages.
- Wood is hard and bark is used as an astringent.

ROSE APPLE
Syzygium jamobs (Linn.) Alston
(Myrtaceae)

- It is a small evergreen tree up to 10 m high with narrow pointed leaves and is indigenous to the Malay Archipelago and Upper Burma, cultivated throughout India and Burma.
- Flowers are seen terminally, about 7 cm in diameter with numerous greenish-white stamens.
- Fruits are rose-scented, ovoid to globose, creamy white or light pink, whitish tinged, usually with 1-2 grey seeds, which are often polyembryonic.
- Tree grows wild in many places and is cultivated in humid and arid tracts in South India up to an altitude of 1300 m.

- They prefer shade and quickly die on exposure to the sun. The most common propagation method adopted is by seeds.
- Air layering, or gootee, is a successful vegetative method in this crop.
- Plant flowers during February-April and fruits ripen in June-August. Fruits drop immediately after ripening. Still, an earlier flowering season is reported in South India.
- Crishp-fleshed and rose-scented fruits are like small apples and tasteless. Fleshy and seedless fruits are produced by the application of growth regulators like NAA or 2,4,5-T during the flowering stage.
- Fruit is eaten fresh or used for making candied fruit, sauces, and jellies. Leaves on steam distillation yield a yellow essential oil (0.18%) with a characteristic aroma that contains ocimene, pinene, camphor and geranyl acetae.
- Fruits contain protein (0.7 g), fat (0.2 g), fibre (1.2 g), minerals like calcium (10 mg), magnesium (4 mg), phosphorus (30 mg) and iron (0.5 mg) per 100 g of pulp.
- It also possesses vitamin A, thiamine, riboflavin, nicotinic acid and vitamin C.

MALAY ROSE APPLE
Syzygium malacense (L.) Mernil & Perry
(Myrtaceae)

- Synonym is Malayan apple is a beautiful tree growing up to 6–20 m high with deep green glossy foliage.
- It is indigenous to the Malay Archipelago and is grown commonly in India.
- Flowers are in cymes produced on older branches, crimson pink, sessile or shortly pedunculate, 5-7 cm in diameter, with numerous long stamens.
- Fruits are oblong or pyriform, about 8 cm long, reddish-pink or white striped crimson pink, the flesh white. The seeds are large and 1-2 in number. Fruit flesh is thick, juicy and fragrant.
- Fruits can be eaten fresh or combined with other fruits to make jams and pickles. Fruits contain vitamins like nicotinic acid, riboflavin, vitamin C and minerals like calcium, phosphorus and iron.
- Trees also possess medicinal properties. Extracts taken from plants (excluding roots) affect the rate of respiration and blood pressure.

- Bark is astringent and is used for making mouthwashes. Root is a diuretic and is used in applications for itching.
- It is a tropical plant that prefers moist soil with drainage facilities.
- The common propagation method is by seeds. Budding during the rainy season on its own rootstock is also successful.
- Seedlings are planted at a spacing of 8–10 m. Trees produce flowers during summer and fruits mature towards the end of the rainy season. Plants are affected by termites, sap feeders, fruit flies and leaf mines.

SURINAM CHERRY
Syzygium iniflora Linn.
(Myrtaceae)

- A small tree native to South America bears dark green glossy foliage, whitish and slightly fragrant flowers, and globose, conspicuously ribbed fruits.
- Fruits are deep crimson red when ripe, with one or two seeds.
- Trees grow well in tropical and subtropical climates and tolerate frost.
- It is usually grown as a garden plant in India. This crop can be propagated by seed or by layering.
- Trees come to flower within 3–4 years after planting and produce two crops per year. Low or zero setting occurs in isolated trees due to pollen incompatibility.
- Fruit has a sub-acid spicy flavour. It is eaten fresh or made into jellies, jams and pickles. Fruit contains 9.3% total solids, 1.02% protein, 0.66% fat and traces of fibre and ash.
- Seeds yield 0.043% essential oil with a yellow colour, peppery flavour and a taste.
- A tea prepared from tender leaves and alcoholic extracts of leaves and fruits is used in South America against disorders of the stomach.
- Leaves yield 0.14% essential oil containing citronellal, geranyl acetate, geraniol and polyterpines. Bark contains 28.5% tannin.

Section 3
Temperate Fruit Crops

50

Temperate Fruits

Temperate Fruit Plants

Temperate fruit plants are requiring specific climate. They can tolerate both diurnal and seasonal wide fluctuations in temperature and are grown only where winter is distinctly cold. They require a specific chilling temperature for a certain period to break bud dormancy and to initiate the bud break. These fruit plants are generally deciduous and suitable for higher elevations as they can withstand frost. Examples are: apple, pear, plum, apricot, almond, peach, strawberry, walnut, pecan nut and cherry.

Classification of temperate fruits: Fruits can be classified on several bases, but their classification on the basis of climatic adaptability and morphological features.

Classification on the Basis of Plant Stature

Temperate tree fruits: Fruits borne on the trees growing in temperate climates, such as apples, pears, stone fruits, etc.

Temperate small fruits: Fruits generally borne on the vines and herbaceous plants grown under temperate climate like strawberry, crane berry, blackberry, blueberry etc.

Temperate nuts: Nuts are characterized by a hard shell on the outside, separating the kernel from the husk of the fruit. Ex. pecan nuts, hazelnuts, and walnuts.

Classification Based on Fruit Morphological

Simple fruits: Simple fruits are derived from a single ovary of one flower. Simple fruits are further classified as fleshy and dry fruits.

Fleshy fruits: These fruits are have pericarp (ovary wall) becomes fleshy and succulent at the time maturity, fleshy fruits may be either pome or drupe.

Pome: The pome is an inferior, two or more celled fleshy, syncarpous fruit surrounded by the thalamus. The fruit is referred to as “false fruit” as the edible fleshy part is not derived from the ovarian tissues but from the external ovarian tissue thalamus. Examples of temperate pome fruits are apples, pears, and quince.

Fleshy fruit: Pome (Leathery carples, edible portion is receptacles)

Drupe (stone): This type of fruit is derived from a single carpel. However, the olive is an exception in that the flower has two carpels and four ovules, but only one carpel develops. Two ovules are borne in most drupes, but only one seed develops. In this type of fruit, the pericarp is differentiated into three distinct layers; the thin exocarp or peel of the fruit; the mesocarp, which is fleshy and hard; and the stony endocarp, enclosing the seed. Examples of temperate drupe fruits are cherry, peach, plum and apricot. At maturity, the exocarp and mesocarp get separated as leathery involucre and are removed before marketing. Only the endocarp containing the edible seed is used hence it is a nut.

Fleshy fruit: Drupes (one seeded, seed within stony endocarp)

Dry fruits: This type of fruit has been classified on the basis of pericarp (ovary wall) at maturity. The entire pericarp becomes dry and often brittle or hard at maturity. They are dehiscent (in which the seeds are dispersed from fruit at maturity) and indehiscent (not split open when ripe) Nuts are typical example of indehiscent dry fruits.

Nut: A fruit in which carpel wall is hard or bony in texture. Fruit is derived from an hypogynous flower (filbert) or an epigynous one (walnut) and is enclosed in dry involucres (husk). It is only one seeded, but in most cases in derived from two carpels. Examples are walnut, almond, chestnut, hazelnut and pecan nut. The dry fruits are not juicy or succulent when mature and ripe. When dry, they may split open and discharge their seeds (called dehiscent fruits) or retain their seeds (called indehiscent fruits).

Achene: A one seeded fruit in which the seed is attached to ovary wall at one point. Example is strawberry.

Aggregate fruits: Aggregate fruits develop from numerous ovaries of the same flower. Individual ovary may be drupe or berry. Ex. Raspberry.

Multiple (composite) fruits: Multiple or composite fruits are produced from the ripened ovaries of several flowers crowded on the same inflorescence. The example of this type is mulberry.

Classification based on bearing habit: The flower bud is either terminal or lateral. Based upon the location of fruit buds and type of flower bearing structure to which they give rise, the temperate fruits are classified as under.

Terminal bearer: Flower buds mixed, flowering shoots with terminal inflorescences. Examples are apple, pear, walnut (pistillate flowers) and pecan (pistillate flowers)

Lateral bearer: Flower bud containing flower parts only e.g peach, apricot, plum, cherry, almond, walnut (staminate catkin) and pecan (staminate catkin). Flower buds mixed, flowering shoot with terminal inflorescences e.g. blackberry, raspberry, blueberry, apple and pear(occasionally). Flower buds mixed, flowering shoot with lateral inflorescences e.g. persimmon, chestnut, pistachio nut, craneberry.

A. Botanical Classification of Fruit Crops

Family	Common Name	Scientific Name	2n	Origin	Fruit type
Actinidiaceae	Kiwi	*Actinidia deliciosa*	58	Central America	Berry
Anacardiaceae	Pistachio nut	*Pistachia vera*	30	Iran/Iraq	Nut
Betulaceae	Filbert/ Hazelnut	*Corylus avellana*	-	-	Nut
Ebenaceae	American persimmon	*Diospyros vixgiana*	-	China	Berry
	Persimmon	*Diospyros kaki*	90	China	Berry
Fagaceae	Chinese chest nut	*Chestua mollissima*	-	-	Nut
	European/sweet chest	*Chestua sativa*	-	-	Nut
Proteaceae	Macadamia nut	*Macadamia ternifolia*	48	Australia	Nut
Rosaceae	Almond	*Prunus communis*	16	Central Asia	Drupe
	Apple	*Malus x domestica*	34	S. W Asia	Pome
	Apricot	*Prunus armeniaca*	16	N. E. China	Drupe
	Quince	*Cydonia oblonga*	34	Caucasus region	Pome
	Peach	*Prunus persica*	16	China	Drupe
	Pear	*Pyrus communis*	34	Western China	Pome
	Plum	*Prunus domestica*	32	China	Drupe
	Sour cherry	*Prunus cerasus*	16	S. E. Europe	Drupe
	Sweet cherry	*Prunus avium*	16	Asia Minor	Drupe
	Strawberry	*Fragaria vesca*	56	France	Etaerio of drupes
Juglandaceae	Walnut	*Juglans regia*	32	Central Asia	Nut
	Peacan nut	*Carya illinoensis*	32	North America	Nut

51

Almond

- B. N.: *Prunus amygdalus*
- Family: Rosaceae
- Sub – family: Prunoideae
- Chromosome No.: 2n = 2x = 16
- Origin: Central Asia (Afghanistan)
- Inflorescence: Cymose/Solitary
- Type of fruit: Drupe
- Edible portion: Kernel or Cotyledon

Introduction

The almond is an important temperate nut fruit with widespread popularity for its kernels throughout the world. It is native to the central Asian mountain areas (India, Iran and Pakistan). Apart from raw consumption as an edible kernel of the drupe, it also has industrial applications as a major ingredient in many manufactured products. The kernels are a rich source of fat (58.9%), protein (20.8 %), carbohydrates (10.5%), and a rich source of Vitamin E. Almond oil contains 49% oil, 62% oleic acid, 24% linoleic acid, and 6% palmitic acid. Its oil, called badamroghan, is considered to have high nutritional, medicinal and industrial value. Almond oil is used in confectionary and also for pharmaceutical and cosmetic preparation. Kernels after blanching, roasting, frying and salting are very delicious and in great demand. Green almond kernels are also consumed in the milky stage.

Area and Production

Almond is cultivated mainly in regions situated between 360 and 450 N latitude. The major almond producing countries are the USA and Spain in world almond production. The USA is the largest producer of almonds in the world. The major almond producing countries are the USA and Spain, which account for about 50 per cent of the total world almond production. Other leading almond producing countries are Italy, Iran, Syria, Morocco, Greece and Turkey. In India, almond cultivation is confined mainly to Jammu and

Kashmir, as well as the high hills of Himachal Pradesh and Uttarakhand. The state of Jammu and Kashmir is the major almond producing state in the country and in Himachal Pradesh, almonds are cultivated in Shimla, Mandi, Kinaaur, Chamba, Kangra and Sirmaur districts.

Morphological Characters

The cultivated almond is a tree of medium height. The branches are glabrous and the one-year-old shoots are pale green to reddish brown in colour. Flowers usually develop laterally on short spurs and have 1–5 flower buds. The flower is perigynous with a single ovary enclosed in the floral cup that bears 30–40 stamens. Flowers are hermaphrodite with white or pink petals, 5 sepals, and a single, unicarpel pistil, which usually contains 2 ovules.

Varieties

- The almonds of seedling origin are classified in four groups viz. (i) paper shelled, (ii) soft shelled, (iii) semi- soft shelled and (iv) hard shelled.
- Almost all cultivars of almond are self-sterile and require pollinizer. Even a few cultivars like IXL and Non Pareil are cross sterile.
- J & K: - Waris, Shalimar, Afghanistan Seedling, IXL, Merced, Rake and Mission
- Himachal Pradesh: Merced, Drake, Katha, Peerless, Texas, Ne Plus Ultra, Drake and Dhebar are pollenizers for almond varieties.
- Drake, Katha, Dhebar are self-fruitful cultivars of almond suitable for low hills.
- **Sloh (Peach x Almond hybrid) is self-fertile.**
- **J K 55: Earliest ripening variety**
- JK 39, JK 57, Hybrid 29, Hybrid 50, NB-258 are improved hybrid cultivars of almond.
- The cross incompatible combinations in almond are Jordanella x HarPariel, Non Pariel x IXL.
- Introduced varieties -**Makhdoom**, Prabhat, Waris, Shalimar, Pranyaj.
- Mutant variety Supernova, late flowering and self-compatible variety.
- Non Pareil and Ne Plus Ultra are highly tolerant to low temperature.

Climate and Soil

Among the various temperate nut and dry fruits, almond is the most exacting in its climatic requirement, favorable environmental conditions are essential for success in the cultivation of almond. Among the various climatic factors, a cool climate during winters plays an important role in flowering and regular sprouting in spring. The chilling requirement for normal bud sprouting depends on the cultivars and is relatively low, ranging from 200 to 700 hours below 7.2°C. However, for successful cultivation, almond require cool winter, frost free spring and warm dry summers. Almond can withstand a temperature of -2.2 to -3°C. Almond requires a very specific climate, cool winter, a frost-free spring and warm, dry summers. The limiting factor in almond cultivation is the spring frost, especially during full bloom or fruit set. South-facing slopes are considered the best for almond growing. Rains during spring and summer cause blossom and fruit infection by brown and green rot organisms, while foggy and rainy weather during summer results in brown strains on the shells (due to stagmini blight) of ripening nuts. Almond can be grown on all types of soils, but well-drained loamy soils having a pH of 5.5 to 6.8 are most suitable for its cultivation.

Rootstocks

Seedling rootstocks: In India seedlings of bitter almond, wild peach, **Nemagaurd (*P. Persia x P. devidiana*) and** behmi (*Prunus nira*) are used as a rootstock. Almond seedling rootstock from the bitter or sweet cultivars are used because of longevity and tolerance to drought, lime soils and iron chlorosis. In winter climate and irrigated areas peach seedling are used.

Clonal rootstocks: The clonal rootstocks like GF 677, GF 557, Marianna 2624, Myrobalan 2032 and Marianna GF 8/1 are being used world over.

Propagation

Bitter or sweet almond seeds are sown either directly in the nursery beds during December or stratified in the moist sand for 50–60 days and then sown in the nursery beds. It is commercially propagated by tongue grafting and shield budding from July to August.

Planting

One-year-old grafted or budded plants having good growth and a well developed root system are planted in a square, contour or terrace system at a spacing of 5 x 5 metres depending upon the rootstocks and soil fertility. The planting should be done in December – January. Since almond is a highly cross-pollinated crop, every third row should be planted with a pollinizer variety to provide 33 % pollinizer.

Training and Pruning

Almond plants are generally trained with an open centre system. After planting, the plant is headed back 70–80 cm above the ground level. In the summer, 3–4 well-spaced branches on the trunk in different direction are selected, while other unwanted branches are pinched off. The lowest branch should be selected not below 0.6 m from the ground level. If these branches are not selected during the summer, then they must be selected during dormant pruning. The primary scaffolds (3-4) developing within 10-15 cm from the top will form the tree crotch. After selecting the primary branches, the leader is removed and the primary branches are headed back to ½ to 1/3 of the growth. Summer pruning is carried out 2-3 times in order to retain shoots for main branches and to eliminate unwanted branches and water sprouts. The summer pruning should also be carried out in the second year to form the secondary framework scaffold. In the third year of dormant pruning, 2-3 secondary branches on each main branch are selected. The shoots growing inward and interfering with the main branches are pruned off. Almonds bear fruit primarily on spurs that last about 5 years and should be renewed by regular pruning every 3–4 years. Pruning should be done in such a manner that one fifth of fruiting wood is removed every year. Unwanted water sprouts and suckers should be removed. Trees with less than 10–12 years of age should make 22–25 cm of annual growth and older trees should produce 15 cm of new shoot growth each year. Pruning is done by heading back new shoots and thinning out unwanted shoots.

Manuring and Fertilizers

Almonds are heavy feeders and thus require proper fertilization. The application of manure and fertilizer depends upon the soil fertility and age of the tree. A proper fertilization schedule is formulated after leaf and soil analysis. The recommended manure and fertilizer schedule for bearing almond trees 7 years or older is 50–60 kg FYM, 500 g N, 350 g P_2O_5, and 700 g K_2O per plant. FYM along with a full dose of P_2O_5 and K_2O are applied in Dec-Jan and half dose N one month before flowering and the remaining half after one month of the first application.

Irrigation

Irrigation is the most important practice to improve the growth and yields of almond trees because summer rainfall is insufficient in the areas where almonds are grown. Almond trees should be irrigated at weekly intervals during April, May and June.

Orchard Soil Management

Sod culture plus mulching of the basin area is the best soil management system for almond orchard. The ground should be cultivated to a depth of 10–15 cm during winter and repeated in spring. After cultivation 10-15 cm of dry grass or black alkathene mulch is used on the basin for weed control and moisture conservation. The weeds can also be controlled with the application of weedicides like glyphosate or paraquat.

Pollination

Most of the cultivars are self-fruitful and require cross pollination for good fruit set. Only a few varieties, like Drake and Dhaber, are self-pollinated and capable of setting fruits with their own pollen. It has been observed that even in self-fruitful cv. Cross pollination increases fruit set. In almond, all the cultivars are not cross-compatible. Usually two pollinizer varieties, possibly of commercial importance are used, one flowering slightly before and the other just after the main cultivars. Good pollination can be achieved by a single row of the main cultivar and pollinizer. Pollination is mainly done by honey bees; usually 4-5 colonies/ha are placed for effective pollination. A spur of almond remains productive for 5 years and hence pruning is done to remove $1/5^{th}$ of the old bearing wood. Incompatibility (self and cross) is the major hindrance in the crop improvement of almond. type of incompatibility: gametophytic incompatibility.

Harvesting

Almonds can be harvested green or dry. Nuts of the thin-shelled variety are also harvested at the green stage for direct consumption. Almonds are ready for harvesting when they change from green to yellowish with cracks or when splitting at the suture starts from the pedicel end. For dry nuts the harvesting is done from August to October. Nuts are harvested by knocking the limbs with long wooden poles. The polythene sheet should be spread beneath the tree prior to harvesting. The nuts should be placed in a shady place for dehulling where they can be dried as well. A yield of 10-12 quintals/ha of shelled almonds is obtained. Dehulling term related to “almond”.

52

Apple

- King of temperate fruits/Symbol of health/Premier fruit of the wolrd/ Patient fruit
- B. N.: *Malus* x *domestica*
- Family: Rosaceae
- Sub-family: Pomoideae
- Chromosome No. 2n = 2x = 34
- Origin: South Western Asia
- Inflorescence: Raceme
- fruit type: Pome
- Edible part: Fleshy thalamus

Introduction

The apple (*Malus x domestica*) is an important temperate fruit and is grown in areas where winters are cold, springs are frost free, and summers are mild. The main ancestor of cultivated apple *M x domestica* is *M. sylvestrii*, Thomas Andrew Knight produced the first apple cultivar of known parentage. Apple is disomic and has complex polyploidy, tetraploid and partially hexaploid. The apple (*Malus x domestica*) is the most widely grown temperate fruit in the world. First in India, the apple was introduced by **Captain Lee** in 1865 in the Kullu Valley of H.P. Later on, red-coloured Delicious group varieties were introduced at Kotgarh in the Shimla district of Himachal Pradesh in 1917 by American missionary **Mr. Satya Nand Stokes.** The region that includes South Western Asia, the Caucasus, Soviet Central Asia, and the Hindu-Kush Himalayan region is thought to be the primary centre of apple origin. The apple plant is deciduous without spiny branches. Climacteric fruit. Dry temperate region is most suitable for apple cultivation in India.

Area and Production

In India, apples are grown on a commercial scale in Jammu and Kashmir, Himachal Pradesh and Uttarakhand. It is also cultivated on a limited scale in North-Eastern states like Arunachal Pradesh, Sikkim, Nagaland, Meghalaya, Manipur and the Nilgiri hills of Tamil Nadu. Apples account 55% of the total area and 75% of the total production of temperate fruits in the country. India ranks 11th in apple production in the world. The apple bowl of India: Himachal Pradesh (H.P.). The largest collection of apple genotypes is from the Plant Genetics Resource Unit, Cornell University, Ithaca, New York.

Uses and Composition

'An apple a day keeps the doctor away'. This is the old proverb that signifies the nutritive importance of apples in the human diet. Apples are a rich source of easily assailable carbohydrates (15 %), protein (0.3 %) and they are also fairly rich in calcium (10 mg/100 g), phosphorus (14 mg/100 g), and potassium (120 mg/100 g). It supplies vitamins B and C. The apple has been under cultivation since time immemorial and today more than 80 % of the world's supply is produced in Europe, where the major producers are Italy, France, and Germany.

Climate and Soil

The apple is a typical temperate fruit and is generally grown in the temperate regions of the world. Apples can also be grown in the subtropical zone at an altitude of 1600–2500 m above MSL also, apples can be grown. Almost all varieties have a chilling requirement of 1000 hours at a temperature below 7.2°C. Most apple varieties require 1000–1600 chilling hours during winter to break the rest period, but some low-chill varieties require only 500–800 chilling hours. less than 7.2°C. Apple seed stratification in moist sand at 4-7°C for 60–90 days. An average summer temperature should be around 21–24°C during the active growth period. Low temperature below 15°C, rains and cloudy weather during bloom restrict the bee activity, which is completely inhibited below 5°C and adversely affect fruit set. The optimum temperature for pollination, pollen germination and fruit setting is 18 to 22°C. The areas with frost free spring and adequate sunshine during summer without wide fluctuation in temperature are most suitable for apple growing. The long drought spells during fruit development and excessive rain and foggy conditions at maturity hamper fruit size and quality. Well distributed rainfall of about 100-125 cm throughout the season is considered most favorable. Apples thrive best in loamy soils, which are rich in organic matter. A soil pH of between 6.0 to 6.5 with good drainage and aeration is considered most suitable.

Species

- Resistance to powdery mildew and codling moth in cultivated apple is transferred from ***M. zumi.***
- Resistant to powdery mildew: *M. zumi* (Pl2): Single dominant gene
- Orthotrophic shoot, rhythmic growth and terminal flowering: Apple
- Orthotrophic, rhythmic growth and lateral flowering: Peach, cherry, apricot
- **Sucker free Species:** *M. siversii, Apomixis species: M. sieboldii*
- **Indigenous to Himalayas:** *M. baccata var. himalacia, M. siddimmensis*

Classification Varaties on Based of Season

1. **Early season:** Irish Peach, Benoni Tydeman's Early Worcestor, Michael, Mollies Delicious, Schlomit, Starkrimson, Anna, Early Shanburry, Fenny, Benoni, Chaubattia Princess.
2. **Mid-season:** American Mother, Razakwar, Jonathan, Cox' Orange Pippin, Queen's Apple, Rome Beauty, Scarlet Siberian, Rich-a-red, Vance Delicious, Top Red, Lord Lambourne Red Chief, Oregon Spur, Red Spur, Red Gold, Silver Spur, Scarlet Gala, Well Spur, Super Chief, Scarlet Spur, Starking Delicious, Red Delicious, Mclntosh, Cortland, Golden Delicious.
3. **Late season**: King Pippin, American Apirouge, Kerry Pippin, Lal Ambri, Sunheri Chamure, Ambri, Baldwin, Yellow Newton, Winter Banana, Granny Smith, Red Fuji, Coe Fuji, Rymer, Buckingham.

Special Characters Varieties

1. **Spur types and colour sports:** Red Chief, Oregon Spur-II, Silver Spur, Well Spur, Red Spur, Super Chief, Starkrimson, Hardi Spur., Schelet Spur, Ace Spur, Royal Red, Vance Delicious
2. **Standard colour mutants:** Vance Delicious, Top Red, Skyline Supreme, Hardiman, Bright-N-Early.
3. **Standard varieties:** Starking Delicious, Red Delicious, Rich-a-red.
4. **Low chilling varieties:** Michal, Schlomit, Anna, Vered, Tamma, Tropical Beauty and Parlins Beauty.
5. **Pollinizing varieties:** Golden Delicious, Red Gold, Tydeman"s Early Worcester, Summer Queen, Golden Spur, Granny Smith, Winter Banana, Mclntosh, Scarlet Gala and flowering crabs like Manchurian, Snow Drift and *Malus floribunda*.
6. **Table purpose varieties-** Golden Delicious, Braeburn, Mongendy, Lonas Gold, Goldstar, Jonathan, Fuji, Pink Lady, Granny Smith, Gala Sales,

Washington, Red Delicious, Empire Liberty, Colapters, Cabarett, Rainette, De Flanders, Mc Instosh and Empire.

7. **Processing-** Spackle Greala, Rome Beauty, York imperial, Stayman winsap, northern Spy, Rhode Island, Greening, Prize, Freedom and Redfree.
8. **Processing and Table purpose-** McInstosh, Cortland, Granny Smith, Redfree, Delcious, Golden Li, Wellspur, Delicious, Gloster, Tonagold, Bolero, Maypole, Polka and Walta.
9. Liberty variety of apple is resistance to all fungal diease.
10. Red Delicious cultivar is mother of all Delicious group of cultivar (important variety in India).
11. Baldwin, Mutsu, Bramlay's, and Winesap Seedling are triploid varieties of apple.
12. Introduced variety from USA - Red Delicious, Starking Delicious.
13. High yielding apple variety- Scarlet Gala, Red Fuzi.
14. Early maturing varieties- Yandik - Ovskoe, Papisovka Canniaga.
15. Natural mutant- Red Elstar (Parent cv. Elstar).
16. **Ambri** - cultivar of apple is said longest shelf life and **indigenously** grown in Kashmir (India). But biennial bearing.
17. Most of the commercial apple varieties are diploid and triploid are rare.
18. The commercial cultivars Red Delicious, Royal Delicious, Top Red, Vance Delicious, Red Chief and Oregon Spur are self-incompatible.
19. Pusa Seb Moolvrint -1 (PSM-1) is a selection of –*Malus baccata* (Shilong).
20. Pink Lady variety developed at Australia.
21. Delicious group of apple varieties are self-incompatible and cross pollinated.
22. Golden delicious, yellow newton, Northern Spy, Mcintosh have good seed viability.
23. Delicious group of varieties are self-incompatible and cross pollinated.
24. English varieties are self-pollinated and act as suitable pollinizer for delicious group.
25. Red Delicious –most popular variety of India.
26. Red Gold –Act as pollinizer for Red Delicious and Starking Delicious.
27. **Golden Delicious:** Commercial varieties of USA and Europe. It is a good pollinizer for all Delicious group.
28. Rymer-Indigenous variety grown in Kashmir.
29. Scarlet gola, Red fuzi are high yielding varieties.

30. **McIntosh** It is a mid-season variety. But this cultivar is scab susceptible and is leading variety of Canada
31. **Bramely Seedling** It is an English cooking apple ideal for stewing especially baking.
32. **Jona gold (Triploid)** Golden Delicious x Jonathan. Developed in New York.
33. **New Jona Gold** A virus free clone of Jona Gold at Japan with super grade having high red colour intensity.
34. **Applethrope Summerdel-** (Delicious x Earliblaze) Developed at Granite Belt Horticultural Research Station, Queensland, Australia.
35. **Richelien**: Apple Developed at Canada Agricultural Research Station by a multiple cross involving Melba, McIntosh, Jonathan, Rome Beauty and *Malus floribunda.*
36. **Romus-2** Released from Romania, resistant to scab. Early maturing (mid-July onwards).
37. **Kodaikanal-1** A clonal selection made from Parlin's Beauty at Horticultural Research Station, Kodaikanal in Tamil Nadu. Adapted very well to warm winter conditions prevailing in South Indian Hills.
38. **Crab Apple varieties**: Red flush, Crimson gold, Yellow drops, and Nodrift
39. **Intergeneric hybrid: Pamapples** (1st successful Pear x Apple hybrid), Developed by Ellis Marks (1952) in John Innes Centre.

Scab Resistant Varieties and Species

1. **Scab resistant varieties:** Prima, Priscilla, Sir Prize, Florina, Flrdous, Shireen, Macfree, Coop-12, Coop-13, Liberty, Freedom, Jonaface, Nova Easy, Red free, Nova Mac, Firdous Shireen. 1st scab resistant Variety-Honey Crisp.
2. Resistant to scab: *M. floribunda, M. atrosanguinea* (VF): Single dominant gene.
3. Scab resistance is controlled partial polygenic.

Hybrid Varieties

1. **Lal Ambri** (Red Delicious x Ambri) and Sunehari (Ambri x Golden Delicious) are developed in 1956 at fruit Research Station, Shalimar, J&K.
2. **Amroyal-** (Starking Delicious x Ambri).
3. **Pusa Gold** – Golden Delicious x Tydmans Early

4. **Pusa Amaratara Pride** - Royal Delicious x Prima) at IARI, Shimla.
5. **Akbar** (Ambri x Coxs Orange Pippin), Tolerant to scab and powdery mildew.
6. **Ambred** (Red Delicious x Ambri), Low incidence of PM, scab and sooty blotch
7. **Ambstarking** (Starking Delicious x Ambri-81), Tolerant to scab.
8. **Ambrich (Richared x Ambri-)** Tolerant to scab and San Jose Scale *(Quadraspidiotus perniciosus)* is the most serious pest, Mashobra, H.P.
9. **Chaubattia Princess** (Red Delicious x Early Shanbury) Early ripening variety.
10. **Chaubattia Anupam** (Red Delicious x Early Shanbury)- Chaubattia, U.K.
11. **Kent** - A hybrid between Cox and Jonathan.
12. **Shamrock A** -cross between irradiated McIntosh x Starkspur strain of Golden Delicious. It is green apple identical in appearance and taste with Granny Smith but matures six weeks earlier.

Rootstocks

Seedling Rootstocks

- Seedling rootstocks are vigorous and not uniform in size.
- Seedling of crab apple (***Malus baccata***) or self-pollinizing varieties like Golden Delicious and Granny Smith are used for raising seedling rootstocks. It is most commonly used rootstock of apple in India

Clonal Rootstocks

Clonal rootstocks are precocious, uniform and resistant to some insect pests and diseases. In recent years' size controlling clonal rootstocks are gaining popularity especially for establishment of high density. The promising clonal rootstocks of apple are:

- Semi-dwarf- M4, M7, M106, M24, Suitable for **HDP**, Resistant to **wooly apple aphid.**
- Semi-vigorous, MM-111 and MM-104, Drought tolerant and Resistant to wooly apple aphid.
- Vigorous- Merton-793, Resistant to wooly apple aphid and collar rot, Early fruiting.
- M9- is a dwarf rootstock, evolved by chance seedling, Suitable for HDP, most widely used.

- **M 27 is a result of cross between M9 x M13. It is ultra-dwarf**.
- East Malling Long Ashton (EMLA) series of rootstock is resistant to viruses.
- East Malling Research Station is located at Kent in collaboration with John Innes Institute.
- MM series (Malling Merton) released from John Innes Institute, England. size controlling and resistant to woolly apple aphid.
- **Northern spy.** These were developed by East Malling Research Station, Kent in collaboration with John Inns Irrigation Institute. resistant to wooly apple aphid (*Eriosoma lanigera*). root stock.
- EMCA series of rootstock is resistant to virus.
- ***Malus sikkimiensis, M. torngoides** and **M. hupehensis areapomictic***root stocks.
- ***Malus floribunda***: Cultivated as ornamental trees for their attractive flowers, Small fruited species and good source of apple scab disease resistance and pollinizer.
- The root-stocks are either related species such as *Malus sylvestris* (crap apple), **M. prunifolia**, *M. sikkimensis* or their hybrid derivatives or seedling progenies of cultivated varieties.
- East Malling Research Station in England developed Malling Stocks (M). rootstocks are size controlling but not resistant to woolly apple aphid.
- M7, MM-106: Tolerant to below freezing (-40°C to – 35°C) resistant to wooly aphids.
- **Robusta-5:** Developed at Ottawa.
- Canada as a selection of Malus robusta – Resistant to wooly aphid. Other rootstocks used are Merton 779, 793.

Propagation

Apple seeds are dormant, which requires stratification treatment (moist chilling) to break dormancy. Seeds are stratified for 60-90 days in alternate layers of moist sand at 4–7°C during December to February. The stratified seeds are sown in nursery beds during March at a spacing of 8-10 cm from seed to seed and 15-20 cm from line to line. After sowing, the nursery beds are mulched with 10 cm thick dry grass and light irrigation is given to avoid desiccation of stratified seeds. Mulch is removed as soon as seeds start germinating. Cultural operations like weeding, hoeing, irrigation and spraying of insecticides and fungicides are done at regular intervals. The seedling rootstock attains a graftable size of 15 mm diameter in a year.

Propagation of Scion

The commercial methods of propagation of apple scion varieties are grafting and chip budding. For grafting and budding, the scion wood should be collected from healthy, disease-free, true to type mother trees during January. The scion wood is collected from one-year-old shoots and packed in moist sphagnum moss after proper labelling of the variety. These bundles of scion wood are stored in cold storage or buried deep in the soil in a shady place until required for grafting. The best time for grafting apples is February to March with tongue and cleft methods. Chip budding can also be done in March and July. common method of propagation of clonal rootstocks of apples: stooling. Root suckering is a major problem in clonal apple rootstocks.

Planting

The best time for planting apples is January - February. The planting distance varies depending on the variety, rootstock, and soil fertility. Before planting, orchard layout should be planned. In flat land square or hexagonal system of layout is adopted, whereas in sloppy land contour and terrace system of layout is done. After the layout of an orchard, the pits of 1 x 1 x 1 m size are dug well in advance of planting. The pits are filled at least one month before planting with soil in which 40-50 kg of well rotten FYM and 1 kg of single super phosphate are mixed. After planting, watering is done and tree basins are mulched with 10 cm of dry grass, which helps in conserving the soil moisture and controlling the weed population. In apple, most of the varieties are self-unfruitful, therefore, at the time of planting proper proportion of pollinizer varieties (25 or 33%) should be planted in the orchard. In India, 11 to 33% of pollinizing trees are recommended for regular cropping. Planting Pits of size 60 x 60 x 60 cm are dug at a spacing of 5 m x 5 m. For effective fruit set (since self-incompatibility exists in most cultivars), pollinating varieties are interplanted. Usually one in six or one in nine plants is used as a pollenizer in self-unfruitful varieties. In an intensive system, planting is done at 4 m x 2.5 m or less. A complete row of pollenizer is used for every two rows of the main variety. For high density planting (HDP) and Medium Density Planting (MDP) for this dwarfing rootstocks like M9, M4, MM106 and M26 have to be used. The grafts are planted at a close spacing of 2 m x 1 m, so that 5000 plants/ha are accommodated. The comparison with M.D.P., where the total number of plants accommodated was 1666 trees/ha (3 m x 2 m), the H.D.P. did not give much yield. An additional yield of 50% more than HDP can be obtained from MDP. The fruit colour and quality are also better in MDP than in HDP. Meadow orcharding: 20000-70000 plants/ha (super high density), popular in European countries.

Training

Training is done to shape or build a strong framework around the trees in order to support a maximum crop when the plant reaches bearing stage. There are several systems of training in apples, especially for dwarf plantations like spindle bush, dwarf pyramid, and cylinder spindle, etc., but the modified central leader system is most suitable for standard plantation.

Modified Central Leader System

Generally, a one-year-old whip without a single branch is planted in the dormant season. Immediately after transplanting, the tip of the plant is headed back 60 to 75 cm above the ground level. Most of the buds on the main branch will sprout the following summer. In order to develop a clean stem up to 45 cm from the ground, the sprouted buds are pinched off soon after their appearance. Three or four well-spaced buds projecting in opposite directions are retained, with the lowest one at 45 cm above ground. If summer pruning is not done, then 3-4 well-spaced primary branches having a wider crotch angle are selected during dormant pruning. The selected branches should be spaced 10–15 cm apart in spiral fashion. The branches emerging below 30 cm from ground level and other undesirable branches are pruned off. The selected branches are headed back to ¼ of their growth to a bud projecting in the outer direction. The leader is also headed back to 30 cm above the last branch. During the second dormant pruning, 2-3 well-spaced primary branches are selected on the leader. From the primary branches selected during the previous year, two secondary branches which are growing in an outward direction should be selected. The selected primary and secondary branches are headed back to 1/3 - ¼ of the growth. The secondary and tertiary branches selected should be spread horizontally and upright or downward-growing branches should be removed. The third year of training entails removing unwanted branches and pruning desirable side branches. The central leader should be headed back to a bud or weak shoot, which will develop in the form of a side branch. By the fourth year, training should be completed.

Pruning

The objective of pruning is to maintain a proper balance between vegetative growth and spur development. Training is completed during the initial 4-5 years after the planting of the plant, but pruning is continued after training throughout the life of the tree. In pruning, thinning out and heading back are two basic components. All upright laterals and those growing inside the trees are thinned out, and leaders and laterals are headed back. In apples, the fruit is obtained not only from spurs but also from fruit buds on young laterals.

Therefore, pruning should be done in such a way that a continuous supply of new, healthy shoots, spurs and branches is maintained. While pruning, some parts of the tree are pruned and some are left unpruned. The pruned parts produce shoot growth and the unpruned parts will produce fruit buds. The laterals which were left unpruned in one year may be either shortened or left unpruned in the next year depending on the growth, spur formation and crop load. Once the laterals have cropped and become weak, they must be severely shortened, leaving them at 5–6 cm long. The severe shortening will promote the production of new growth from these stubs, which in two years will give a crop. The whole cycle is accordingly repeated every year to ensure regular growth, spur formation and cropping. At the time of pruning, dead, diseased and broken branches are removed and on the cut surface, Chaubattia or Bordeaux paste is applied to avoid any fungal infection. The best time for pruning is during the dormant season (December to January).

Morphological Characters and Flowering

The apple plant is deciduous without spiny branches. Buds are ovoid with imbricate scales, leaves are serrate or lobed, folded in bud and stipulate. Floral buds are mixed buds borne terminally on spurs and terminally or laterally on long shoots, depending upon the cultivar, age and vigour of the tree. The initiation of flower primordia starts about 3-6 weeks after full bloom (June). The inflorescence is determinate, having five flowers. Flowers are white or pink or carmine in cymes. Most cultivars have flowers that are epigynous and hermaphrodite. The ovary is inferior. A flower consists of five petals, five sepals, 15-20 stamens and a pistil, which is divided into five carpals, each containing two ovules. Pollinizers facilitate cross-pollination. Pollination is carried out by insects. The flower colour is white to pink. The diploids are self-fruitful and the triploids are self-unfruitful.

Pollination and Pollinizers

Most apple varieties are self-fruitful and will not produce fruit if fertilized by their own pollen. They thus require some compatible cultivars for cross pollination and good fruit set. Apple trees produce abundant bloom but fail to set fruit, especially under adverse climatic conditions due to lack of pollination Inadequate fruit set often results from a failure during the pollination period, which is associated with pollen production, transfer and germination, pollen tube development, or fertility of the ovule. Sterility and incompatibility are two main causes of unfruitfulness in apples. Self-incompatibility is most common, although cases of cross-incompatibility are also known. The commercial cultivars Red Delicious, Royal Delicious, Top Red, Vance Delicious, Red Chief and Oregon Spur are self-incompatible. Various climatic and edaphic factors are also responsible for poor pollination and fruit set. Low temperature, rainfall and cloudy weather at flowering time adversely affect the bee activity, transfer of pollen to stigma, pollen germination, and ultimately result in poor fruit set. Planting at least 25% to 33% pollinizers can improve apple cross-pollination and fruit set. Placement of 5 to 6 honey bee colonies per hectare, top working of 2-4 shoots of commercial varieties with pollinizers.

Manure and Fertilizers

The manure and fertilizer requirements depend upon the soil fertility, the age of the tree, cultural practices and crop load. To judge the actual nutrient requirement, a leaf and soil analysis should be done to know the status of nutrients in the tree and soil. The manure and fertilizer schedule for any fruit crop is suggested after long term fertilizers trial. In the orchard of optimal fertility, nitrogen, phosphorus and potassium are applied in the ratio of 70:35:70 g per year of apple tree age, and doses of these NPK fertilizers are stabilized at the age of 10 years. For ten or more than ten-year old apple trees, 100 kg FYM, 700 g N, 350 g P_2O_5 and 700 g K_2O should be given annually. The N, P and K are applied in the form of calcium ammonium nitrate, single super phosphate and muriate of potash fertilizers, respectively.

Time and Method of Application

FYM, along with a full dose of P_2O_5 and K_2O, is applied during December-January. Nitrogen is applied in two split doses. One-half dose of N is applied one month before flowering (March) and the remaining half dose one month after fruit set. These fertilizers should be applied when there is enough moisture in the soil. The fertilizers should be broadcast in tree basins 30 cm away from the trunk. However, in very steep slopes and heavy rainfall areas band application of fertilizers is recommended to avoid leaching and runoff losses of fertilizers. For higher fertilizer efficiency, soluble fertilizers can also be applied through drip irrigation.

Foliar Application of Nutrients

The micro-nutrients, particularly boron, zinc, iron and manganese, are essential for the normal growth of plants and their deficiency is frequently observed in apples. It can be corrected by two foliar sprays of $ZnSO_4$ (0.5%) and boric acid (0.01%) before flowering and in May-June.

Weed Management

Weeds, if allowed to grow in the orchard, exhaust nutrients and moisture from the soil. Thus, it is suggested that apple orchards should be managed as clean basin management. In this system, the basin of trees is kept clean and free from weeds either by hand weeding, use of mulches and weedicides. In the initial years of plantation, the intercrops like peas, beans, cabbage, cauliflower and ginger were grown in the vacant area in between the trees but not in the basin area. Some green manuring crops like beans, peas and gram can also be grown, which helps in improving soil texture and nutrient status. In a bearing orchard, mulching of the basin area of trees with 10 cm thick layers of hay or black alkathene is a good floor management practice for efficient weed control and in situ moisture conservation. Sod grasses like white clover, red clover, orchard grass and rye grass are grown in the vacant area between the trees. For herbicidal weed control, pre-emergence and post-emergence weedicides can also be used. Sprays of simazine at the rate of 4 kg/ha in March, followed by two sprays of glyphosate at 800 ml/ha at monthly intervals in July and August, have been found very effective to control the weeds in apple orchards. Clean basin management is the common practice of floor management in apples.

Irrigation

Apples require optimum soil moisture, particularly during the period of fruit growth and development to get good size and better quality fruits. The cultivation of apples in India is mainly in hilly areas, where land is sloppy and water for irrigation is also inadequate. Moreover, in these areas, rainfall is also very low or no rainfall occurs, particularly during a critical period of growth. From April to July, when flowering, fruit set, fruit growth and development occur, the most critical periods of water requirement are from April to July. Therefore, water management is very important in apple orchards Water management includes rainwater harvesting, in-situ soil moisture conservation and efficient use of water for irrigation through drip irrigation. The rain water should be harvested and collected in water storage tanks. The harvested water should be used for irrigation, particularly during critical periods of water requirements. The drip irrigation method saves more than 50 percent of irrigation water and can be adopted in water-scarce areas. About 6-8 irrigations at 7-10 day intervals should be given during April to July in Apple. Besides irrigation, tree basins should be mulched with hay or black polythene in the month of March. The most critical period for apple water requirements is from April to August.

Fruit Thinning

Heavy bearing in apples during the year results in small-sized and poor quality fruits. Therefore, judicious thinning at the proper stage of fruit development (pea stage) is a must to regulate cropping and improve fruit size and quality. The thinning can be achieved either manually or with the use of growth regulators. Removal of fruit lets by hand thinning is very laborious and uneconomical. Therefore, chemical thinning with a foliar spray of 20 ppm Nepthalene acetic acid (NAA) at petal fall results in optimum fruit thinning.

Fruit Drop

Although apples bloom profusely, yet a small percentage of flowers will mature into fruit. Most of the flowers fall soon after full bloom, with a small amount dropping later. Fruit drop is a serious problem in apples. In most commercial varieties, 40-60% of fruit drop occurs in three stages.

- **The early drop** is a natural and occurs due to a lack of pollination after petal fall and up to three weeks later. This drop can be controlled by assuring effective pollination, placement of bee hives, and planting of more pollinizer varieties in the orchard.
- **The June Drop (major drop)** which is caused by moisture stress. This drop can be controlled with the application of irrigation water and mulching of tree basins.
- **The Pre-Harvest Drop** occurs before harvesting of fruits and caused economic loss to farmers, which is due to reduction in levels of auxins and increase in ethylene in the fruit. Application of 10 ppm NAA 20-25 days before harvest checks this drop.

Use of Growth Regulators

1. Plant propagation: GA_3 at 100-200 ppm can overcome physiological dormancy and stimulate germination in seeds. The use of IBA at 2000 ppm and NAA at 200 ppm aids in the rooting of clonal rootstock cuttings.
2. Effect on growth: Growth regulators like auxin and GA_3 help in increasing the vegetative growth. Growth retardants like P_{333} retard the growth of trees and help in flower bud formation. Foliar spray of P_{333} @ 500 to 1000 ppm reduces the vegetative growth of apple trees. Application of 250 ppm GA_3 stimulates the vegetative growth but also enhances the alternate bearing phenomenon.
3. Effect on fruit set and yield: Application of triacontanol (20 ppm), miraculan (0.6 ml/L), Paras 0.6 ml/L or Biozyme at 2ml/ L sprayed at bud swell and petal fall stages helps in improving fruit set and yield in Delicious apple.
4. Fruit drop: The problem of pre-harvest fruit drop is more severe in early maturing cultivars where 40–60% of the fruit drops. The pre-harvest dropping is mainly because of hormonal imbalance and can be controlled

with the application of 10 ppm NAA (1 ml of Planofix in 4.5 L of water) one week before the expected fruit drop.

5. Fruit thinning: The application of 10-20 ppm NAA, 7–15 days after petal fall, is most effective for fruit thinning and to maintain regular production of good-sized and quality fruits. Carbarly or sevin @ 750-1000 ppm or NAA @ 10-20 ppm,7-15 days after petal fall is most effective for fruit thinning.
6. Improvement of fruit shape: Application of promalin at 30-60 ppm (GA3+7 + cytokinin) at pea stage helps to improve the shape of the apple fruits.
7. Improvement of fruit colour and maturity: Application of ethrel (2-chloro ethyl phosphonic acid) @ 1200 ppm a.i. or 4.5 ml ethrel/litre of water + 25 ppm NAA improves surface red colour in apples.

Maturity Indices and Harvesting

An apple is a climacteric fruit in which the maturity of the fruit does not coincide with ripening. The fruits do not attain ripe edible quality on the tree at harvest. If the fruits are picked immature, they lack flavour and taste, which shrivel in storage. The harvesting of over-ripe fruit develops soft scald and internal breakdown and has poor shelf life. To judge the optimum harvesting time, several maturity indices are adopted. Maturity indices: The starch index should be 1 to 2. The maturity indices like days from full bloom to harvest, change of ground colour of fruit from green to yellow to pale, TSS of fruit pulp, ease of separation of fruit from spur, change of seed colour to light brown and fruit firmness are used singly or in combination. The entire fruit does not attain full colour and mature on the tree at one time; therefore, fruits should be picked in 2-3 pickings. Harvesting of fruit is done in such a way that bruising and stem puncture are avoided and the pedicel must be retained with the fruit. Apple fruits are grasped between the index finger, middle finger, and thumb, and a quick twist of the wrist easily removes the fruit and its pedicel. Fruits are placed softly in the picking bag or basket and transported to packing houses for grading and packing.

Post-harvest Management

After the harvesting of fruits, proper handling, grading, and storage are important. In apple fruit, about 30% post-harvest losses have been recorded, which can be minimized by proper handling of the fruits.

a) Precooling

Immediately after picking, the fruits should be placed in a cool and ventilated place to remove field heat before packing. For the removal of field heat, methods like the use of air coolers, cold water sprinkling, fruit washing and keeping of fruits overnight in a cool place are adopted.

b) Grading

The fruits are graded according to fruit size and fruit appearance, like colour and shape. On the basis of size, the apple fruits are graded manually or by mechanical grades as:

On the basis of fruit colour, shape, quality, and appearance, fruits are graded as AAA, AA, A, and Tatoo. The quality of fruits during transportation, storage, and marketing is influenced by the packing. Usually, apples are packed in telescopic corrugated fiberboard (CFB) cartons. The usual dimensions of a telescopic CFB carton with trays are 50.4 x 30.3 x 28.2 cm (outer jacket) and 50 x 30 x 28.2 cm (inner case). Each layer of fruit in a carton is separated by a fruit tray, which is different for different size grades. Early varieties losses about 40–60%.

Storage

Apple fruit has a long storage life, and the shelf life of fruits can be prolonged by providing optimal storage conditions. The cold storage retards fruit deterioration and reduces decay from pathogens and helps prevent shrivelness resulting from water loss. Apple fruits can be stored for 4-6 months after harvest in cold storage at a temperature of -1.1 to 0 °C with 80-90 per cent relative humidity. Among the fruits, apples have a long storage life. Stone cells are absent in the flesh, Major sugar: Sorbitol.

Processing

About 30 per cent of apple fruits are rejected as cull fruits before packaging for the fresh fruit market. The culled fruits can be processed and preserved as fruit pulp for making various value-added products like jam, fruit leather, fruit toffees, and fermented products like wine, which can be prepared from preserved pulp. Cider is a fermented wine prepared from apples. from which fresh fruit juice can be made. Fruit colour is controlled by 3 independent genes: A, B, and C. The cream yellow flesh colour is controlled by a single dominant gene.

Insect-pests and Diseases

1. **San Jose Scale (*Quadraspidiotus perniciosus*)**

   It is the polyphagus pest which feeds on apple plants. The insect is covered with a deep grey armature. The San Jose scale got its entry into India from France in 1906. The San Jose scale got its entry into India in Kashmir from

France in 1906, caused by *Venturia inaequalis*- most serious disease in the world. By lifting the armature, a yellow-colored insect underneath can be seen. The nymphs and adults suckered sap from the aerial parts. Heavily infested trees have bark covered with deep grey overlapping scales. The plant's vigour is reduced, which results in poor fruit setting and quality.

Control

Spray 2 per cent dormant oil (Servo orchard spray oil/Hindustan petroleum spray oil) or 1.5 per cent summer oil like orchaks 796/IPOL/shelter909 at half-leaf to tight cluster stage. If oil spray is not applied, then spray with 0.04 percent chlorpyriphos (200 ml durmet in 100 L water) after petal fall to kill the crawlers and newly settled scale.

1. **The woolly apple aphid (*Eriosoma lanigerum*)**

   It feeds on apples and lives in colonies on the aerial parts and roots of the plant. In the aerial parts, it is seen as a white woolly mass. Damage is caused by sucking of sap from stems, twigs, and roots, resulting in gall formation, plant remains stunted. Fruit set and quality are also reduced under severe infestation. The Wooly apple aphid is the most devastating pest in the world.

Control

Spray infested trees with 0.04% chorpyriphos (200 ml durmet in 100 L water) during May-June and again in October. Aphid infestation on roots can be reduced by drenching the collar region of the tree with chlorpyriphos (0.1%) in October-November using a 10-15 litre solution per tree. Use Malling's Merton (MM) series clonal rootstocks for raising nursery plants that are resistant to woolly aphid.

Aphilenus mali* is a predator of WAA. Resistance to WAA: *Malus hupehensis

1. **European red mite (*Panonychus ulmi*)**

   This is a serious pest in apples and causes damage by feeding on the green matter of leaves. The leaves turn bronze in colour and cup upward, followed by leaf drop and the weakening of fruit bud. The maximum population is observed during May-July. The mite completes 5-7 generations in a year.

Spray 2% dormant oil (Servo orchard spray oil/Hindustan petroleum spray oil) from half leaf to tight cluster stage to control. Dormant oil spray, as suggested for scale, is effective against mites. Spray 1% summer oils like Orchaks 796/IPOL/shelter 909 @ 1% at petal fall and again at fruit walnut stage. If population is high, spray with fenazaquin (25 ml Magister 10 EC/100 L) or propargite (100 ml Omite/ 100 L water) twice at 20 days' interval in June- July.

Defoliating Caterpillars

They feed on newly emerged leaves and defoliate the trees. The growth of plants is retarded.

Control

Spray 0.05% Endosulfan (150 ml Thiodon or endocil 35 EC in 100 L water)15-20 days before flowering or when caterpillars appear.

A. **Apple scab** (*Venturia inaequalis*)

   This is caused by fungus. Light brown or olive green spots which soon turn musty black appear on either or both sides of the young leaves in spring. Young lesions are velvety brown to olive green becoming more distant with age, leading to curling of leaves. On the fruits, small lesions develop and slowly increase leading to misshapening and cracking of fruits. Scab epidemic in Jammu and Kashmir-1772-73 and Himachal Pradesh-1778-79.

Control

Follow the spray schedule of dodine (0.1%) or mancozeb (0.3%) at silver tip to green tip, mancozeb (0.3%) + carbendazim (0.05%) at pink bud, benmy/ carbandazim (0.05%) at petal fall, zineb (0.3%) or dodine (0.075%) at pea size fruit, mancozeb + carbendazim in June-July and urea 5% spray after fruit harvest to control the disease.

1. **Powdery mildew (*Podosphaera leucotricha*):** This disease is caused by the fungus, which survives as mycelium on dormant buds. The young leaves show white mildew growth on their surface and also on twigs and look silvery white.

   Control

   Prune off the affected twigs. Spray during dormancy, green tip, petal fall and two weeks after petal fall with fungicides like wettable sulphur (200-300 g/100 L) or contaf (50 g/100 L) or Baycor (50 g/100 L).

2. **Canker:** Many fungi have been reported to be involved in canker complexes. Symptoms appear on the trunk and branches, resulting in the production of wounds which develop length-wise more rapidly. These are normally elliptical and the wound may increase to up to a metre in length. The bark beneath the rough exterior becomes hard, dry and tough.

Control

Cut and burn the badly canked portion of the tree. Scarify the cankered portions up to healthy portions and paint with Chaubattia paint. Immediately after pruning, apply chaubattia paste or copper oxychloride paint to the cut portions of the shoot. Spray copper oxychloride (300 g) or captan (200 g) in 100 litres of water after fruit harvest.

3. **White root rot** : This is caused by Dematophora necatrix fungi. Affected trees show sparse foliage, slow growth, bronzing or yellowing of leaves. Such trees ultimately die. The roots turn brown and remain covered with white cottony mycelial fungi in the rainy season.

   Control

   Improve the drainage of an orchard. After leaf fall, remove the infected roots and apply Chaubattia paste (Red lead, copper carbonate and linseed oil (1:1:1.25) or copper oxychloride (300 g/100L) in Nov- December on the cut ends of the roots. Give at least four drenches of carbendazim (100 g) along with mancozeb (300 g) in 100 L of water during April, June, July and September in infected trees.

Low Productivity in Apple

The decreased productivity of apple orchards in recent years has become a serious concern for the growers in all the apple-growing areas. Apple productivity has been fluctuating year to year between 2 MT and 8 MT/ha. The factors which influence yields are climate, soil, cultivar, rootstocks, and cultural management practices. Most of the factors influencing yield are manageable to a large extent, but the climatic factors are beyond the control. The appearance of certain diseases and pests in epidemic form has also adversely affected yield in recent years. The outbreak of red spider mite attack and premature defoliation of apples in the past 9–10 years has remained a persistent problem in apple orchards.

Causes of Low Productivity

- Climatic factors
- Varietals factors
- Inadequate pollinizer
- Lack of pollinators
- Inadequate nutrition
- Poor soil conditions
- Poor canopy management

- Senile orchards
- Pathological factors
- Entomological factors

Climatic Factors

The low temperature at the time of flowering and fruit setting adversely affects the production of fruits. The area's most vulnerable to the effects of low temperatures are located between 5000 to 6000 feet elevation, where a good spring season with enough sunlight promotes apple flowering from mid-March to mid-April.Fluctuating temperature during this period, particularly rain accompanied by low temperature inhibits the cross-pollination due to restricted bee activity and washing off pollen and poor pollen tube growth. It is well established that the flowers are killed below 2.2°C and bee activity is completely inhibited below 4.4°C.

Varietals Factors

In Himachal Pradesh, the delicious group of apple varieties constitutes more than 80 per cent of the total production of apples. The predominant varieties like Starking Delicious, Red Delicious, Rich-a-red and certain improved bud sports and spur types are self-unfruitful and require cross pollination for fruitfulness. Moreover, these varieties have a strong tendency to alternate bearing, which is also one of the reasons for low production during the off years.

Inadequate Pollinizer

In Himachal Pradesh, a 25 to 33 per cent proportion of pollinizing varieties in orchards is recommended for adequate fruit set. However, the actual proportions of these varieties are only 5–10 %. The problem is further compounded due to the predominance of Golden Delicious as a pollinizing variety, which does not synchronize the flowering of the Delicious varieties in many agro-climatic situations and the strong tendency of alternate bearing.

Lack of Pollinators

Honey bees are the major agent, besides other wild pollinators, for effective pollination in apples. Over the years, the population of honey bees and other pollinators has declined due to the indiscriminate use of pesticides. Placement of honeybees in the orchards has also not picked up due to the scarcity of beehives.

Inadequate Nutrition

Apple cultivation is mostly done on the slopes, which poses a serious problem of water and nutrient losses. Even if nutrients are applied adequately in the soil, frequent dry spells in April-June and September-November render them unavailable to the plants. Contrarily, leaching of the nutrients during the rainy season from July –August further affects the health of the trees. It has also been noticed that the fertilizers are not applied according to the requirements of the trees.

Poor Soil Conditions

In orchards which are planted on the slopes, run-off losses render the soils nutritionally and structurally poor. In many orchards, on soils that do not have adequate drainage, temporary water logging conditions develop during the rainy season, killing feeder roots and temporarily restricting the uptake of nutrients. All these factors adversely affect the plant's health and productivity.

Poor Canopy Management

Dependence of the orchardists on hired pruners is increasing day by day. The plants are not properly trained and pruned by these untrained pruners, resulting in poor canopy development. At lower elevations where the vegetative growth is excessively more due to warm conditions, hard pruning promotes more vegetative growth and reduces reproductive growth, which is considered a wrong orchard practice. In such conditions lesser heading back and more thinning out of shoots as per tree behaviour is required to balance cropping and growth.

Senile Orchards

Orchards more than 40 years of age face the problem of unfruitfulness more seriously than younger orchards. Such orchards do not produce adequate annual extension growth and usually have small-sized foliage. The old orchards have also been planted under traditional systems of planting at a spacing of 20–25 feet, which takes 15-20 years to come to commercial fruiting after planting. Use of ethephon for early maturity and colour improvement at lower elevations has also proved counterproductive. Continuous use of ethephon combined with poor orchard management practices is a cause of senility and poor shelf life and quality of the fruits.

Pathological Factors

The number of disease has been found affecting the apple orchards. The most serious among these is apple scab. Besides apple scab, other diseases are premature leaf fall, root rot, colour rot, and replanting problems. Powdery mildew, cankers and viruses. Most of the pollinizing cultivars like Golden Delicious and Red Gold have been found to be higher susceptible to scab.

Entomological Factors

The magnititude of pest incidence varies from region to region and orchard to orchard. The most common is aphid, which affects 82% of orchards, followed by San Jose Scale (71 %), blossom trips (70 %), European mite (62 %), apple leaf roller (43 %), root borer (26 %), stem borer (9 %), defoliating beetle (6 %), and hairy caterpillar (5 %).

Suggested Remedial Measures

Among the standard varieties, there should be more proportion of regular-bearing varieties than the Delicious cultivars. An adequate proportion of pollinizing varieties should be compensated for what has not been provided by top working of the trees with pollinizing varieties. Proper orchard soil and canopy management practices should be given adequate priority. There is an urgent need to go for high density plantation in different temperate fruit crops. Indiscriminate use of insecticides, pesticides, and fungicides should be avoided in order to maintain the population of natural predators and avoid the problem of tolerance by pests.

Premature Leaf Fall

For the last few years, apple plantations in Himachal Pradesh have been ravaged by a unique phenomenon of leaf shedding in mid-summer. The problem starts in the month of June-july and by mid-August the majority of orchards are affected. In severe cases, only fruits are seen hanging on the defoliated branches near maturation. The disease, which was first noticed in 1995 in some orchards, has now spread to all apple-growing districts of the state. All the commercially delicious cultivars are susceptible. Premature leaf shedding has also been reported recently in Kashmir, neighbouring Uttaranchal, and Bhutan.

Symptoms

Disease symptoms first appear as dark green circular patches on the upper surface of the mature leaves, giving rise to 5-10 mm in size brown leaf spots, especially in the months of June and July, which turn dark brown in due course. When lesions are numerous, they coalesce into larger dark brown blotches and the surrounding areas turn yellow. In the weeks that follow, severe leaf shedding occurs, resulting in a large amount of natural leaf fall in autumn.

In affected orchards, fruits nearing maturity are commonly seen hanging on the defoliated branches. Symptoms also appear on the fruit as clear brown spots, which are initially circular (3-5 mm in diameter) and become oval, depressed and dark brown later. Numerous small black colour pinhead specks, the acervuli, are visible in the affected tissues.

Causes Organism and Disease Development

This disease is caused by *Marssonina coronaria*, but alternaria leaf blight is also involved in this malady. This fungus is reported to perennate in the fallen leaf litter on the orchard floor. The seeds of the fungus are mature by the time of blooming in the spring and they are liberated in the orchard for a quite long period. In Himachal Pradesh, the perfect stage of this fungus is not frequently intercepted. This fungus was found to perennate in the infected leaf litter in the form of acervuli, which produce fresh conidia in early summer to start primary infections. Frequent rains are helpful for disease development. Infections first appear on mature leaves, turning yellow and abscising prematurely. Countless conidia and microconidia are formed on diseased leaves, which cause secondary infections, leading to epiphytitic development in favourable humid conditions.

Management

This disease can be controlled effectively by integrating different technologies like field sanitation, proper pruning and judicious use of fungicides as follows, the orchardists are advised to collect and destroy the fallen leaves from the orchard floor in winter. Urea (5%) spray on the leaf litter can also be helpful in reducing the primary inoculums by enhancing the decomposition of leaves. Proper pruning allows adequate air circulation in the tree canopy, thereby modifying the microclimate and reducing disease development. Protective 3-4 sprays of fungicides like mancozeb (0.3%), carbendazim (0.05%), thiophanate methyl (0.05%), benomyl (0.05%), propineb (0.3%), dodine (0.075%), ziram (0.3%), dithianon (0.05%) and zineb (0.3%) are effective in controlling the disease.

Replant Problem

Replantation pertains to the plantation of new plants of apple in the fields vacated by the removal of old and declining trees. In India, old trees in many apple orchards have either outlived their economic bearing life or declined due to the adverse effects of non-curable insect pests' or diseases' problems and/or natural calamities. Moreover, many growers want to introduce new improved and highly productive varieties in their old orchards, that too, sometimes under high density plantations. Because of land limitations, the growers the growers are mostly compelled to plant new apple trees on the old apple sites. There has been increasing concern about poor growth, delayed fruiting and the short life of apple trees planted in the old apple sites. The growers' problem is referred to as the "replant problem. It is also, sometimes, termed as 'Replant Disease' when only biotic causes are involved to develop such situation.

Causes of Replant Problem

There are many causes of poor growth of young trees in apples, and these vary from region to region or even orchard to orchard in a particular region. These can include weak or diseased nursery stock, poor planting and management techniques, water deficiencies or excessive, spray injuries or damage from insects, diseases, rodents etc. But our concern here is only with the soil-related causes, which include biotic (harmful microorganisms) and abiotic (nutritional deficiencies or excessive soil pH, phytotoxins) factors in the soil. Various types of micro-organisms like fungi, bacteria, actinomycetes, nematodes and their interactions cause replant problems. When we remove the old plants, some of the root systems (mainly fine roots) are left behind at the old site. The soil adhered to such roots consists of a good population of micro-organisms which later almost rob the newly planted trees of their vital elements, thereby adversely affecting their growth. These microorganisms can also have a direct impact on newly planted trees by causing diseases in their root systems. A number of fungi belonging to oomycetes, hypomycetes, and basidiomycetes have been reported as causal agents of replant disease. In England, Canada and the U.S.A., many investigators have shown that species of Phytophthora and Phythium are the primary causes of replant disease, Phythium sylvaticum has been identified as the cause associated with apple replant disease in Canada.

Management of Replant Problem

Liming has been found effective in soils where Dermatophthora and Phytophthora are the causes of replant problems. The elements N and P have been reported to suppress the growth of replant disease caused by fungi and bacteria and subsequently to promote the growth of bacteria antagonistic to these usual organisms. Replacing the soil at a replantation site with the steamed/fumigated old apple soil or some nonapple soil or potting soil mixture (containing peat, sand or soil, dolomite, lime, NPK fertilizer, minor elements) has also been reported as effective in attaining good growth of replanted apple trees. Inter-cropping with herbaceous crops and growing mustard and radish before planting greatly improved the growth of newly planted apple seedlings. Antagonistic crops like marigold (Tagetes pastula) successfully reduced the population of nematodes, Pratylenchus penatrans and the fungus, Phythium spp. in replanted soil. Cultivation of red fescue (Festuca rubra) and red top (Agrostis alba) also reduces the P. penetrans population. Soil sterilisation by fumigation, steaming, or even solarization checks the population of soil-borne

Pathogens. Generally, pre-plant treatments are more useful than post-planting treatments.

The recommendations to manage the replant problem in Himachal Pradesh include:

i) Dig out and destroy the stumps and roots of old/dead apple trees.

ii) The fresh layout of the orchard be done by avoiding frequently/possibly the old pit sites.

iii) New pits of a bigger size (5 x 3 ft) should be dug and kept open for exposure to sunlight for about one month.

iv) Pre-plant soil sterilization by fumigation (early winter) or solarization (during summer) to check the population of soil-borne micro-organisms. For fumigation, make a 9-inch-high heap of pit soil outside the pit, drench with formalin solution (1 litre of commercial formalin in 9 litres of water) and cover with a transparent thin polyethylene sheet for at least 48 hours. Remove the sheet and turn the soil daily for about a week so that formalin fumes escape from the treated soil. In the case of solarization, a 9-inch-high heap of pit soil is covered with a transparent polythene sheet for 2 months during the summer before filling the pits.

v) Incorporate well rotten FYM along with 1 kg SSP fertilizer and 200 g Kanadane dust in treated pit soil and fill the pit upto one foot above ground level.

vi) Always use the healthy and strong rooted plants for replantation. In high problem areas preferably use tolerant rootstocks of apple like Merton793.

vii) To improve the growth of newly planted apple seedlings, grow mustard, radish, marigold, and red fescue (*Festuca rubra*) and red top (*Agrostis alba*) as decoy/biofumigation crops.

Unfruitfulness in Temperate Fruits

Unfruitfulness is a serious problem in apple, cherry, almond and walnut and is associated with both internal and external factors. The sterility is mainly due to

i) Impotence

ii) incompatibility

iii) abortion of embryo.

1. **Impotence:** It relates to the condition when either one or both the sex organs fail to develop to required stage and fails to form flowers or abortion of male and female flower organs occur.
2. **Incompatibility:** It relates to the condition, where both male and female flowers develop and their organs are functional but sterility is due to the incompatibility.
3. In some cases both the male and female flowers are formed and function but embryo abortion occurs. Sterility associated with internal functions may be related to evolutionary tendencies due to factors associated with constitution of protoplasm, genetic influence and physiological reasons.

Evolutionary Tendencies

A. Defective Flowers

Self-fertilization is not possible in many temperate fruit species due to imperfect flowers, heterostyly, dichogamy and pollen impotence. In most temperate fruit plants like apple, pear, stone and nut fruits, flowers are perfect and present on the same plant (monoecious), but problems of unfruitfulness still occur due to incompatibility (apple, cherry and almond) and dichogamy (walnut, pecan nut and chestnut). In kiwifruit, the male and female flowers are present on different plants, which restricts self-pollination. In apples, cherries, and almonds, the cross incompatibility results in unfruitfulness. Some varieties of persimmon are staminate constant (bear staminate flowers every year) and pistillate constant. The presence of short styles with long filaments or long styles with short filaments is dimorphism, a type of heterostyly and the basal gap between filaments is more which allows the bees to enter without touching the stigma to collect nectar (in some apple varieties), which results in unfruitfulness.

Abortiveness Leading to Impotence

Interference either in the development of flowers or in the full development of sex elements and their functions may lead to unfruitfulness. In certain varieties of plum, the pistil is degenerated and unfruitfulness results. In pecan and walnut, the terminal clusters consisting of pistillate flowers fall off before pollination, which leads to unfruitfulness. In strawberries, too, late flowers are abortive, and no fruit set occurs in these flowers. In some temperate fruits, there are some varieties having defective pistils (plum), defective embryos, and embryo-sacs (apple), which also lead to unfruitfulness.

Non-viable Pollen

When the pollen is not viable, the production of fruit is not possible. Pollen of some varieties of plum, peach, pear and cherry are non-viable due to their abortion leads to unfruitfulness in these fruits.

Genetic Factors

- Hybridity is associated with sterility as well as unfruitfulness.
- Hybridity is also responsible for seedlessness in some varieties of temperate fruits.
- Hybridity and incompatibility are two type of sterility which are directly due to genetic factors.
- Self-sterility depends on inheritance but its development is controlled by environment. Peach- plum hybrids known as Blackman are completely sterile. Similarly, Peach- sour cherry hybrid Kamdesa is completely sterile. Pyronia (*Pear x quince*) flowers and fruits freely but is always seedless.
- Incompatibility between pollen and ovule is one of the causes of unfruitfulness in apple, cherry, almond, pear, apricot, plum.

Physiological Causes

It is known that the poor growth rate of the pollen tubes in the styles, possibly due to hormonal or chemotropic control (apple, pear, cherries), is known. The pollen tube growth is slow due to the low temperature. Unfruitfulness can also result from a difference in the stage of maturity of pollen grains vis the pistil and embryo. The proper development of the flowers, their maturity and fertilization leading to the development of a fruit with viable seeds is controlled by the abortive condition within the plant existing at the time of pre-blooming and post-blooming stages. The nutritive status of the plant determines the time taken by the pollen to fertilize. Fertilization takes a shorter time in strongly vegetative conditions than in poorly vegetative conditions. Defective pistils are formed in exhausted or weakned trees caused by overbearing, drought and poor nutrition.

External Factors

Environment: The climatic conditions at the time of flowering affect pollen germination, pistil formation and transfer of pollen, pollen tube growth and ultimately fruit set. Low temperature and rainfall at the time of flowering adversely affect the transfer of pollen and pollen tube growth, thereby resulting in poor fruit set in most temperate fruits.

Nutrition: Proper nutrient supply affects fruit setting as well as unfruitfulness. Rich soils having an optimum level of nutrients in the soil will produce a self-fruitful Jonathan apple. The high nitrogen content in the plant at the time of flowering encourages fruitfulness.

Pruning: Moderate to heavy pruning induces good setting and yield.

Age and vigour of plant: Very young and vigorous tree set less fruits than the moderate vigorous and old trees.

Water relations: Moisture stress promotes the formation of abscission layers, leading to flower and fruit drop.

Control

- Pollination management by planting pollinizing varieties, placement of bee-hives.
- Proper nutrient management, irrigation and proper pruning

Physiological Disorders

1. **Bitter pit-** in apple is due to Ca deficiency and susceptible varieties- Yellow Newton, Golden Delicious, Gravenstein
2. **Internal browning:** Browning streaks radiating into flesh from the core, susceptible variety: Yellow Newton
3. **Scald:** mottling on greener surface of fruits-immature fruits are most susceptible
4. **Boron deficiency in apple**: Hardy corky tissue, fruit cracking, blossom blast
5. **Zinc deficiency:** Blind bud, little leaf,
6. **Alternate bearing:**

- It can be corrected by two foliar sprays of $ZnSO_4$ (0.5%) and boric acid (0.01%) before flowering and in May-June.
- Foliar spray of P_{333} @ 500 to 1000 ppm reduces the vegetative growth of apple trees.
- Application of 250 ppm GA_3 stimulate the vegetative growth but also enhances alternate bearing phenomenon.

53

Apricot

- Apricot/Coppery fruit/Amerian plum
- *B. N=Prunus armeniaca,*
- Family = Rosaceae
- Sub-family = Prunoideae
- Chromosome No.: 2n = 2x = 16
- Origin: West China
- Inflorescence: Cymose/Solitary
- Fruit type: Drupe (Stone)
- Edible part: Monocarp and Endocarp

Introduction

Its fruits are attractive, delicious and highly nutritious. This is a temperate crop that is drought-resistant and salt-tolerant. largest Producer-Turkey. It is a rich source of protein and oil (40-45 %). Use it for tumer and ulcer treatment. highest antioxidant content (carotenoids) Laetrile, a purported alternative treatment for cancer, is extracted from apricot seeds. Moor Park one of the best apricots for outdoor cultivation in a small garden. It is very rich in Vitamin-A and also contains more carbohydrates, proteins, and phosphorus than the majority of common fruits. They are mostly used as dessert, and because of their perishable nature, they can be canned, frozen and dried. The fruits are also processed into jam, nectar and squash. The kernels, which are either sweet or bitter depending upon the variety, are a valuable by-product. Sweet kernels are used in confectionary, while bitter kernels are used for oil extraction and raising rootstocks. It is very rich in Vitamin-A: 3600 IU and Thiamine: 217 mg/100 g.

Origin and Distribution

Cultivated apricots have their origins in Western China. The apricot fruit moved westward from central Asia through Iran and the Transcaucasus region and reached Italy during the first century, England in the 13th century, and North America by 1720. The major apricot producing countries are Turkey,

Uzbekistan, Iran, Algeria, Italy, China, Spain, Greece, France and the USA. Commercial cultivation of apricots in India is of recent origin and was started by European settlers and missionaries after 1870. Presently, apricots are grown commercially in the hills of HP, J & K, Uttarakhand and to a limited extent in the north-eastern hills.

Climate and Soil

Apricot can be successfully grown at an altitude of between 900-2000 m above mean sea level. Apricot is an important fruit crop in the mid-hills and dry temperate regions of the country. The white-fleshed, sweet-keeled apricots require cooler climate and are grown in dry temperate regions up to 3000 m mean sea level. The long, cool winter (300-900 chilling hours below 7°C), frost-free winter, and warm spring are favourable for fruiting. An average summer temperature between 16.6 to 32.2°C is suitable for growth and quality fruit production. In general, the sites located on the north-eastern aspect at lower elevation and the south-western aspect at higher elevation are suitable for its cultivation. Spring frost causes extensive damage to the blossoms, which are killed when the temperature falls below 4°C.An annual rainfall of about 100 mm, well distributed throughout the season, is good for normal growth and fruiting. Apricot is quite hardy and can grow in most soils, but deep, fertile, well-drained loam soils are much more suitable for its growth and development. The pH of the soil should be 6.0-6.8. In Ladakh and Kinnaur, large wild apricot trees are found growing in sandy well drained soil which is not very fertile. In India, apricot is grown in mid and high hills and dry temperate areas having variable climatic conditions. Varieties which are suitable for mid-hills are not suitable for high hills or dry temperate regions.

Varieties

The promising varieties recommended for different regions of the country are as under:

1. **Himachal Pradesh-** New castle, Early Shipley, Shakarpara, Nugget, Royal Suffaida, Charmagz, St. Ambroise, Moorpark. Nari, Charmagz, Suffaida, Kaisha, Halman, Nari and Bebeco.
2. **Uttarakhand** Charmagz, Kaisha, Moorpark, Turkey, St. Ambroise, Shipleys Early, Chaubattia Alankar, Chaubattia Madhu, Chaubattia Kesri
3. **Jammu and Kashmir:** Halman, Rakcha Karpu, Tokpopa, Narmu and Khante, Turkey, Australian, Charmagz, Rogan and Shakarpara.

4. **Early maturing** - Baiti, Beladi.
5. Late maturing - Farmingdale, Alfred.
6. Chaubattia Alankar (Kaishu x Charmagz)- Low chilling and early ripening.
7. Chaubattia Madhu (Turkey x Charmagz) -Early ripening.
8. Chaubattia Kesari (St. Ambroise x Charmagz)- Mid-season variety.
9. Low chilling subtropical varieties: New Castle, Early Shipley, St. Ambroise. Kaisha

Rootstocks

Wild apricot-Zardalu (Chulli) originated from India and wild peach seedlings are commercially used as a rootstock. However, wild apricot seedlings are generally preferred as rootstocks because the graft union is good and the trees are more vigorous than on plum and wild peach seedlings. Peach is a satisfactory rootstock for light soils and dry conditions, but sometimes the graft union is enlarged or rough. Different clonal rootstocks have been recommended in different regions. *i.e.,* Myrobalan 29 C and Marianna 2624 for California; GF 31, GF 8-1 for France and Marianna 7/7 for South Africa. For raising the seedling rootstocks, the seeds are collected from fully ripe wild apricot fruits. Seeds are stratified for 45–50 days at 4°C to break dormancy. Pre-stratified seeds are sown in nursery beds, which attain a graftable size after one year. Under excessive soil moisture conditions, apricots on Myrobalan plums make better growth.

Propagation

It is propagated by shield budding on wild apricot or peach rootstocks. The seedlings which attain pencil thickness in one year are grafted with the tongue method in the month of February, while the seedlings of less thickness are T-budded in the months of May- June. Chip budding performed in the month of September also gave good success.

Planting and Planting Density

The apricot plants are planted during the dormant season, *i.e.,* the end of December to February, but early planting gives better establishment of plants. Undesirable trees and shrubs should be removed from the land during its initial preparations by digging and ploughing. On the flat land, a regular planting layout system such as square and triangular is followed, while on the hill slopes, a contour system is generally practiced. The spacing of plants varies with the soil, climate and vigour of varieties and rootstocks. The plants are generally planted at a spacing of 6m x 6 m. The pits of 1 x 1 x 1 m dimension are dug about a month before planting and are filled with a mixture of soil and 50–60 kg of well-decomposed FYM. About 1 kg of SSP and 10 L. of Chlrophyriphos

(4 ml/ 1 litre of water) are also added to each pit. In comparison to other temperate fruits, high density planting in apricot has begun rather late as there are very less dwarfing rootstocks. A density of 7200 trees/ha has been reported in cv. Canino.

Training and Pruning

The apricot is trained in both the open vase and modified centre leader systems, with the open vase system being more popular in hilly areas. One-year plants are headed back at about 60–70 cm above the ground and 3-5 well-spaced shoots are allowed to grow in all directions, while undesirable shoots are pinched off during summer. In the first dormant season, 3-5 primary scaffold branches arising at a proper angle of 45 degrees, well-spaced (10–15 cm apart), and spirally arranged around the tree trunk are selected. The lowest branch should be 40–45 cm above ground level. All the primary scaffold branches are headed back to half their growth to get the secondary branches on them. During the second dormant pruning, 4-5 well-spaced secondary branches are selected on each primary branch and the others are removed. At the end of the third year, the pruning is confined to the thinning of branches which are crossing or crowding each other. Apricot bears on spurs and laterally on one-year-old shoots. The spurs have a short life of 3–4 years and many of them get broken during fruit plucking. The production of young growth is therefore essential for the initiation of new spurs, which generally takes place at the base of the growing laterals. In young bearing trees, pruning should be light and of corrective type, but in older trees, heavy pruning should be done to maintain balance between growth and fruiting. In new castle apricot, 25-30 percent thinning of one-year-old shoots and $1/3^{rd}$ heading back is recommended to improve size and fruit quality. After pruning, mashobra or Chaubattia paste is applied on the cut end of the shoots.

Flowering

In apricot, usually three buds develop in the axil of a leaf at each node on a shoot and spur. The central one being a vegetative bud, the two side buds are floral. The time of flowering and its duration varies with the variety and the prevailing weather conditions. Under mid hills condition the flowering in apricot comes in the month of March and higher hills at the end of March and April.

Pollination and Fruit Set

Most of the commercial as an apricot are self-fruitful and set fruits without pollinizer. However, varieties like Charmagz and Perfection have been reported as self-incompatible. There is generally good fruit set in the apricot cultivars growing in appropriate climatic conditions. There is 40-60 % fruit set in the cultivars commercially grown in the mid-hills, but fruit drop is to the extent of 79 % in these cultivars, which occurs mostly in the second week after fruit set. A spray of 10 ppm NAA at the beginning of pit hardening reduced the pre-harvest drop. Apricot bears on spurs and laterally on 1-year-old shoots. Spurs' life: 3–4 years. It bears on spurs and shoots.

Manures and Fertilizers

Apricot trees remove large amounts of nutrients from the soil and require organic manures as well as chemical fertilizers for normal growth and fruit production. The manurial requirements depend upon the age of the tree, type of soil, climate conditions, and cultural practices, which vary from region to region. The FYM is applied during December-January along with a full dose of P and K by the broadcasting method. Nitrogen is applied in 2 doses via first half dose of N is spring 2-3 weeks before flowering and remaining half Nitrogen a month later, if irrigation facilities are available. Under rain-fed conditions, the second half dose of N should be applied at the onset of monsoon rains or through one or two foliar sprays of urea 0.05 % after fruit set. Fertilizers should be broadcast on the soil surface under the spread of the trees and mixed with the soil. It should not be applied to wet or too dry soil. In high rainfall areas with steep slopes, the band application of nitrogenous fertilizers should be preferred over broadcasting.

Thinning

Fruit set in apricot is rather heavy, which results in undersized fruits and also increases the tendency of biennial bearing. Fruit thinning improves fruit size, promotes regular bearing, decreases limb breakage due to heavy crop load and maintains the tree's vigour. Fruit thinning should be done within forty days after full bloom, i.e., during the last week of April or first week of May, because this is the effective period influencing fruit bud formation. Both hand and chemical thinning methods are employed. Depending upon the crop load, the fruit may be thinned till the fruits are 6-10 cm apart. A spur should have no more than two fruits. Foliar spray of 25 to 50 ppm NAA 20 days after fruit set is best for thinning.

Orchard Soil Management and Weed Management

Sod culture plus mulching of basins with dry grass or black polythene is the common orchard floor management practice followed in apricot orchards. In apricot orchards, atrazine or diuron @ 4.0 kg/ha as pre-emergence and gramoxone @ 2 kg/L or glyphosate @ 800 ml/ha as post-emergence have been found effective and economical in controlling weeds. Mulching of trees basin with 10-15 cm thick dry grass also checks the weed growth. During the initial 3–4 years of orchard life when the plants are young, intercropping with leguminous crops like peas, beans, soyabeans, cowpeas, and also tomatoes and strawberries is recommended as they enrich the soil and also give economic returns.

Irrigation

Apricots are tolerant of dry conditions but require irrigation water during critical periods of fruit growth and development. Water requirements varies with the soil, tree age, climatic conditions and irrigation method. The peak water use period is from the end of April to the middle of June, which coincides with the fruit development period. Irrigation at 20 per cent depletion of soil moisture from field capacity improves fruit size and yield. The irrigation interval should be 10 days during May and 6-8 days during June. In all, eight irrigations in a season are sufficient for apricots in the mid-hills of Himachal Pradesh.

Maturity Indices, Harvesting and Yield

Change of surface colour, days from full bloom to harvest and T.S.S. are considered indices of maturity (May to June). For the fresh market, the fruits are plucked when the surface colour turns green to yellow. Fully ripe fruits are, however, harvested for freezing, canning and drying. In Himachal Pradesh, days from full bloom to harvest and fruit T.S.S. have been standardized for different varieties to judge the optimum time of harvest. Since apricot fruits are very perishable, due care is required during harvesting, packing and transportation. The fruits should be harvested in the morning hours and direct exposure of fruits to the sun should be avoided during grading and packing. Apricot trees start fruiting at the age of 5 years and give an economic yield of up to 30-35 years. Apricots attain full bearing age at about 8–10 years of age and yield about 50–80 kg of fruit per tree.

Post Harvest Management

Before packing, the fruits are graded according to their size of the fruit. Fruits are packed in wooden boxes or CFB cartons. Each box is lined inside with newspaper sheets, keeping the margins overhanging the flaps. The boxes are initially packed with pine needles at the bottom to avoid brushing the fruits. In apricot, there is no wrapping of individual fruits. Then fruits are arranged in layers and the top layer is covered with paper by bringing together overhanging flaps. The top of the box is nailed in the case of wooden boxes, whereas CFB cartons are sealed with adhesive tape. Apricot fruits can be stored at 0 degrees Celsius and 85-95 percent relative humidity for 1-2 weeks.

54

Cherries

- Cherry: *Prunus avium* (Sweet cherry), *P. cerasus* (Sour cherry)
- Family: Rosaceae
- Chromosome No.: 2n = 16 (Sweet cherry), 2n = 32 (Sour cherry)
- Origin: South East Europe
- Inflorescence: Racemose
- Fruit type: Drupe
- Edible part: Mesocarp and epicarp

Introduction

Cherries occupy an important position among temperate fruits all over the world. Cultivated cherries are divided into two main groups, *i.e.,* sweet cherries (*Prunus avium*) and sour cherries (*P. cerasus*). Cherries are rich in protein, sugar, potassium, calcium, iron and zinc. Sweet cherries are believed to have originated in an area between the Black and Caspian Seas in Southern Europe. The earliest records indicate that it was first domesticated in Greece around 300 BC. It spread to Italy from Greece where it was established as a fruit crop by 37 BC. Early setter's brought seeds to North America and parts of South America. In India, it was introduced by British setter's in Kashmir, Kullu and Shimla hills during the pre-independence era. World leading breeding of cherry: John Innes Horticultural Institution, Merton. A putative ancestor of cultivated sweet cherries: Anthocynins (antioxidants) are the red pigment found in cherries.

Uses and Composition

The sweet cherry is mainly used for table purposes, and sour cherries for processing. Cherries are rich in protein, sugar, potassium, calcium, iron and zinc. It was first domesticated in Greece around 300 BC. It spread to Italy from Greece, where it was established as a fruit crop by 37 BC. Early setter's brought seeds to North America and part of South America. Cherries are extensively grown in all the temperate countries. The leading cherry producing countries are the USA, West Germany, Italy and France. In Europe, cherry wine, Dirschwasser, is distlled from the pulp of cherries. Maraschinos.Cherry fruits reached first in the market among temperate fruits. Cherries have more caloric value than apples. **Methyl anthrinilate and methyl salicylate** give the flavour of cherry. colouring compound present in cherries: keracyanin chloride.

Area and Production

Cherries are extensively grown in all the temperate countries. The leading cherry producing countries are the Turkey, USA, Chile, Uzbekistan, Iran, West Germany, Italy and France. In India, it is extensively grown in the states of Jammu and Kashmir and HP.

Morphological Characters

Sweet cherry is a tall tree with erect branches and large, thin leaves that are pubescent beneath and serrated. The petioles are long having two or more swollen glands. The fruit is cordate in shape, has a deep cavity and the apex is rounded or pointed. The floral buds are borne on 2-year-old shoots or at the base of 1-year-old shoots and are found only on lateral, simple flower buds. Flowers are white in colour and have 5 petals, numerous stamens, a single style and an ovary with a single carpel containing two ovules. The fruit colour of cherries varies with varieties and may be yellow, red, or purplish black. The texture of the flesh is tender or firm, sweet and yellow or red or dark purple in colour. On the basis of flesh, sweet cherries have been divided into two pomological groups: The Heart Group and the Bigarreau Group. The Heart cherry varieties. have soft and tender flesh and heart shaped fruits. The fruit colour varies from dark with reddish juice to light coloured with less colour. The Bigarreau Group of cherries is usually roundish. The colour of the fruit and juice also varies from dark red to light red.

Climate and Soil

Sweet cherry requires a cool climate. The highest chilling requirement It is grown successfully in areas between 2,000 and 2,700 m above mean sea-level, requiring 1,000- 1,500hr chilling period during winters. which is highest among temperate fruits. Cherry blossom is very sensitive to spring frost; therefore, frost-free sites on hill slopes and valley areas with a drainage of cold air are preferred. Since the southern and south-western aspects are warmer, they should not be selected for cultivation at lower elevations. Generally, north-east or north-west aspects are the most suitable for cherry cultivation. An annual rainfall of 100–120 cm, well-distributed throughout the year, is desirable, but high rainfall during flowering results in heavy blossom wilt. At the time of fruit ripening, heavy rains cause fruit cracking. Therefore, the weather should be dry at the time of fruit ripening and locations having an early onset of monsoon should be avoided. A well-drained deep sandy loam soil with pH 6.5-7.0, which can hold moisture during summer, is most suitable. The cherry plant is very sensitive to waterlogging and, therefore, heavy soil should be avoided. Cherries occupy an important position among temperate fruits all over the world. sensitive to frost and waterlogging.

Species

- Sweet cherries (Prunus *avium*), mostly used for table purpose
- Sour cherries *(P. cerasus = P. avium x P. fruticosa)* mostly used for canning or cooking
- Duke cherry *(Prunus gudouini)*- interspecific hybrid *(Prunus avium)* x *(Prunus cerasus)*

Varieties

About 120 varieties are available in germplasm repository. Most of them belong to sweet cherry group. All varieties are divided into 2 groups.

1. **Heart Group:** Fruit is heart-shaped having soft and tender flesh. Colour of fruit varies from dark with reddish juice to light coloured with colorless juice. Red Heart, Black Heart, Victor, Chico Beauty, Waterloo, Ebony, Lambert are important varieties of this group., Bing, Van, Jubilee are cultivars of sweet cherry.
2. **Bigarreau group:** Fruit roundish, colour of fruit and juice varies from dark to light red. Sam, Summit, Sue, Sunbrust, Lapins, Napoleon white, Yellow Spanish. Tradescant's Heart, Compact Stella and hybrid (13-17-40) are new promising varieties.
3. **Amarelles group:** Early Richmond and Montmorency
4. **Morellous group:** Flemish Red, English Morello and Louis Phillipe
5. **Duke cherries:** Ronald's Duke, Royal Duke, May Duke and Arch Duke
6. **Others varieties** –Pink Early, Black Republican,7 White heat, Early Rivers, Governors wood, Sunbrust, Compact stele, Smmit,
7. Donar varieties – Stella, Vista, Vic, Seneca and Vega
8. Universal donor varieties: Stella, Vista, Vic, Seneca and Vega universal
9. Sweet cherry varieties: Royal Ann (Napoleon) is most popular variety
10. Resistant to fruit cracking: Sam, Sue, Windsor, Victor

11. While Bing Cherry is susceptible and Sam, Sue, Windsor, Victor and Van are resistant to fruit cracking.
12. Early Richmond and Montmorency are self fruitful cultivars of sour cherry.

Rootstocks

Cherry is propagated by grafting on seedlings as well as on clonal rootstocks.

1. Seedling rootstock: Seedling rootstocks of cherry are paja (*Prunus cerasoides*), wild birld cherry (*Prunus paddum*), mahaleb is also known as *St. Lucie Cherry* and mazzard.
2. **Clonal rootstocks :** Clonal rootstocks are commercially recommended for raising its plants as trees on **paja** show symptoms of delayed incompatibility.
   - Clonal rootstocks are **Colt, Giesela, Charger, SL64**.
   - Colt is semi-dwarf, compatible with almost all varieties of sweet cherry, has good anchorage, and is tolerant to gummosis, crown-rot, moderately resistant to stem-pitting virus and bacterial canker but susceptible to oak-root fungus.
   - **Mazzard, F 12/1** is semi-vigorous.

Propagation of Seedling Rootstocks

Paja seeds do not require chilling treatment to break dormancy. It has now been discouraged as a rootstock due to delays in compatibility, although most of the earlier plantations existed on paja rootstock only. Seeds of mahaleb and mazzard require stratification before sowing. Seeds are soaked in 500 ppm GA_3 for about 24 hours, then they are stratified by placing them between the layers of sand in a cool place at 2-4°C for 80–120 days for mahaleb and 120–50 days for mazzard to break seed dormancy. During stratification, the medium is kept moist. As the embryonic root comes out of the seed coat, these are transplanted 6 cm deep and 10-15 cm apart in rows spaced at 20–25 cm in nursery beds. The nursery beds are mulched with 10–15 cm of thick hay and irrigated lightly. Mulch material is removed when seedlings attain 5–6 cm in height. The nursery should be regularly watered and kept free from weeds. The seedlings attain graftable size in a year. The common method of clonal rootstock multiplication: Mound or stool layering.

Propagation of Clonal Rootstock

Mound layering, or trench layering, is the common method of clonal rootstock multiplication. The stool beds are established during December by planting healthy mother plants 30-45 cm apart in a row spaced at 60-70 cm. Before new growth starts, the mother plants are cut back to 2.5 cm above the ground level. New shoots develop on the stub in the spring. When the shoots are 25-30 cm long, their bases are covered with a mound of soil or sawdust, building the mound to a height of 20–25 cm as the shoots grow during spring. The suckers are ringed at the base and then covered with soil to encourage rooting. In the difficult-to-root Mazzard F12/1 root stock, IBA (7,500 ppm) is applied to the ringed portion of the shoots during summer. The shoots are separated in winter and then lined out in the nursery beds. If the suckers are well-rooted and more than 0.8 cm in diameter, they are grafted in the spring; otherwise, they are kept in the bed for a year to produce strong plants for grafting. Colt rootstock is easy to root and can also be multiplied through cuttings. Hardwood cuttings of 30–45 cm in length and of pencil thickness are taken in February. Cuttings are treated with IBA (2,500 ppm) for 10 seconds and planted in nursery beds for rooting. Rooted cuttings are lined out in December and grafted with scion varieties in March.

Propagation of Scion

Cherry plants are propagated mainly through grafting. Tongue (whip) grafting during February-March is recommended, which gives a bud-take of more than 90%. For grafting, the scion wood is collected during winter when the buds are dormant. Scion wood is packed in moss grass and then wrapped in moist gunny bags. These packed bundles of scion wood are stored at 2-4 C till they are used for grafting.

Planting and Planting Density

In India, cherry cultivation is confined to hilly areas on sloppy land, so the planting is done on contours or terrace systems. However, in valley areas, a square system of layout is recommended for the establishment of an orchard. The planting distance depends upon the soil fertility and the rootstock used. Spacing: In the rectangular system: Sour cherry measures 10m x 14m and sweet cherry measures 7m x 10 m. A spacing of 6 m x 6 m is recommended for plants raised on seedling rootstock in Himachal Pradesh. For semi-dwarfing rootstock like colt, spacing can be reduced to 4.5 x 4.5 m. Pits of 1 m x 1 m x 1 m size are dug and filled with a mixture of 35-40 kg farmyard manure and half kg super phosphate one month before planting. The pits are filled at last, up to a height of 15 cm above the ground level. The best time for planting is December-January.

Pollination

Therefore, since most of the cherry varieties are self-incompatible as well as cross-incompatible, compatibility of varieties is required for cross-pollination. There are many cross-incompatible groups, and cross-compatible varieties within a group should be planted in the orchard for proper fruit set. Care should be taken to see that not only do their flowering periods overlap, but they also do not have the same sterility alleles. The varieties that have the same sterility alleles when planted together will not set fruit and require at least one other variety having a different sterility allele for better fruit set. The universal donor varieties like Stella, Vista, Vic, Seneca, and Vega are good cherries as pollinizer to planted with any variety to get good fruit set provided their flowering period overlaps. For good pollination, planting should be done in such a manner that plants of one variety adjoin each other in the planting arrangement. The placement of 2-3 bee hives/ha is required in a cherry orchard at balloon stage for better pollination and fruit set. Type of incompatibility: gametophytic incompatibility. Most commercial varieties of cherry are self-sterile. Most of the cherry varieties are self-sterile; they need cross pollination. Most of the sour cherry varieties are self-compatible. Incompatibility (both self and cross) is one of the cherry production problems.

Training

Cherry trees are trained on modified leader systems. Plants are headed back to about 60-80 cm at the time of planting. The central leader is retained and 3-5 wide-angled branches, 20-25 cm apart, spirally around the tree are selected in the first dormant pruning. The lowest branch should be 40–60 cm above the ground level. The selected scaffold branches are headed back to a minimum and only one-fourth of the growth is pruned off. In second dormant pruning, 3-4 well-spaced main branches are selected whose one-fourth growth is pruned off and on each main scaffold, well-spaced 3-4 secondary branches are selected. After 3–4 years, the central leader is headed back and lateral branches are allowed to grow, resulting in the development of a strong and moderately spreading tree.

Pruning

Cherry plants require more corrective pruning rather than too much heading back of the branches. Bearing trees need some pruning to keep the centre of the tree open. The top is kept fairly low to generate new growth. Pruning is restricted to eliminating dead, diseased and intercrossing branches. Fruits are borne laterally on spurs of one-year-old shoots. The average productive life of fruiting spurs is 10–12 years, requiring less super renewal pruning.

Manuring and Fertilization

Cherries require all the essential nutrients for better growth and quality fruit. Since fruit development and vegetative growth occur simultaneously, there is a high demand for mineral nutrients. The amount of manure and fertilizer to be applied is influenced by the age or size of the tree, soil type and fertility, cultural practices and anticipated fruit yield. Farmyard manure should be applied in December along with a full dose of super phosphate and muriate of potash. One half dose of N is applied in spring before flowering and the other half dose of N is applied in spring before flowering and the other half one month later. Fertilizers are broadcast in tree basins about 30 cm away from the tree trunk.

Aftercare

Cherry orchards are maintained under permanent sod with clean basin management. The basins are kept clean by hand-weeding or using weedicides. Application of Diuron at 4 kg/ha as pre-emergence and Paraquat (0.5%) as post-emergence is recommended to suppress the growth of weeds for 4-5 months. Mulching tree basins in April with 10-15 cm of thick hay also helps control weeds and conserve soil moisture. Green manuring crops-bean, pea, red clover and white clover-can also are grown in tree basins to improve soil texture and fertility. The most important cultural operation in cherry, apple, and pear is sod culture. The most commonly used sod crops are rye, vetch, buckwheat, oats and millets.

Irrigation

Due to sloppy lands and non-availability of irrigation water, cherries are grown under rain-fed conditions in our country. The distribution of rainfall throughout the year is uneven and owing to less rainfall during April-May, the plantations should be irrigated frequently. Irrigating cherry trees at weekly intervals during fruit growth and development is recommended for better fruit size and quality.

Fruit Cracking

Cherry fruit cracking is a serious issue that causes 50 to 80 percent losses. Sweet cherry cultivars differ in the susceptibility in cracking because of differences in the rate of water absorption and the capacity for expansion of the peripheral tissues to accommodate the increased fruit volume that results when water is absorbed. The cultivars with a rapid rate of absorption and a low capacity for expansion tend to be immune. Any treatment that decreases the rate of water absorption or increases the capacity of fruit tissues to stretch without rupturing reduces the amount of cracking. At 25-30 days before harvest, a spray of calcium chloride at 300 g per 100 litre of water, GA_3 at 2000 ppm, or NAA at 10 ppm prevents fruit cracking. cracking.

Maturity Indices and Harvesting

The yield and quality of cherries are appreciably affected by the stage of maturity at which the fruits are harvested. Early-picking results in flat fruits with less yield as cherries usually develop rapidly in the last few days before maturity is reached. Harvesting of over-ripe fruits results in a loss of weight, volume and quality. Colour development, TSS and flavour are the best standards for judging the optimum time of harvesting. The fruits are packed in boxes lined with paper. Generally, 1 kg cardboard boxes are used for packing nowadays.

Common Disorder

- Common disorder in cherry: Sunscald.
- Abnormal reduction in fruit size: Zinc and Interveinal chlorosis is due to Mg.

55

Chest Nut

- B N: *Castanea dentata*
- Family: Fagaceae
- Chromosome No.: 2n = 24
- Origin: Europe

Introduction

The chestnut is similar to the acorn of oak. Chestnut trees have been growing in the temperate climate of the world for more than 4,000 years for beauty, fuel and shelter. The sweet chestnut is some nutritious low-fat food rich in vitamin B. Freshly harvested nuts contain about 50 per cent moisture, 40-42 per cent carbohydrates, 2.9 per cent proteins and about 1.5 per cent fat. In addition to providing a good food source for humans, chestnuts also provide food for a number of wild life species. Its wood is durable and used for timber and furniture work. Chestnuts and chinquapins are exceptions to all other nut trees in that they contain little oil and are high in carbohydrates, particularly in starch.

Area and Production

There is no organized plantation of chestnut in India. Only stray plantations exist and some plants grow wild in the forests of Himachal Pradesh, Darjeeling and Khasi hills. The leading countries in chestnut production are the USA, China, Turkey, Korea, Italy, Japan and Spain.

Flower and Fruits

The Chinese chestnut is the latest tree to bloom. The flowers are produced in two kinds of catkins borne on current season shoots near the terminal portion of the shoot. The fruit of the Chinese chestnut is borne in a spiny involucre known as bur. Three nuts are usually produced in each bur (upper, left and right). The bur is a vegetative structure that encloses the nuts or fruits. Botanically, each nut is a complete fruit. The shell of the nut develops from the ovary wall. The kernel of the nut is a young embryo plant that develops from the fertilized eggs of an ovule. The edible portion of the kernel is made up of two fleshy cotyledons and minute internal growing points of shoots and roots. The kernel is enclosed in a membranous covering called the pellicle, which originates from the integument of the ovule.

Climate and Soil

Chestnut can be grown in a wide range of climates in temperate areas. It is as hardy as a peach and can withstand temperatures as low as -29 °C in deep dormancy. It requires less chilling to break bud dormancy in spring. Buds respond quickly to warm temperature and thus become subject to damage by late spring frosts. Chestnut can be grown in all types of soil, but it grows best in well-drained sandy or sandy loam soil. The soil should be moderately to slightly acidic. The chestnut trees can withstand moderate drought after well establishment.

Rootstocks

Some species of chestnut are used as rootstock for propagation. Chestnut is highly cross-pollinated and hybrid seedlings used as rootstock are a possible cause of graft union failure in chestnut. Therefore, mixed hybrid strains should not be used as rootstock.

Propagation

Chestnut seeds planted immediately after harvest showed poor germination. The seeds are stratified for 50–60 days in moist sand at 0–2.2 ^{0}C to break the dormancy of chestnut embryos and ensure uniform germination. The chestnut stratification in moist sand kept in wooden boxes at cool shady place proved better than in refrigerator because of fluctuating temperature under field condition help in promoting the seed germination. The stratified seeds are sown in nursery beds in the month of March. The chestnut has only one bud at each node and if it is killed by spring frost, bud break fails. Early grafting should be avoided. Splice grafting and tongue grafting done in March are generally used for the propagation of chestnut. Chip budding also performed well.

Varieties

The cultivars which are of commercial importance are: Abundance, Crane, Kuling, Meiling, Nanking, Orrin, Colby and Hemming.

Planting Density

The planting is done during the winter. Before planting, the site should be properly laid out with contour or terrace systems. The pit should be prepared well in advance and refilled with soil mixed with 60 kg of well-rotted FYM. Plants should be spaced about 10 m apart as trees have very large canopies.

Pollination

All cultivars and species of chestnut are self-sterile. Two or more cultivars and seedlings must be planted in the orchard to ensure cross pollination. Young orchard trees may not set nuts when the first flowers are formed because of lack of pollination. When young trees start flowering, the few staminate catkins apparently do not produce enough pollen to fertilize the pistillate flowers. Chestnut is wind pollinated as well as insect pollinated.

Training and Pruining

The Chinese chestnut forms a low-headed tree if left unpruned. Unpruned trees start bearing fruit earlier than severely pruned ones. Early pruning stimulates vegetative growth. Hence, for early nut production, orchard trees should not be pruned for a few years. An ideal practice is to allow the trees to come into bearing before any pruning is done, then remove only a few of the lowest branches each year until the tree is properly headed. Lower limbs are removed gradually from the tree and the practice is continued. The lowest branch should be kept at a height of one metre from the ground and therefore, branches should be spaced spirally at a distance of 40 cm from one another.

Manures and Fertilizers

The chestnut trees are mostly planted on eroded upland soils, so it is important to fertilize trees regularly. At least ½ kg of 15:15:15 NPK mixture per year of tree age should be applied before sprouting or in early spring. The full-bearing trees should be supplied with 100 kg of FYM and 6-8 Kg of NPK mixture during December–January.

Irrigation

Well established trees can withstand a moderate amount of drought. Chestnut is generally grown under rain-fed conditions but needs adequate moisture for at least 2 months after blooming. Irrigation at fortnightly intervals after blooming is desirable for better fruit size, yield and nut quality. Row crops such as corn, soyabean or other pulses can be grown between the trees for the first few years. Filler trees of stone fruits can be planted for supplementing income in the early years and should be removed later on after chestnut plants start bearing full crops.

Maturity and Harvesting

The chestnut matures in the first fortnight of October in Himachal conditions. The bur colour changes from green to light brownish and splits open during maturity, releasing the nuts. Chestnuts are very perishable crops that require prompt harvesting every three days. Traditionally, chestnuts are hand gathered from the ground after falling naturally. In the USA and other developed countries, the chestnut is harvested mechanically by shaking the burs from the trees and using a mechanical pick-up device together the nuts. The harvested chestnuts are treated with fungicides to prevent spoilage. These nuts are then cured for 5 days at 2°C. On an average, the harvest period of each tree is 23 days as maturity is not uniform in chestnuts. A seedling tree yielded 26 kg of nuts at 12 years of age.

Storage

Fresh chestnut contains 40–45 per cent carbohydrates, mostly in the form of starch, about 5 per cent oil and about 5 per cent moisture. These are highly perishable because the nut loose moisture rapidly at room temperature, causing the kernel to become hard and inedible. For storage, the moisture of nuts must be less than 10 per cent and relative humidity of storage atmosphere must be 70 per cent or lower and storage temperature must be 0°C or lower. Mold on chestnut can be destroyed with a hot bath (57.7°C for one hour). The best way of drying chestnut is to put them in bags at 4.4°C with well circulated air at 70 per cent relative humidity. Under these conditions, the nuts will cure and dry to the optimum moisture content.

56

Hazelnut

- B N: *Corylus avellana*
- Family: Betulaceae
- Chromosome No.: 2n = 22
- Origin: Europe and Western Asia
- It is typically a temperate zone nut crop
- Chilling requirement: 1500 hours below 7.2°C

Introduction and Origin

It is typically a temperate zone nut crop and is mostly grown in Turkey, Italy, Spain, Germany, France and England. Turkish hazelnut production of 625,000 tons' accounts for approximately 75 % of worldwide production. The common hazel (*Corylus avellana*) is native to Europe and Western Asia. In Himachal Pradesh, it is found growing wild in the Pangi region of Chamba district, where it is locally known as Thangi. Hazelnut is also known as cobnut and filbert. In the UK, a distinction is made between filbert, which has a husk longer than the nut, and cobnut, in which the husks are shorter than the nut. Hazelnuts are extensively used in confectionary to make praline and are also used in combination with chocolate truffles.

Climate and Soil

The hazelnut tree is quite hardy but only produces satisfactory crops under moderate climate conditions. A temperature of minus 10°C is critical, especially if accompanied by wind, which may kill both pistillate and staminate flowers. The chilling requirement of hazelnut is about the same as that of most commercial cultivars of apple and thus it can be grown successfully in the apple growing regions of Himachal Pradesh, Jammu and Kashmir, Uttarakhand, and the North-Eastern Himalayan regions of India. Hazelnuts are more shallow-rooted than most fruit and nut trees, and do not tolerate wet soils. However, moisture retention in the soil is important since the tree can not tolerate excessive dry summer heat and hot winds. Soils must be moderately fertile and heavy clay soils should be avoided. Hazels will grow in pH ranges from 4.5 to 8.5, but around pH 7 is ideal.

Varieties

1. **Important varieties are** – Tonda Romana, Barcelona, Negret, Tonda Giffoni, Tonda Gentiledelle, Langhe, Pauetet, Tombul.

Planting and Planting Density

The area for planting should be marked and cleared or sprayed with herbicide. Tree spacing is highly variable in the different countries, as they depend on the fertility of the soil, rainfall and variety vigour. A planting density of 860 trees/ha is recommended, with rows 4–5m apart (to allow machinery access) and 2–3 m within row spacing. To ensure adequate pollination it is advisable to plant atleast 10% of other varieties, evenly distributed throughout the stand. Planting is usually done during the winter months. In the orchards of Oregon (USA), trained in vase, tree density normally varies from 278 to 400 trees/ha (6 m x 6 m or 5 m x 5 m), while in the South-West of France they oscillate between 666 (5 m x 3 m) and 800 trees /ha (5 m x 2.5 m).

Manures and Fertilization

Prior to planting and up to bearing age, organic and inorganic fertilizers should be applied according to soil analysis as follows: Organic fertilizers like FYM should be applied at around 30 tons/ha if the soil organic matter is below 2 per cent. Where the soil pH is around 5.5, it should be raised to 6.5 by liming, but not more than 5 t/ha should be given in a single dressing. Fertilizer application to mature trees should be based on leaf and soil analysis. The fertilizer recommendations for hazelnut are 120 to 150 kg/ha of N, 60 to 70 kg/ha of P and 100 kg/ha of K.

Training and Pruning

The traditional training system in hazelnut orchards in the main production areas (Turkey, Italy and Spain) has been a multistem bush, according to its normal tendency of bushing growth. However, the training system used in the new orchards of the United States, France, Italy and Spain is a vase with only one stem.

Pruning Systems

Hazelnut bears its fruits laterally and terminally on wood of the previous season's growth and the pruning after the tree has come into bearing should be such as to stimulate a moderate amount of new growth each year.

Weed Control

Weeds compete with the crop for moisture and nutrition and adversely affect the nut yield and quality. Therefore, planting should be kept free of weeds either manually or through the use of herbicides. The most commonly used herbicides in mature orchards, according to their application time, are: Simazine, Diuron, Napromide, Oxifluorphen, Propyzamide, Trifluraline in pre-emergency and Paraquat, 2,4-D, Aminotrizol, Glyphosate, Ammonium gluphosinate, Terbutilazine + Terbumeton etc. in post-emergency.

Harvesting and Yield

When nuts change from green to brown and the abscission starts, it is the best time for harvesting. Usually it comes in the mid-autumn. Hazelnut trees reach maximum production between fifteen and twenty. Five years of age and the average yield per plant ranged from 2-2.5 kg /Tree / year.

57

Kiwi Fruit

- Kiwi fruit/China's miracle/Horticultural wonder of New Zealand/National symbol of New Zealand/Chinese gooseberry/Peach of the macaque Monkey
- B.N =*Actinidia deliciosa*
- Family = Actinidiaceae
- Chromosome No.: 2n = 58
- Origin: South China
- Fruit type: Berry
- Fruit growth follows triploid growth curve
- Flower colour: Creamy white to yellowish

Introduction

This dioecious fruiting vine is native to the Yangtze River valley in south and central China. This dioecious fruiting vine. Seed requires stratification. Although this fruit vine is originated in China, yet its full economic potential was exploited by the New Zealand which accounts for over 70 per cent of world trade. In other countries of the world, its cultivation gained momentum after 1960, and now it is cultivated on a commercial scale in the China, New Zealand, Italy, USA, Japan, France, Germany and Australia. In India it is grown in mid hills of Himachal Pradesh, J&K, Uttarkhand, Sikkim, Arunachal Pradesh, Meghalaya, Nagaland and Nilgiri hills of south India. It was first introduced at Lal Bagh garden, Banglore, but the plant did not come to bear due to a lack of chilling requirement during winter for its bud break. Later on, in 1963, it was introduced in the Shimla hills, where the plant came to bear in 1969. The kiwifruit is rusty brown with a hairy surface and looks like a sapota fruit. Fruit has a refreshing and delicate flavour, a pleasing aroma and high nutritive and medicinal value. It is high in vitamin C, has more potassium, phosphorus, and iron than any other food, and has a low calorie count.

Climate and Soil

Although kiwifruit is a very hardy deciduous vine that can withstand a wide range of climatic conditions, for the production of a good crop of quality fruits it has specific climatic requirements. It can be grown successfully in warm temperate to sub-tropical regions lying between 3000 to 5500 feet a.m.s.l, which provide 600–800 chilling hours to break dormancy. Low temperature (-2.5°C or below) and frost during spring and autumn are very injurious, which kill immature shoots and fruit buds. In summer, high temperature >38°C accompanied by high insulation (100) and low humidity may cause scorching of leaves, sunburn of fruits and even death of the plants. A rainfall of about 120–150 cm, well distributed throughout the growing period, is sufficient for proper growth and development. High velocity of wind during April and May cause breakage of young tender flowering shoots, whereas low temperature, hails and rainfall during flowering hamper the fruit set. Kiwifruit can be grown in almost all types of soil, provided adequate soil moisture is available. Deep, well-drained, sandy-loam soil with a good amount of organic matter is ideal for its cultivation. Heavy clay soil with poor drainage is not suitable. A soil pH of 5.5 to 6.5 is considered ideal for vine growth and fruit production.

Varietal Status

Kiwifruit is a dioecious plant, bears staminate and pistillate flowers on separate plants.

A. Pistillate varieties: Hayward, Allison, Abbott, Monty, Bruno

B. Staminate varieties: Allison, Tomuri, Matua

- Abott, Allison, Bruno, Hayward, Skelton and Monty are commercial cultivars of kiwifruit.
- Tamuri, Motus, Allison, Monty and Motua are male cultivars often used as pollinizers.
- Kiwi is functionally dioceious. Female flowers have infertile pollen
- **Brono**: Rich source of Vitamin-C variety
- **Hayward**: Shy bearing and late maturing cultivar and more suitable to high hill
- **Abott**: Early flowering cultivar

Rootstock

The Bruno and Abbott varieties are commonly used as rootstocks in kiwifruit. Seeds of these two cultivars are preferred because of their good germination and strong seedling vigour. cold-tolerant species: *Actinidia arguat.*

Propagation

Commercial propagation: stem cuttings: soft wood cuttings Softwood, semi-hard, and hard wood cuttings are used to raise their own rooted plants. The ideal cutting is 0.5 to 1.0 cm thick with relatively short internodes, about 10–15 cm in length having at least 4-5 buds. Cuttings should be taken from the middle portion of the current season's growth shoot in June-July in the case of semi-hard cuttings and from one-year-old shoots during January-February in the case of hard wood cuttings. After preparation, the cuttings are dipped in 4000-5000 ppm IBA solution for 10-15 seconds and then planted in mist propagation chambers or in open field conditions for rooting.

Propagation through Grafting and Budding

Kiwifruit vines are also propagated by grafting and budding on seedling rootstock. This method takes two years to develop a nursery plant, but it is the easiest, most economical and used for large scale multiplication. Before sowing, seeds are stratified in alternate layers of moist sand for 30–35 days at 0–5°C to break dormancy. Stratified seeds are sown in the seed beds during February. Germinating and baby seedlings are very sensitive to direct sunlight, so they must be protected by creating shade or using shade nets. During July and August, seedlings are transplanted in polybags and then into nursery beds at the three-leaf stage. The seedlings attain a graftable size within a year. Tongue grafting and chip budding done in the last week of January to the end of February gave 90–95 % bud-take success. Nursery management practices like weeding, irrigation, and staking of grafted plants are done at regular intervals.

Planting

Flat land with a gentle slope is ideal for the planting of kiwifruit. Plant spacing varies according to the cultivars being grown and the training system adopted. In general, planting is done at a spacing of 6 m from plant to plant and 4 m from row to row in varieties like Allison, Abbott, and Monty trained on the T-bar trellis system. whereas Hayward is less vigorous and is planted at a spacing of 5 m x 5 m. In the pergola system of training, a spacing of 6 m x 6 m is recommended for getting better fruit production. Planting is done during the dormant season (December to January). While planting, the male plants are spread throughout the block, with every female adjacent to a male. This is achieved by a 1:8 or 1:9 males to female ratio.

Training

Kiwi fruit is a vine-like grape and requires a similar training structure but is stronger than grapes. A number of training structures are used for training vines, but T-bar trellis and pergola systems are more popular. In the T-bar trellis system, the pillars of iron and concrete, about 1.8m in height above the ground level, are erected at a distance of 6m from each other in a straight line. A cross arm (1.5 m) is fixed on each pole, which carries five outrigger wires at a distance of 45 cm each. Vines are trained up to the wire as single stems, then two leaders in opposite directions along the centre wire are selected or developed. From these permanent leaders, temporary fruiting arms 25–30 cm apart are selected at the right angle along both sides of each leader. Training of vines on a pergola system is similar to that of a T-bar. A flat-topped network of criss-cross wire is prepared on the erected pillars. This system is costly to prepare, but vines trained on this system give higher yields.

Pruning

The main objectives of pruning are to obtain a high yield of quality fruits and to manage excessive vegetative growth. The following principles should be kept in mind at the time of pruning. The fruit is developed on the current season's growth, arising from a one-year shoot. Only the basal buds of the nodes 4–12 in the current season of growth are productive. Vines grow 2-3 m every year, which becomes overcrowded if not controlled by summer and winter pruning. The shoots developed on older wood by heading back will not fruit normally in the first season. Keeping in mind these principles, the vine pruning is carried out in such a way that the fruiting areas are available every year, requiring the wood to be young. This is achieved by a 3–4-year lateral replacement system. In dormant pruning, the fruiting lateral is cut back to 2 vegetative buds beyond the last fruit. In the second year, these vegetative buds produce the fruiting shoots, which are pruned again. The arms on lateral shoots are pruned and allowed to fruit for 3–4 years. After this, the lateral is removed from the main branches and other laterals are selected and pruned accordingly so that the balance between vegetative and reproductive growth is maintained for the continuity of the fruit production. In summer, the pruning shoot is cut beyond 6-8 buds from the last fruit during June-July.

Pollination and Pollinators

The Kiwifruit crop is highly dependent on pollination as the plants are functionally dioecious. For effective pollination, one male plant is planted for every 9 pistillate plants. Wind and insects play a very significant role in pollination. However, the introduction of honeybees in the orchard further increases the fruit set and size of the fruit. For effective pollination, kiwifruit colonies of about 8–9 per hectare are required in kiwifruit orchards. Besides insect pollination, hand pollination is essential to get fruits of better size and quality. Gender change due to bud mutation in a mature male vine.

Orchard Floor Management

Clean cultivation with mulching of the tree basin area with 15 cm thick hay grass or black polythene mulch is recommended for kiwifruit orchards. Intercrops such as strawberries, peas, beans, cowpeas, and vegetable crops such as tomatoes were planted during the first two years. Ginger, etc., can also be grown in the vacant area between the trees.

Manures and Fertilizers

The vines make much vegetative growth, yield heavily, thus requires adequate amount of manures and fertilizers for normal vine growth and fruit production. The quantity of fertilizers to be applied varies with soil fertility and the age of the vine. In general, a basal dose of 20 Kg FYM along with 0.5 Kg of NPK fertilizers mixture containing 15 per cent N be applied each year of age. For fully bearing vines (8 years and above), 60-80 kg FYM, 800 g N, 560 g P_2O_5 and 1200 g K_2O/vine are applied every year per vine. FYM along with a full dose of P_2O_5 and K_2O is applied during December-January, while N fertilizer is applied in two equal dressing, half before bud-burst and the remaining full dose at the onset of monsoon, *i.e.,* in July. Application of a 2/3rd dose of recommended NPK fertilizers through drip irrigation in eight equal split doses from bud burst at 10 day intervals is recommended, which results in a 25 per cent saving of fertilizers and an increase in yield of quality fruits over the traditional method of fertilizer application.

Irrigation

The water requirement of kiwifruit is very high because of vigrous vegetative growth and a larger leaf surface area. In general, fully grown vines require 80-100 L of water for total daily transpiration from 16–17 m of canopy area during summer. Supplemental irrigation is important for improving the productivity of fruits due to the irregular distribution of rainfall in most kiwifruit growing areas. Young vines should be irrigated at 2-3 days' intervals, while bearing vines should be irrigated at 20 % depletion of soil moisture from field capacity (5–6 days' intervals) during summer to get better size fruits. Drip irrigation at 100 per cent etc gave a higher yield of quality fruits.

Crop and Quality Regulation

All the cultivars of kiwifruit except Hayward bear heavily every year. This heavy crop creates severe competition between the fruits for water, nutrients and photosynthates, which leads to the production of small-sized fruits. Therefore, to harvest a quality crop of good size, hand thinning is essential as chemical thinning is ineffective. In a study on thinning in kiwifruit, hand thinning (20% thinning) to the extent of retaining 4-6 flower buds or fruits/ flowering shoots produced more fruits of A grade without any adverse effect on total yield, as chemical thinning was found ineffective in kiwifruit. In hand thinning, only lateral flowers or fruits are removed. Flowers borne: current season shoots. Flowering period: from the last week of April to the third week of May.

Harvesting and Marketing

In kiwifruit, determination of optimum fruit maturity is difficult because no change takes place in the skin or flesh colour or size of the fruit at fruit maturity. A maturity index of 6.2°B TSS is considered satisfactory for fruit harvest. Days from full bloom to harvest, along with TSS, were found to be the best index to judge the optimum harvesting time. Besides this, at maturity, the hairs present on the fruit skin are removed very easily and can be used to judge the harvest or maturity. Harvesting is done manually by giving a gentle twist to the fruit, and fruits are plucked without the stalk.

58

Macadamia Nut

- Macadamia nut/Australian nut/Queensland nut
- B. N.: *Macadamia ternifolia*
- Family: Proteaceae
- Chromosome No.: 2n = 48
- Origin: Australia
- Fruit type: Nut

Introduction

The macadamia nut (*Macadamia ternifolia*) is a fruit from Australia that was recently introduced into India. The macadamia nut is a unique, world's finest dessert nut characterized by its creamy colour, delicate flavour and crunchy nature. Having two important species, *M. integrifolia* and *M. tetraphylla* where the former one is adapted to a subtropical climate and the latter to temperate regions. Another species, *M. ternifolia*, produces bitter, inedible nuts. Hawaii is the world's largest producer, followed by Australia, South Africa, Guatemala and Kenya.

Use and Composition

Raw and roasted nuts can be eaten and the quality of the kernel is related to its oil content and composition. The kernel contains 75-80 g of oil per 100 g of edible portion, out of which 89% is with unsaturated fatty acids. It also possesses carbohydrates (10 g), protein (9.2 g), potassium (0.37 g), phosphorus (0.17 g) and magnesium (0.12 g) and minerals like calcium, iron and zinc. It is used in many confectionaries, ice creams and desserts. Important cultivars of this nut tree fall into Hawaii an and Australian selections. Keauhou, Kakea.

Climate and Soil

Macadamia is a subtropical plant and temperature is a major factor influencing vegetative flushing, flower initiation and nut development. It can be grown up to an elevation of 1000 m and above this altitude, the growth of the plant and nut size are retarded. Seeds are common propagating materials but these trees are slow in growth habit. This tree comes up well in a wide range of soils, but soils with good drainage and high organic matter content are the most preferred ones.

Cultivars

Ikaika, Keaau, Mauka and Makai are Hawaiian cultivars. Own Choice and Hinde (H_2) are important Australian cultivars.

Propagation

Seedlings are common propagating materials, but these trees are slow in growth habit, require longer time to bear fruit and are not true to type. Vegetative propagation methods like cuttings (using girdled branches treated with IBA-5000 ppm), layering and grafting are successful in macadamia.

Planting

High density orchards require selective mechanical pruning to allow light penetration. Spacing adopted is 10 x 5 m, but recently high density planting (5 x 3 m) is also practiced. Tree is trained to central leader system. Irrigation is given in low rainfall areas. Mulching tree basins helps to conserve soil moisture and to reduce weed growth.

Manures and Fertilizers

Fertilizer application of 120 g urea, 50 g super phosphate, and 80 g muriate of potash per plant per year increased growth and yield. Tissue and soil analysis are to be performed to recommend the exact dosage of nutrients. Flower initiation in Macadamia nuts takes place during shortened days where the night temperature is between 11 to 15°C. After anthesis, both ovules increase in size, but normally one gets fertilised and it inhibits the development of the other.

Flowering and Fruiting

Nuts can be left on the ground for 4 to 6 weeks without deterioration, except in wet conditions, which favour germination. Hence, nut set was enhanced by cross-pollination, mainly by introducing honeybees in orchards. Less than 1 to 44% of flowers set and a large number of these abscise 5 to 6 weeks after anthesis. A flower pistil has an ovary with two ovules. Flowers are borne on long pendulous racemes and are partly self-incompatible.

Harvesting and Grading

Roasted kernels with a specific gravity (SG) of less than one first grade, SG of 1.0 to 1.025 second grade and SG of more than 1.025 third grade. From flowering to maturity, it takes about 7 to 8 months. At maturity, nuts fall on the ground and are picked from the orchard floor. Third grade kernels are commercially unacceptable. Mechanical harvesters are used in many orchards. The yield of nut varies from 40 to 50 kg per tree and 3 to 4 tonnes per hectare. After harvest, nuts are dehusked using machines. These nuts are then dried and the moisture level reduced to 1–2%. They can be lightly roasted in coconut oil and salted or packed raw into vacuum-filled-foil-laminate bags. Nuts can be kept at a cool temperature for about one year.

Harvesting and Grading

59

Peach

- B. N.: *Prunus persica*
- Family: Rosaceae
- Sub-family: Prunoideae
- Chromosome No.: 2n = 2x = 16
- Origin: China
- Inflorescence: Solitary
- Fruit type: Drupe
- Edible part: Mesocarp and epicarp

Introduction

It comes to the market early in the season, particularly the low chilling peaches and growers get remunerative prices due to the non-availability of other fresh fruits in the market. It has high nutritive value, being rich in protein, essential amino acids, minerals, and vitamins. Peaches originated in China, where its culture dates back to 3000 years. Although it is not exactly known how and when the peach reached India, it is opined that some traders might have carried over the stones of peach from China to Kashmir. Peaches were introduced by Mr. Alexander Coutts in Himachal Pradesh in 1870. Later on, its cultivation was initiated in the mid-hills of Himachal Pradesh on the advice of an American horticulturist, Prof. R. W. Hodgson. Some low-chilling varieties of peach were also introduced at Punjab Agricultural University, Ludhiana in 1968 from California. peach with a strong flavour and aroma.

Nectarine: (*Prunus persica* var- nucipersica)

Nectarines are smooth-skinned mutants allied to peaches. It is a non-pubescent peach of smaller size. Nectarines have evolved as a result of a single dominant gene mutation in peach for fuzziness. Sun Red Neutered, Sun Grand, Sun Light, Sun Red, and Sun Rose Ripe are cultivars of nectarine. Nectarine cultivars are mostly preferred for table purpose. Nectarine is a temperate, juicy fruit of excellent appearance and quality. Fruits are fuzzy with free or clinging stone.

However, the nectarines are fuzzless peaches. Peaches and nectarines are the same species.

Uses and Composition

It is a good source of sugar, vitamins and minerals. TSS - 8-13°**Brix, 8** % total sugars Ascorbic acid - 6-13 mg/100 g, Carotenoids - 0.19-0.53 % - white flesh cultivars, 0.75-0.79 % - yellow flesh cultivars. Peach kernel contains 39-55 % Fat, 23-30 % Proteins, 14.8 % Crude fibre,2.7 % Minerals. Red blush skins due to anthocyanins. The yellow colour of peaches is due to xanthophylls. predominant acid present in peach-malic acid. Amygdalin is present in the seeds of peaches. Peach seeds contain 40-50 % oil. Prunacin is the principle glycoside present in the pulp of a peach.

Area and Production

Peach is basically a temperate zone plant and its commercial production is confined between the latitude of 30 and 40° N and S, although it is now grown almost all over the world. The major peach producing countries are China, Spain, Italy, Turkey Greece, USA, France, and Japan. In India, peach is grown on a commercial scale in the mid-hills of Uttarakhand, Himachal Pradesh, Jharkhand, Jammu and Kashmir, and as well as on a limited scale in the north-eastern states, viz., Arunachal Pradesh, Meghalaya, and Manipur. In Himachal Pradesh, peaches are commercially grown in the Rajgarh area of District Sirmour, which is also known as the peach bowl of India. Low chill varieties of peaches are commercially grown in Punjab, Haryana and Eastern U.P.

Morphological Characters

The peach is a small to medium-sized, upright-spreading, open-topped deciduous tree. The trunk bark is dark brown and rough, and the young shoots are smooth and pinkish in colour. The leaves are simple, large, oblong-lanceolate, glabrous above, pubescent beneath. Vegetative and flower buds are borne in the axils of leaves. Flowers are numerous, sessile, white or pink, appearing before leaves. The flower is of the perigynous type as the perianth surrounds the pistil but is not fused to it. The floral configuration is five sepals, five petals, 30 stamens and a single ovary. Fruits are fuzzy with free or clinging stone. However, the nectarines are fuzzless peaches. Self-compatible in nature and flower pink colour.

Climate and Soil

Peaches require a humid climate with a cold winter and a dry summer. It is moderately winter hardy and sensitive to low temperature injury. Swelling buds are injured at -6.5°C. Sites free from early spring frost are more suitable as peaches bloom early in the season. Peaches need about 500 to 800 hours of chilling during winter to break dormancy. The chilling requirement is easily met in areas located between 900 to 1600 m above mean sea level. There are some low-chilling varieties that require 250 to 300 chilling hours. These varieties are grown successfully in sub-mountain tracts of Punjab, Haryana, UP and Jammu, where the summer temperature reaches 45°C with very low humidity. The limiting factors in peach cultivation are the minimum winter temperature, chilling hours, spring frosts, hailstorms, high humidity and desiccating winds during summer. Deep sandy loam soil rich in organic matter is best for its successful cultivation. Peaches are highly susceptible to waterlogging and prefer perfect drainage. The pH of the soil should be between 5.8 and 6.8. Among temperate fruits, peach has lowest chilling requirement and earliest in flowering. Chilling hours at 7.2°C more effective, Temperature < 2-3°C does not break dormancy. High temperatures in December – January, bloom abnormalities and failures of fruit set is noticed. High temperatures in winter prolong dormancy and multiply chilling requirements. Peach with a low chilling requirement: 500 hours.

Varieties

Peaches are used for the table as well as for canning purposes. The choice of suitable varieties for any region depends on the type of market to be served and their adaptability to the local soil and climatic conditions. The varieties used for table purposes should be yellow-fleshed, freestone, and relatively free from fuzz. For canning purposes, the fruits should have yellow flesh, clingstone, non-splitting pit, and good symmetrical size.

Early season varieties: Alton, World's earliest, Red Heaven, Stark Red gold, Early Candler, Quetta, Sharanpur Prabhat and Florida Prince.

1. **Mid-season varieties:** Julie Elberta, Alexandar, Co Smith, Earligrande, Stark earlyglo, white giant, candor.
2. **Late season varieties:** J.H. Hale, Parrot deluxe, Peregrine.
3. Kanto-5, Shimizo Hakuto, Matchless and Suffeda are improved cultivars of peach.
4. Saharanpur Prabhat - (Sharbati x Florda Sun) is an improved cultivar of peach developed at Saharanpur (U.P.).
5. The varieties used for table and canning purposes should be yellow fleshed, freestone and relatively free from fuzz- Sharbati, Prabhat, Florida Sun and Shane-e-Punjab.

6. For canning purpose, the fruits should have yellow flesh, clingstone, non-splitting pit, good symmetrical size.
7. Sharbati: Developed through clonal selection.
8. Recent hybrids: Gala, Glory and (Regal)-Resistance to cold and bacterial spot disease.
9. Commercial cultivation of plains of Punjab: Florida, Florida Prince, Early Grande.
10. J.H. Hale, June Elberta and Halberta are male sterile cultivars of peach.
11. Sharbati, Saharanpur Prabhat, Florda Prince, Florda Sun and Shan-e-Punjab, Florda red, Sun Red, Sun Gold, Dawn Ramber and Dawn rose and Early Grande are low chilling cultivars of peach.

Rootstock

Peach seedlings are generally used as rootstock, though plum, apricot and almond seedlings can also be used. Since wild species produce more vigorous and hardy seedlings than cultivated varieties, wild peach seedlings are preferred in the hills. In the plains, seeds of Sharbati, Safeda and wild apricot are used to grow the rootstock for peach propagation. Clonal rootstocks are precocious, size-controlling, very productive and resistant to insect pests and diseases. Due to these plus points, clonal rootstocks are preferred over seedling rootstocks for raising nursery plants of these fruits.

- Sloh (peach x almond) hybrid and GF 677 clonal rootstock is commercially used. This rootstock is useful on alkaline soil due to resistance to chlorosis.
- **The clonal rootstocks are Siberian C (cold hardy), St. Julien hybrid No.1 and 2 and Damas GF-677 (resistant to water logging and alkaline soil),** Nemagaurd (*P. persica* x *P. davididasa)* and Shalil (nematode resistant).
- Peach itself is a successful rootstock Nematode attack is common. But 'Nenaguard and Okinoura'
- Apricot (*P. armeniaca*)– Root knot nematode rest., Almond, *P. amygdalis* – dwarfing rootstock,
- Western sand cherry – *P. berseyi*, *P. tomentosa* – dwarfing. Namking cherry – *P salicina* – medicinal to large trees.
- Damas (*P. domestica* x *p. spinose)* better rootstocks for heavy soil
- GF 557 and G 677 are rootstocks of peachs. GF stands for Grande Ferrade.
- GF series of rootstocks are developed by Bernhard in 1965 by crossing Peach and Almond.
- *Prunus behmi* is a natural hybrid of Almond and Peach

- Myran (*Prunus helsiana* x *Yunnan)*- Tolerant to drought and root knot nematode.
- Marianna (*P. domestica* x *P. munsoniana)* better rootstocks for heavy soil.
- Radhar (*Halehaven* x *Kalhaven)*.
- Sharbati, Sufeda and wild apricot is used to grow the rootstock for peach propagation

Propagation

The commercial methods of propagating peach are tongue or cleft grafting and T budding. For raising seedling rootstocks, seeds of wild peach are used. Before sowing, seed is first stratified at 4-5°C or below for 10–12 weeks in the moist sand. Pre-sowing treatment of seed with certain agro-chemicals and plant bio-regulators viz., thiourea (0.5%), GA (200 ppm) or BA (600 ppm) not only reduces the stratification period required and enhances seed germination but also improves the seedling growth. The stratified seeds are sown in well-prepared beds about 5 cm deep and 15 cm apart, at a row spacing of 20 cm. Seed beds are mulched with 6–10 cm thick dry grass and light irrigation is a must after sowing to avoid desiccation of seeds. The seeds start germinating in March and the seedlings become graftable in one year. The clonal rootstocks are multiplied through mound and trench layering. In the hills, tongue grafting during February and T budding during May - June are recommended. In the plains, grafting is performed during November - January and budding occurs during April - June and in September.

Planting

The plantation operation is carried out in the winter season. Before planting, the site of an orchard should be leveled, bushes and weeds should be cut down, and proper planning of the layout should be adopted. In the plains, a square planting system is common, whereas in the hills, an orchard is laid out using a contour and terrace system. Pit size 1 x 1 x 1 m at a distance of 4.5 m in the hills and 6.5 m in the plains, are dug in the months of September – October, which are refilled with soil and 40–50 kg of well-rotted farmyard manure and 1 kg of single super phosphate. High density planting: 3 m x 3 m, Tatura trellis system: 5 m x 1 m (2000 plants/ha), Meadow system: 2 m x1 m (5000 plants/ha).

Training and Pruning

Pruning of young plants for frame work in the initial 3 – 4 years of planting is termed as training. Training is mainly done to give a proper shape and to build a strong framework of trees. Another benefit is that we can utilize the available space and sunlight to the maximum extent for the production of quality fruits. Peaches are generally trained in the form of open center. However, in the plain area where plenty of sunlight is available, trees are trained on a modified central leader system with 4-5 scaffold branches. In high density, peach trees are trained with the Tatura trellis system. The main objective of pruning is to maintain a balance between vegetative growth and fruiting. Bearing peach requires heavy and regular pruning because it bears fruit laterally on the previous season's growth. It is known that once a plant has fruited, it will never bear fruit again in its life. Therefore, pruning is done to remove the unproductive parts, which in turn will form new fruiting branches in the following season. In peach pruning, thinning and heading back of shoots are two basic components. Pruning should be done so as to produce 30–70 cm of growth under subtropical conditions and 25–30 cm under mid-hill conditions, which is sufficient for optimum fruit production. For good quality fruit production, 40-50 per cent of thinning out and 75% heading back of shoots is suggested under mid hills conditions. At the time of pruning, dead, diseased, and broken branches should be pruned off. December-January is the best time for pruning peaches. Fruiting takes place laterally on the previous season's growth. Pruning intensity is the most severe. It is trained to open the centre/Vase system in the hills. Peaches require regular and heavy pruning. Mid-winter (December to January) is the best time for pruning. Pruning intensity is most severe because fruiting occurs laterally only on the previous season's growth. Regular pruning is required to get fruits.

Open Centre System

After planting, the plant is cut back to 40 – 60 cm above the ground level. During the growing season, about 3 to 6 laterals, in addition to the central leader, are produced on the tree. In the first winter pruning, 3 to 4 scaffold branches that are well located and have a wide angle should be selected and the remaining unwanted branches are removed. The central leader is also completely removed. The selected branches are headed back to ¼ to ½ of the growth. During the second dormant pruning, 2-3 secondary branches are selected on the primary branches. The major consideration in selecting secondary branches should be their location so that after pruning, the tips of primary and secondary leaders are about 30 – 40 cm apart from each other. The height of secondary branches is staggered in different years by pruning all branch leaders more severely. The vertical ones are pruned more severely. This will produce branch leaders at different heights and prevent overcrowding when the tree is mature. In the following years, the head should be fully formed and the selection of secondary branches completed.

Manures and Fertilizers

The nutritional requirement of peaches is comparatively higher than other stone fruits because it bears on one year's growth, making the production of annual growth imperative for fruit production. Recoupement of the wood removed every year in pruning is also necessary. Fertilizers are therefore applied every year. A large number of factors, like the type of soil, its fertility, climate, cultivar, planting density, shoot growth and irrigation facilities, determine the amount of fertilizers to be applied. The farmyard manure, along with a full dose of P and K, should be applied during December and January. A half dose of N is applied in spring before flowering and the remaining half dose a month later. Peaches are very susceptible to Fe deficiency. Peaches require high N and K and bear fruit after two years.

Irrigation

Irrigation is very essential for harvesting peaches of better size and quality. Sufficient moisture in the soil before the emergence of leaves and flowers is required for proper fruit-set and growth. Frequent irrigations are needed during the fruit's development. Lack of irrigation, particularly during the dry and hot summer, results in fruit drop, reduced fruit size and quality. At least two-three irrigations and, in the plains, weekly irrigation should be given during the fruit development period. In general, for quality fruit production, irrigation at 80 % of field capacity is recommended. Orchard soil management and weed management. During initial 3-4 years after planting, the intercrops like peas, beans, tomatoes, cabbage, zinger and colocacia are grown in between the peach trees and the basin area is mulched with hay or alkathene mulch. In fully grown tees, sod grasses are grown in vacant areas and the basin area is mulched with suitable mulch materials. Weedicides like simazine and atrazin at 2.0 kg/ha, terbacil at 0.8 kg/ha as pre-emergence, paraquat at 4.0 litres/ha and glyphosate at 4.32 kg/ha as post-emergence herbicide proved to be the most effective in controlling the weeds in peach orchards.

Crop Regulation

Heavy flowering and fruiting are the characteristics features of peach trees, resulting in small-sized, poor-quality fruits and a reduction in flowering in the subsequent season. Hence, for the production of quality fruits, crop regulation through thinning is essential in peach. The criteria for fruit thinning in peach trees are based on leaf to fruit ratio, spacing between fruits per tree. Generally, 30–40 leaves per fruit is the appropriate ratio. Application of Ethephon (300 ppm) wih in 20 at petal fall in July Elberta is recommended for optimum fruit thinning. However, in Redhaven peach, Ethephon (600 ppm), 20–30 days after fruit set, when the fruitlets are 20–25 cm in diameter, should be used for thinning. Hand thinning at 5-7.5 cm fruit spacing before the pit hardening stage is equally effective. Fruiting takes place laterally on the previous season's growth. The blooming period can be delayed by the application of GA_3 (200 ppm) before leaf fall or by the application of ethephon to avoid the risk of spring frost.

Maturity

Harvesting of peaches at the proper stage of maturity is essential as the post-harvest quality and storage life of fruits are controlled by maturity. The number of days required from flowering to maturity varies by cultivar and ranges from 78 to 127 days. Early season varieties like Flordasun take 81 days, Alexander 86 days, mid-season July Elberta 101 days and late season cv. Elberta takes 127 days from full bloom to harvest. Ground colour variation in conjunction with flesh firmness is one of the best maturity indices in peaches.

Maturity Indices

- Days until full bloom
- Fruit size
- Fruit firmness
- Pit discolouration
- Freeness of pit
- Ground colour
- The sugar-TSS-arid ratio.

Apple is a climacteric fruit the use of the growth regulators Alar (SADH) and ethrel improves ripening, increases flesh colour and color, and improves fruit quality. Uniform ripening, Alar @ 500 ppm at pit hardening, Ethrel – 100, 700 ppm before harvest - effective. Ethrel 50-100 ppm effective in increasing fruit size, fruit weight increased. The pulp to stone ratio increased and the quality increased.

Harvesting and Yield

Peach fruits do not mature uniformly and several pickings are needed during harvesting. The peaches came into bearing after 2 years of planting in the field. The economic bearing life of a peach plant is about 20–30 years. The yielding capacity increases with the age of the plant. Hand picking is the standard method for harvesting fruits. The picking containers are lined with cushion materials to avoid cuts and bruises. Immediately after harvesting, fruits are stored in a cool place or marketed. Pre-harvest application of calcium nitrate at the rate of 1.5 per cent increases the storage life of peaches. The average yield of fully grown trees of different varieties varies from 50 to 125 kg in the hills. In conventional plantation, 7–10 tones/ha and under high density with the Tatura Trellis system of training, about 23 tones per ha of yield have been obtained.

Storage

Peaches have a shorter storage life than most other temperate fruits. Cold storage temperatures of 0-0.3°C and relative humidity levels of 85-90% are recommended. In these conditions, free stone peaches and nectarines can be kept for two weeks and clingstone for 4 weeks. Pre-cooled peaches can be stored for 28–36 days. Peaches can be stored for up to 42 days in controlled atmosphere storage containing 5%. Soft textured high moisture is highly perishable and spoils within 2-3 days. Pre-cooling – reduces field heat. Hydrocooling with running H2O @ 100C. Storage diseases Brown rot, Whiskers rot, grey mould, green mold, blue mould, black mould.

Diseases

Peach leaf curl cause by (*Taphrina deformans), transmit* aphid (*Brachycaudus helichrysi*): This is most serious pest of peach. They also infest plum plants. The aphid sucks sap from the buds and sprouting foliage causing curling, yellowing and thickening of leaves. The activity of aphid is seen with the emergence of new growth during March. Floral buds also become weak and result in poor setting.

Control: The pest is controlled by spray of 0.025 % methyl demeton (200 ml metasystox 25 EC) or dimethoate 0.03% (200 ml Rogar 30 EC) in 200 litres of water 7-10 days before flowering. The spray should be repeated after 5 days.

- Peach short life syndrome (PSLS) is caused by ring nematode
- Splitting and gumming is due to prolonged dry period and sudden rain.
- Peach is very susceptible to Fe deficiency.
- Major problem: Sun Scald

60

Pear

- Pear/Gift of god
- B. N.: *Pyrus communis*
- Family: Rosaceae
- Sub-family: Pomoideae
- Chromosome No.: 2n = 2x = 34
- Origin: Western China
- Inflorescence: Corymbo
- Fruit type: Pome (With girt cell)
- Edible part: Fleshy thalamus

Introduction

Pears are next only to apples in importance, acreage, production and varietal diversity amongst the temperate fruits. The genus *Pyrus* probably originated in the mountainous regions of Western China. It is a rich source of carbohydrate as sugars, starch and cellulose and minerals like calcium (8 mg/100 g) phosphorous (15 mg/100 g) and sulphur (14 mg/100 g).

Vavilov (1951) three centres of Origin

1. **Chinese centre:** It is primary gene center comprising the regions of north and central China, Japan and Korea. The important species which originated in this track are *P. pyrifolia, P. ussuriensis, P betulifolia* and *P calleryana*.
2. **Central Asiatic centre:** Western Tian-shan, Uzbekistan, North-west India and Afghanistan are included under this centre. The indigenous *species P communis, P. paishia, P. salicifolia.* originated under this centre.
3. **Near Eastern centre:** Asia minor and Caucasus mountains are the tracks where great variability exists in *P. communis, P. syriaca and P. caucasica* are the important species found in this centre.

- Common pear/French pear/English pear/European pear *P. communis* Derived from *Pyrus causasia*, fruits are pyriform in shape without any grit cells
- Japanese pear/Oriental pear/Asian pea/Hard pear P. *pyrifolia* Derived from *Pyrus ussuriensis)* are roundish in shape with grit cells are pear species grown in India.
- **Chinese/Sand Pear *P. sinensis* commonly grown in North Indian plains**
- **Wild Pear/Kainth/Mahal (*P. pashia)*** *is* common rootstock in the Northern India
- **Snow pear *(P. nivalis)*** is suitable for perry preparation
- Willow leaf pear (*Pyrus salicifolia*) is frost, draught and salt resistant root stock
- Intergeneric sterile hybrids: Mule (Troth Early Peach x Wild goose Plum).
- Intergeneric sterile hybrids: Kamdesa (Peach x Sand Cherry),
- Intergeneric sterile hybrids: Pyronia (Pear x Quince).
- Classified into 4 groups on the basis of geographical distribution.
- European group: ***P. communis***, *P nivalis, P cordata, P. caucasica*
- Asian group*: P. calleryana*, P. *beautifolia, P. dimorbhophylla, P Koehnei*
- North Africa group: *P. longipes, P. memorensis, P. gharbiana*
- East Asian group: *P. pyrifolia, P. ussuriensis, P. hodoensis*

Area and Production

Both European and Asian pears were domesticated by hybridization and selection from local wild species in prehistoric times. Chinese and Japanese migrants took Asian pears to the USA as seed and scion material. After this, many varieties of pear were introduced from England, France, Italy and Japan. Pears in temperate areas of the state are mostly grown as intercrops in apple orchards. However, in recent years, some compact orchards of pear have been established in the Kullu and Shimla districts of Himachal Pradesh and Kashmir valley. Due to wide genetic variability, pears are cultivated in many temperate and tropical countries of the world. Uttrakhand and Jharkhand has the maximum area and production in India. In India, various English missionaries introduced pears along with other fruits some times during 1970 and planted them in Kullu, Shimla in Himachal Pradesh and Kumaon Hills in Uttarakhand.

Plant Description

A plant is a deciduous tree or shrub, with leaves serrated, crenate, or rarely lobed. Flowers are with or before leaves, in an umbel like raceme, white, rarely pinkish, with sepals reflexed or spreading. Petals suborbicular to broad oblong, stamen 20-30, anthers red to purple, style 25, free, closely constricted at base. Ovules, 2 per locule. It is more tolerant of wet soils but less tolerant of drought than apples. Italy is the largest producer of pears in the world. Browning of pears is due to polygalacturonase activity. Major acid: Malic acid, Flower colour: White, Perry wine prepared from pear. Pear genome sequencing was done by China, 2012: The size of the genome: 51Mb.

Climate and Soil

Pear can be grown in wide range of climatic conditions, as it can tolerate as low as –26°C temperature during dormant season and as high as 45°C during growth period. A large number of pear cultivars require about 1200 hours below 7°C during winter to complete their chilling requirements to flower and fruit satisfactorily. However, Bartlett needs about 1500 hours compared to other temperate pears. The pears of variety Patharnakh need only 150 hours of chilling and can also stand high temperature and hot winds during summer. The medium-chilling-required pear cultivars like Le Conte, Kieffer, and Gola perform well in areas experiencing mild winters. Spring frosts are detrimental to pear production and temperature at –3.3°C or below kill the open blossoms or flowers. As a result, low-lying areas should be avoided for planting. The north-eastern slopes are better for pear plantations because warmer aspects have a problem with spring frost, which kills the bloom. Pears grow best in deep, well-drained, fertile, medium-textured loamy soil. It is more tolerant of wet soils but less tolerant of drought than apples. Pears even do well on poorly aerated heavy soil with a high water table, which is heavy in texture for most deciduous fruits. A pH range of 6.0 to 7.5 is desirable because Fe deficiency appears in alkaline soils. For active growth of pears, soil depth should be 180 cm. Premature ripening due to night temperature < 7.1°C and day temperature < 21°C. Most pear cultivars require chilling for 1200 hrs. below 7°C. The chilling requirement of European pear is 1200-1500 hrs below 7°C.

Varieties

- **Pear varieties belongs to 2 groups:** European Pears and Asian Pears.

1. **European Pears:** Bartlett, Doynne du Comice, Max Red Bartlett, Red Bartlett. Starkrimson, Laxton"s Superb, Baggughosha, Clapp's Favourite, Anjou, Conference, Winter Nellis, Flame Beauty (self-fertile, good pollinizer), Fertility, Dr. jule, Guyot.

2. **Asian Pears:** Shinseiki, Chojuro, Kosui, China Pear, Nijisseiki, Kikisu, Yali, self-fruitful and suitable for processing purpose.

- Chilling requirement of *P. communis* is more than *P. pyrifolia.*
- Red Blush, Punjab Gold and Punjab Nectar are improved pear selections made in Punjab.
- Bud sports in pear include Max Red Bartlett from Bartlett in Belgium; Flemish beauty a chance seedling;
- Eldorado from Bartlett in California; Starkrimson from Clapp's Favourite and Red Anjou from Anjou.
- **Anjou: Originated in France.** It is fairly resistant to very low temperature and fire blight. Fruits have high dessert quality and very long keeping quality.
- **Flemish Beauty:** It is a self-fruitful variety. It can also be used a good pollenizing cultivar and free from grit cell.
- **Max Red Bartlett:** A bud mutant of Bartlett; plants and fruits resemble parent except that the fruit colour is dark cranberry red and shoots and leaves have a reddish tinge.
- **Moonglow'** and '**Magness**' are two fire blight resistant varieties evolved in USA, Flesh is free from grit cells.
- **Jorgonelle:** It is a European type with superior quality and adapted to South Indian Hills like Kodaikanal (warm winter conditions).
- **Starcrimson:** Trees are medium sized fairly upright and spreading, Fruits medium sized, oblong ovate, pyriform, dark red change to attractive Crimson red in cold storage. Flesh cream white, moderately juicy, aromatic, high in TSS, sweet with excellent eating quality. Early China Trees are upright and compact. Fruits round, small, greenish with red blush and very attractive. Asian types and Hybrids (common pears)
- **Kieffer:** It is well adapted to different climatic conditions and moderately resistant to fire blight. The fruit is brownish, gritty and hard. It is a self unfruitful variety.
- **Gola:** It is found to be suitable for lower altitude. Fruits are large, round and possess excellent keeping quality. Hence it is suitable for long distance transport.
- **Le Conte:** Suitable for lower are mid hills as its chilling requirement is low. But it is a blight susceptible variety.
- **Patharnakh:** (sand pear) self-pollinated, this is another low chilling variety. Tolerant to very high temperature and hot winds. Fruits have tough skin and hence suited for long distance transport.

- Pear exhibits vegetative parthenocarpy and produces seedless fruits.
- Waite and Magness are pollen sterile cultivars of pear.
- *P. communis* is susceptible to fire blight of pear *P. pyrifolia* is resistant.
- *P. communis* is resistant to oak root fungus *Armillaria mellea.*
- Most of the pear cultivars grown in hills are partially self-fruitful.
- Punjab Agricultural University (PAU): Red Blush, Punjab Gold, Punjab Nectar, Punjab Soft.
- Most popular varieties in North India: Nash and Patharnakh.
- Most popular varieties in temperate region: Bartlett.
- Most popular varieties in Tamil Nadu: Kieffer (Kodaikanal).
- Spontaneous mutation (Bud sports): Starkrimson, Clapp's Favourite(Red coloured pear).
- Red colour and flavoured variety: Starkrimson Delicious.
- Prabhat: Sharbati x Florda Sun (Early maturing hybrid).
- Interspecific hybrid: Le Conte and Kieffer European (*P. communis) x* Oriental pear (*Pyrus serotine).*
- Soft fleshed selection: Red Blush, Punjab Gold, Punjab Nectar.
- Fire blight of pear (Most serious disease) is caused by bacteria (*Erwinia amylovora).*
- Resistant to fire blight: *Pyrus calleryana.*
- Used rootstock in pear but susceptible to oak root fungus, fire bight and excess lime and cold.
- Double grafting/Double working is practiced in pear to graft Bartlett pear on Quince rootstock.
- **Low chilling varieties:** Kieffer, Le-Conte, Patharnakh, Gola, Punjab Nectar.
- **High chilling varieties:** Anjou, Bartlett, Conference, Flemish Beauty
- Graft incompatibility is overcome by double grafting with Old Home or Hardy varieties
- Barltlett variety is famous for dry pear slices.

Rootstock

1. Pear is commercially propagated through grafting on seedling or clonal rootstocks.
2. Seedling rootstocks are seedlings of Kainth (*P. pashia*) and Shiara (*P. serotina*).
3. In Punjab, root suckers of cultivated pear are also employed as a rootstock.

4. Clonal rootstocks are Quince A (Vigours), Quince B (Intermediate) and Quince C (Dwarf) and BA 29, OH x F-230 (Semi-dwarf), Oregon 211 (dwarfing), Oregon 260 have been recently released.
5. Commonly used stock pear in the Southern India: Country pear (*Pyrus pyrifolia).*
6. Most commonly used clonal rootstock: Quince A (*Cydonia oblonga),* origin South Europe.
7. Quince is commercially propagation by cuttings.
8. Pear & quince are mostly incompatible.
9. Old Home is the better interstock for compatible.
10. Vigorous rootstock in pear: *Pyrus pashia.*
11. *Pyrus* is differentiated from *Malus* due to the presence of gritt cells.
12. Lenticels are very prominent in pear fruit.

Propagation

Pears are commercially propagated by shield or "T" budding and also by whip and tongue grafting. Commercial propagation by budding or tongue grafting. The root-stocks A number of F1 hybrids of Pyrus communis, such as Old Home x Farmingdale, are multiplied clonally and used as rootstock. Quince (*Cydonia oblonga*) is another related genus which shows very good compatibility with a number of commercial cultivars and it is resistant to woolly aphids, nematodes but susceptible to Oak root fungus, fire blight, cold and excess of lime in soil. Quince clones are easily propagated by semi-hardwood and softwood cuttings. Some of the commercially used clones of quince are QA, QB, and QC. But commercial cultivars of pear like pear Bartlett, Bosc, and Eldorado need "Old Home" as interstock as they don't have compatibility with Quince. The chilled scions of Doyenne du Comice grafted at 1.0 m height on 1.0 m long shoots of Pant pear –18 stock were found to be the best for higher fruit yield and quality. For a crop on its own rootstock (pear), an initial spacing of 3 m x 2 m is given, which is changed to 6 m x 4 m after 4-5 years. For pear on quince, a planting distance of 3.5 m x 1.1 m is enough since quince has the effect of dwarfing the trees. Tongue grafting is a commercial method of pear propagation.

Propagation of Seedling Rootstock

For raising seedling rootstocks, seeds are extracted from fully mature fruits of Kainth and Shiara. For good germination, the seeds are stratified for 30–40 days in alternate layers of moist sand at a low temperature (4-5 °C) during December- January. The pre-stratified seeds are sown in the nursery beds at a distance of 6-10 cm seed to seed and 10-15 cm in rows at a depth of 3-4 cm. In very cold areas, the seeds are directly sown in the nursery beds during December. After sowing of the seeds in the nursery beds, the beds are mulched with grass mulch and light irrigation is given. When the seeds start germinating and attain 2-3-inch growth, the mulch material is removed. Regular weeding, hoeing and irrigation are required for the proper growth of the seedlings. About 80–90 % of seedlings are ready for grafting within a year.

Clonal Rootstocks

Clonal rootstocks are multiplied through mound or trench layering as discussed under apple. Raising rootstocks from root suckers on plains, the root suckers from pear trees are separated during October-December along with their roots. These suckers are planted in the nursery beds at 15-20 cm a part in rows 30 cm apart.

Own rooted: Hard-wood cuttings are taken during the dormant season from one-year-old shoots. Before planting, the cuttings are treated with 100 ppm IBA for 24 hours.

Propagation of scion: The seedling and clonal rootstocks are grafted with the tongue or cleft method of grafting during February. The clonal rootstock Quince is not compatible with most varieties, particularly the Bartlett group, so Beurre Hardy and Old Home are used as interstock. T-budding during April-May also gives good bud-take success.

Planting

The land selected for pear plantation should be cleared of all the old stems and roots of earlier growing trees and shrubs need to be removed. After cultivating the soil thoroughly, green manure crops are grown to improve the soil fertility. A planting plan is prepared by adopting a particular layout system before actual planting. The layout system depends on the plant density to be adopted and the topography of the land. Generally, square and rectangular systems of planting are followed on flat land, while terrace and contour systems are on sloppy land. Under high density, a hedge row and bed system are used for pear plantation. The planting distance ultimately depends upon the soil fertility, cultivars, rootstock, and training system. Planting is done during December-January. About one month before planting, pits of 1 x 1 x 1 m size are dug and filled with a mixture of soil, 50-60 kg FYM and one kg of single super phosphate. High density pear orchard: 1000-4000 trees/ha. Cultivars grown on hills are partially self-fruitful and require pollinizer planting every fourth tree in every 4th row as pollinisers are adequate.

Training and Pruning

Pears are usually trained according to a modified central leader system. Modified central leader system generally, a one-year-old whip without a single branch is planted in the dormant season. Immediately after transplanting, the tip of the plant is headed back to 60 to 75 cm above the ground level. In the following summer, most of the buds on the main branch will sprout the following summer. In order to develop a clean stem up to 30 cm from the ground, the sprouted buds are pinched off soon after their appearance. Three or four well-spaced buds projecting spirally around the trunk are retained. If summer pruning is not done, then 3-4 well-spaced primary branches having a wider crotch angle are selected during dormant pruning. The lowest branch should be selected at least 30 cm above the ground and the other branches spaced vertically 10–15 cm apart in spiral fashion. The branches emerging below 30 cm from ground level and other undesirable branches are pruned off. The selected branches are headed back to ¼ of their growth to a bud projecting in the outer direction. The leader is also headed back to 30 cm above the last branch. During the second dormant pruning, 2-3 well-spaced primary branches are selected on the leader. On the primary branches selected during the previous year, two secondary branches which are growing in an outward direction should be selected. The selected primary and secondary branches are headed back to 1/3 or ¼ of the growth. The third year training consists of thinning out of unwanted branches and heading back of desirable side branches. The central leader should be headed back to a bud or weak shoot, which will develop in the form of a side branch. By the fourth year, training should be completed. Pears are trained in a number of systems like pine shape, pyramid, spindle, palmette and trellis. Among these, the palette system and tatura trellis are found to be commercially superior. In a tatura trellis, the rows are oriented north-south. Pear is trained in the central leader system. Pear bears both on spurs and shoots.

Manures and Fertilizers

The fertilizer dose depends upon the soil fertility, type of soil, kind and age of trees, cultural practices, climate and crop load. The dose of manure and fertilizer should be determined on the basis of soil and leaf analysis. In an orchard of optimal fertility, nitrogen, phosphorus and potassium are applied in the ratio of 70:35:70g per year of tree age, which is stabilized after 10 years of age (700: 350: 700/g N: P: K per tree). These applications may be supplemented by FYM at the rate of 10 kg per year depending on the age of the tree, with a maximum of 100 kg for a 10-year-old tree. N, P and K are applied in the form of calcium ammonium nitrate, single super phosphate and muriate of potash, respectively. Since the response to phosphorus application is poor in phosphorus rich acidic hilly soils, it would be better to apply phosphorus every 2-3 years or reduce the dose to half annually in such conditions. In bearing trees, FYM along with phosphorus and potash should be applied during December-January. Nitrogen is applied in the months of February-March in two split doses. Where the irrigation facilities are available, half of N is applied 2-3 weeks before bud break and the second half dose one month after flowering. The fertilizers should be broadcast in the tree basins 30 cm away from the tree trunk to the canopy drip line and mixed well in the soil. A band application of nitrogen is preferred in high rainfall areas with steep slopes having small basins.

Use of Growth Regulators

1. Paclobutrazole (P_{333}) @ 500 to 1000 ppm restrict the vegetative growth, increased fruit set in pear. cv. Flamish Beauty.
2. GA_3 at 10-20 ppm applied 10-14 days after full increased fruit set, fruit retention but it reduced flower bud differentiation in the following year.
3. NAA 5-10 ppm control fruit drop in pear.

Irrigation

Most of the pear orchards in India are established in rain-fed sloppy areas where irrigation facilities are inadequate. During the summer months, there is very less and low rainfall, resulting in drought-like conditions. Soil moisture stress affects fruit growth and development. Therefore, irrigation is essential in pear orchards. The most critical period of water requirement in pear is April to June, and the peak requirement is after fruit set. Irrigation applied at 80 % of field capacity increased fruit set and yield, tree growth and reduced fruit drop in the Flamish Beauty cultivar. In lighter soils, more frequent irrigation and heavier application of irrigation water are required. During the summer months, trees should be irrigated at 8–10 day intervals, while after harvesting, irrigation at 20 day intervals is required till October. No irrigation is required during the winter months as plants under goes dormancy.

Maturity

For local consumption, fruits are picked at a slightly later stage because fruits hanging on trees make a considerable gain in size, weight, and overall fruit quality. Fruit weight may increase by up to 20 % in delayed picking. However, it reduces the storage life and the ease with which the stem can be separated from the spur by an upward twist is used as an index of maturity. Fully mature fruits are harvested while still firm and green for distant markets. The number of days from full bloom to maturity is fairly consistent in a region but varies between regions. In Washington (USA), harvest maturity for pear varieties varies from 110 to 115 days for Bartlett; 130–135 days for Bosc; and 145–150 days for Anjou. Besides this, the change in surface colour from green to yellow is also taken as a maturity index for harvesting. Generally, Bartlett pears are harvested at 19-pound pressure. After harvesting, pre-cooling treatments are given to fruits to remove field heat and arrest ripening. Pear fruits are picked individually by giving them a gentle twist rather than a direct pull. Harvesting should be done in 2–3 pickings at 3–4 day intervals rather than single picking. In India, no specific grading and packing standards have been fixed for pear, but the farmers grade pear fruits according to fruit size as under.

Harvesting

The pear exhibits vegetative parthenocarpy and produces seedless fruits. Harvesting for canning and the distant market: Fully mature but firm and green. Pears can be stored for 120–180 days at 1°C and 85 % relative humidity in cold storage. From a well-maintained orchard, a yield of 30–40 tonnes/ha/ year can be expected. Ripening can be accomplished by keeping it at 15 to 21 and 21 to 25°C and 80-85 % RH for 3-6 days depending on the cultivar. Most commercial cultivars require this post-harvest chilling treatment for proper ripening.

Disorders

1. Pear decline is caused by MLOs which is transmitted by pear psylla.
2. Free from storage disorders: Anjou.
3. High N_2 is not suitable because incidence of pear psylla and fire blight is more.
4. Premature ripening –pink coloration near blossom end. it is due to Night temperature < 7.1°C and day temperature < 21°C.
5. Boron deficiency: Corky tissue, calyx end rot and blossom blast, black end and cork spot is due to deficiency of calcium.
6. Calyx end rot and blossom blast are pear disorders due to boron deficiency, core break down or brown heart is due to abnormal cool season.
7. Hard end of pear is due to unfavorable water conditions.
8. European pear is highly susceptible to fire blight of pear than oriental pear.

61

Pecan Nut

- Pecan nut/Queen of nut
- B. N.: *Carya illinoinensis*
- Family: Juglandaceae
- Chromosome No.: 2n = 32
- Origin: North America (USA)
- Inflorescence: Catkin (male) and spike (Female)
- Fruit type: Nut
- Edible part: Fleshy thalamus

Introduction

The pecan is a valuable horticultural gift from North America to the world. In the USA, it is considered the "Queen of nuts" because of its value both as a wild and cultivated nut. As compared to other nut fruits, which have been grown since time immemorial, pecans have been under cultivation only for about one hundred years but have gained enormous popularity. The pecan occupies fifth rank among the leading tree nuts because of its excellent nutty flavor. Pecan nuts are a rich source of fat (72 g), protein (9 g), carbohydrates (15 g) and minerals. Pecan is usually taken roasted or salted to supplement normal diet. Almost 90 % of the nuts are sold shelled and the rest in shell. The nuts are commonly used to add aroma, flavor and crispness, a rich color, or to garnish a large number of dishes. However, most commonly, it is used in baking dishes and in ice cream. Pecan shell, a bay product, is also used to manufacture tannin, charcoal and abrasives in hand soap. The pecan tree has value as a timber too, due to its strength and hardiness. There is a great demand for its veneer and lumber in decorative paneling, fine furniture, and flooring, as well as in pallet manufacturing. good source of phosphoric acid. The most important nut fruit in the world, 5th in production.

Origin and Distribution

It is said to be native to North America, but it is also opined that it originated during the Cretaceous period in Canada and moved south in recent geological times. Spanish explorers found pecans growing along the Mississippi river and its tributaries. The native habitat of pecan ranges from Texas in the southwest to Alabama in the southeast, extending up to Southern Illinois. Besides the USA, the worldwide distribution of pecan is confined to Australia, Canada, Egypt, India, Israel, Mexico, Peru, Turkey and South Africa. In India, it is grown in Himachal Pradesh, Jammu and Kashmir, Uttarakhand and in the Nillgri hills, where it was introduced in the mid-thirties. In Himachal Pradesh, its plantation remained confined to the areas of Kangra, Mandi, Solan and Kullu districts, which constitute an approximately 700 ha area.

Climate and Soil

The climate requires a warm temperate. The climate is probably the main constraint on the expansion of pecan cultivation. It needs a warm temperate climate, meaning that the extremes of both temperate and subtropical climates are undesirable. The pecan requires a long frost-free period from the time when growth starts in the spring until the nut matures in the fall. It is influenced by the minimum temperature and the number of chilling hours during winter. An ideal climatic environment for pecan cultivation consists of a growing season of 2240–280 days, an average temperature above 26.7°C, heat unit accumulation (base 10°C) of 5000 degree days for 7 months, a nut growing period, and a mean temperature for the three coldest months between 7.2°C and 12.8°C with at least 400 hours of chilling. High humidity affects adversely the growth, fruiting and regularity of bearing in plants by preventing pollination and increasing the incidence of diseases on leaves and nuts. Hailstorms and winds of high velocity are harmful to trees. Pecan trees can be grown on varied soil conditions ranging from the sandy loam to clay loam and alluvial soil of rivers and small streams, but they all have certain common characteristics. The soil should be deep, loose, well drained and well aerated to a great depth. Soil that is deficient in any of these requisites may support young trees but fails when the trees attain greater age and size, with corresponding greater requirements. Pecan can be grown successfully in soils with pH levels ranging from 5 to 8.

Varieties

- In pecan nut size, shell thickness and cracking attributes are desirable for selecting a variety.
- Some important pecan varieties grown in country and other parts of the world are Mahan, Nellis, Stuart, Western Schley, Mohawk, Cheyenne, Chicksaw, Desirable.

- Desirable and Cheyenne varieties are suitable for high density planting.

Rootstocks and Propagation

- Seedlings of various cultivars like Burkett, Nellis and western Schley are generally used as rootstock for pecan nut as there is no clonal rootstock.
- The bitter pecan (*C. aquatica)* is adapted to poor drainage, flooding condition and low pH but gives low yield.
- For raising the seedlings rootstocks, the seeds are stratified at 4°C for 70-90 days.
- Cold stratification of the nuts for 70 days followed by soaking in GA_3 (500 ppm) for 48 hrs gives good germination and subsequent seedling growth.
- After stratification, the nuts are sown in well prepared nursery beds at a spacing of 15 cm from seed to seed and 20 cm apart in rows.
- After sowing of seeds, the nursery beds are mulched with 10 cm dry grass and light irrigation is given.
- As the seeds start germinating, mulch is removed and proper weeding, irrigation and hoeing is done at regular intervals so that seedlings attain graftable and buddable size in a year.
- Pecan is commercially propagated by budding and grafting onto seedling rootstocks
- The seedling rootstocks are budded with patch and annular method in July and tongue grafted in Feb - March.

Planting

The best time for planting pecans is during the dormant season, late winter or early spring, just before bud swelling. On flat land, the pecan nut is planted in a square system at 10-12 m spacing. On sloppy land, contour and terrace layouts of planting systems are adopted. The plant spacing depends on the cultivar and the fertility of the soil. The distance can be reduced to 8-10 m on soil with low fertility.

Training and Pruning

The training in pecan begins with the planting when 1/3 of the top is removed and branches are allowed to emerge as high as 1-1.5 m from the ground level. The pecan trees are trained in a central leader system and, from the second year onwards, the subsequent branches should be spaced spirally at 30–35 cm from one another. As the pecan trees advance in age, they become larger and more crowded, making pruning, spraying and harvesting operations difficult.

Pruning of pecan nut trees is, therefore, desirable but is often neglected. Once the framework is established, very little pruning is done to mature trees. Dried and broken and overcrowded branches are removed.

Aftercare

Pecan trees have a very long growing season and require proper management throughout to ensure good plant growth and productivity. Mulching the basins with dry grass helps in conserving the soil moisture, controlling weeds and adding organic matter through decomposition. Rats pose big problems for pecan plantations and require protection by baiting against rats, keeping basins free from weeds and cultivation regularly. Preemergence and contact herbicides are generally used to control weeds. Diuron and simazine (2.5 kg/ha) are widely used herbicides. Besides, paraquat at 1 kg/ha can also be used as a contact herbicide. Cover crops can also be grown during summer and winter, especially on hilly slopes, to check soil erosion and leaching of nutrients. Crops like soyabean, beans, cowpea and clover can also be grown in summer in young orchards before they come into bearing. Since pecan is planted at greater spacing and comes into bearing late, it is advisable to grow intercrops till it starts bearing. Besides the seasonal crops, filler trees such as peaches can also be grown as an intercrop in the orchard.

Pollination

The pecan is a monoecious tree. The staminate and distillate flowers are borne separately on the same tree and organized into catkins and spikes, respectively. The pecan has compound buds enclosing floral and mixed buds in separate bud scales but with a common outer scale. The floral bud develops into a catkin, whereas the mixed bud develops into either a vegetative shoot or a single pistillate inflorescence. There is a complete or less complete dichotomy in pecan, which often poses the main difficulty in pollination, especially in isolated plantings. Most pecan cultivars require cross-pollination for good fruit set. Wind is a pollinating agent which carries pollen for about 900 meters. For good fruit set in pecan, planting should include 3-4 cultivars evenly distributed throughout the orchard, or at least 10% pollinizer cultivars. Male and female flowers are borne in mixed flower buds. One major problem: biennial bearing. Pecans exhibit heterodichogamy (protogynous and protoandrous). rarely self-pollinated. Pecan bean mixed flower buds have female flowers (spike) and male flowers (catkins). Pecan shows a tendency to biennial bearing.

Manures and Fertilizers

Pecan responds to fertilizer applications very slowly and effects on growth and yield are observed after two to three years. Thus, the leaf analysis may not hold good in determining the nutritional status of the plant. Every pecan tree should be manured with 100 kg of farmyard every year in the month of December. In addition, apply 500 g N: P: K mixture (15:15:15) per year age of the tree up to 16 years. The full-bearing trees of 16 years and above should be given 8 kg of NPK fertilizer mixture every year. Pecan trees are prone to zinc deficiency, which can be corrected with foliar application of zinc sulphate @ 0.5 percent.

Irrigation

Adequate moisture is required to improve the growth and productivity of pecans. Irrigation improves kernel yield, nut weight and diameter, appearance and oil content. Pecans require adequate soil moisture all year and require irrigation even before shucking to reduce stickiness and viviparous nuts. The pecan trees are commonly irrigated through flood, basin, and drip irrigation methods at 6-7 day intervals during critical periods, i.e., flowering, fruit set, fruit and nut growth, and water requirements.

Maturity and Harvesting

The nuts are harvested when the husk or hull covering the shell becomes fairly loose. A single shake will bring down the bulk of the matured nuts, which can be collected on a plastic sheet. To enhance splitting, the hulled nuts may be dipped in water to moisten the shell and spread out in the sun to dry. One method of salting the split nuts is to boil them in a salt solution for a few minutes, then re-dry and store them. If stored in plastic bags, nuts will last for at least 4 to 6 weeks in the refrigerator. Frozen nuts will last for several months.

62

Persimmon

- Persimmon/National fruit of Japan /Persimmon/ Ebony tree/ Chines date plum
- B. N.: *Diospyros kaki*
- Family: Ebenacace
- Chromosome No.: 2n = 90
- Origin: China
- Inflorescence: Cymose solitary
- Fruit type: Berry
- Edible portion: Epicarp and mesocarp

Introduction

It was introduced from China into Japan in ancient times, where most of its improvement work has been done. The term "kaki" is Japanese. In the nineteen century, it was introduced from China to many other countries, namely in the USA, Brazil, Italy, New Zealand, Israel, Australia and South East Asia. In India, the European settlers introduced persimmon in hilly states, namely Himachal Pradesh and Jammu and Kashmir in 1921. In India, it is mostly cultivated in the Kullu valley, where it was introduced. At present, it is grown on a limited scale in Jammu and Kashmir, Himachal Pradesh and Uttarakhand. Ripe fruit is delicious, flesh is sweet and jelly-like. The entire fruit is edible except the seeds and calyx. Hexaploid fruit crops, deciduous and monoecious fruit crops. Fruits are highly astringent due to their tannin content. Flowers appear in spring on new growth. **Types of flowers: male, female and bisexual.** Cross-pollination, Mode of pollination: Insects.

Uses and Composition

Persimmons contain high quality vitamin C (40 mg/100 g) and more citric acid. Dry persimmon has four times more vitamin A. reducing sugar type in Hirantanenasi. High malic acid in Fuyu, and citric acid Hirantanenasi. It is rich in vitamin A (2710 IU/100 g).

Climate and Soil

Persimmon can be grown in a wide range of subtropical and warm-tempered climates. The areas of 1000 to 1500 m a MSL that have a moderate winter and a relatively mild summer are ideal for its cultivation. It is a deciduous tree and can tolerate a low temperature of -15°C during dormancy, but its chilling requirement is less than 200 hrs. Bud sprouting occurs early in the spring. These buds are damaged by the spring frost. During the maturation period, temperature is the most important factor for obtaining quality fruits. Non-astringent cultivars require warmer conditions for fruit maturation than the astringent types. Although persimmon can grow on a wide range of soil types, it thrives best on deep, well-drained loamy soil. A pH range of 6.5 to 7.5 is ideal for itsgrowing.

Varieties

- The fruit of Japanese persimmon shows wide variations in size, shape, and colour, and are broadly classified into two major groups.
- 1. Non astringent
- 2. Astringent group Varieties are also divided on their response to pollination.
- i) Non-astringent and pollination constant (PCMA)
- ii) Non-astringent and pollination variant (PVHA)- Fuyu, Jiro, Hana Fuyu, Suruga, Gosho, Wase fuyu, 20th century, Mastumoto
- iii) Astringent and pollination constant- - Eureka, Hachiya, Honan Red, Triumph, Saijo
- iv) Astringent and pollination variant - Chocolate, Gailey, Hayakume, Maru
- Indian Persimmon –*D. embryopteris*-Dai Dai Maru, Fuyu, Hachiya (Seedless)
- It more than 0.5 % of soluble tannins exist in persimmon they cannot be eaten
- Upright growing species is :(D. *Virginiana)* Varieties-Early Golden, Goldner, Ruby, Queen, Miller
- Cheongdobansi, Sago ku, Taruga, Hirantanashi have strong Parthenocarpy ability.
- Pollinizer cultivar are Fuyu, Jiro, Atago and Hamagosho
- Imperfect sweet persimmon –Nishmura wase

- PCA-Perfect astringent persimmon-Saijo, Atago
- PVA-Imperfect astringent persimmon-Hiratanshi, Tonewase
- The fruit of Japanese persimmon shows wide variations in size, shape, and colour, and are broadly classified into two major groups.
- Astringent cultivars: Hachiya, Fuji, Tonewase, Nightingale, Flat Seedless, Triumph, Hyakuma, Eureka, Honan Red, Saijo and Hiratanshi
- Sujika: Formation of white lines on the rind of the fruit
- Most Preferred fruit: PCNA type cultivars, Fuyu and Jiro
- Commercially grown in India and leading commercial cultivar in California: Hachiya

Rootstock

- ***D. kaki, D. lotus* and *D. virginiana* are used as a rootstock for persimmon**
- The seeds are extracted from fully ripened fruits during Late-October.
- The stratification of seeds for 60-90 days improved seed germination.
- The stratified seeds are sown in the nursery beds.
- The best seed germination is obtained at 28°C. The seeds at this temperature take about 2-3 weeks to germinate.
- Young seedlings usually take a year to be of suitable size for grafting.
- Temperature 8-11°C for 888 hours is sufficient to complete dormancy of persimmon.
- Japanese or Oriental Persimmon: *Diospyros kaki.*
- Inarching on its own rootstock on: Date plum (*D. lotus).*
- *Diospyros lotus* is used as the rootstock in India.
- D. *Virginiana* (American Persimmon/ American Plum) is used as rootstock for Japanese persimmon in Israel and USA. *D. Virginiana* is native place USA.

Propagation

The seeds are extracted from fully ripened fruits during Late-October. The stratification of seeds for 60-90 days improved seed germination. The best seed germination is obtained at 28°C. The seeds at this temperature take about 2-3 weeks to germinate. Propagation is done by grafting the scion on seedling rootstocks, using the tongue graft for smaller diameter stocks and the cleft and veneer grafting on larger stocks. Veneer grafting is generally more successful than budding and should be carried out in September with the start of sap movement. Tongue grafting is also done with a success rate of 60–65%.

Planting

In India, winter planting during January-February is recommended when trees are dormant. Planting after bud sprout generally results in poor survival. The trees are planted at a distance of 5.5–6.0 m. The pits are 1 m x 1 x 1 m. Well-rotted farmyard manure is mixed while filling the pits in the same manner as used for other deciduous trees. At the time of planting, the bud union should be kept 75 mm above the ground level. The planting density for persimmon depends on the cultivar, rootstock and soil type. Dwarfing cultivars (Jiro) can be closely planted at 5 m x 2.5 m (800 trees/ha), semi-dwarf cultivars (Fuyu) at 5 m x 3 m (660 trees/ha) and vigorous cultivars at 6 m x 4.5 m (370 trees/ha). Pollinizer interplanting is used at 10-15% of the total population.

Training and Pruning

The trees should be trained to form a low head. To achieve this, the plants are headed back 60 cm from the ground at the time of planting to develop a frame work of strong branches. The dwarf and semi-dwarf cultivars are best trained on a modified central leader system. The more vigorous, upright, narrow crotch angle types should be trained to a vase or palmate system. The pruning is done during the winter when the trees are dormant in January. Because *D. kaki* flowers are produced on current season wood, heavy pruning reduces crop setting by forcing excessive vegetative growth. Summer pruning of mature trees may thicken laterals and increase fruit size and colour. With more vigorous and less precocious cultivars, techniques such as cincturing and limb spreading may be beneficial to achieve higher and earlier fruit production. In the case of grown-up trees, practically no pruning is done except for removing weak, interfering, discarded or insect-damaged shoots and branches. Training system: Dwarf and semi-dwarf: modified central systems, Vigorous types: palmette or vase system.

Manures and Fertilizers

Persimmon does not require high fertilizer doses. An application of balance fertilizer (10:10:10) at a rate of one pound per inch of trunk diameter at ground level is recommended before bud-break.

Aftercare

After planting, young plants need continuous care for their survival. The plants need staking to keep them in a straight position, which helps in selecting the well-spaced laterals in the coming growing season. Watering at a 7–10 day interval is essential. The plant basins are kept free of weeds. During hot summers when the evapo-transpiration rate is high, the plants require mulching with dry grass/dry leaves. Mulching needs to be done in the first week of March. Flowering time: February, Fruiting time: August.

Irrigation

Irrigation is considered essential for the successful production of persimmon. Dry periods during fruit growth reduce the size, quality and number of fruits carried to maturity. Early summer moisture deficiency may cause fruit drop. The peak water requirement periods are in mid-summer. Due to high evaporation during the summer months, irrigation is very essential to maintain uniform soil moisture. The young and bearing trees are irrigated at 6-7 day intervals during the summer months.

Maturity Indices

Persimmon fruits are harvested when they have attained a yellow to reddish colour but are still firm. The fruits are clipped from the trees with shears, leaving the calyx attached to the fruit together with a short stem. More care is needed to avoid bruising. Persimmon fruits mature in mid-September, although the period of maturity varies among the different varieties. If fruits are harvested too early, they develop poor colour, sweetness and flavor. Fruits after harvesting should be wrapped individually in paper and packed in a single-layer crate. Persimmons soften at room temperature. Ripe persimmons are delicious. Flesh is sweet and jelly-like. The Brix level at maturity in different varieties varies between 14- 17.

Maturity

The persimmon trees start bearing fruit 4-5 years after planting. However, dwarf and semi-dwarf cultivars start bearing 2-3 years after planting. Mature trees of Fuyu are capable of producing 50 kg of fruit/plant. The Jiro cultivar has recorded over 80 kg/plant, whereas in the Hachiya, the yield is over 100 kg/plant. The decrease in astringency during growth and maturation of astringent cultivars and the disappearance of astringency from non-astringent varieties are most striking.

Post-Harvest Management

Various methods have been suggested to remove the astringency from the astringent cultivars. However, most of these result in partial softening of the fruits. Treatment of the fruit with carbon dioxide has been the most successfully developed technique till date. Dip the fruits for 2 minutes in 500 ppm Ethephon solution to remove the astringency in cultivar Hachiya, and the fruits are ready for consumption after 2-3 days of storage. The persimmons are also allowed to sweeten naturally after harvesting from the plants at room temperature, although they can be held in form conditions for 2-3 months at 30–32°C and 85–90% relative humidity. The average freezing point of flesh is 28.20°C. Persimmons are graded by size and quality. During grading and packing, handing should be kept to a minimum to avoid bruising. The most popular packages for persimmons are single-layer trays commonly used for stone fruits.

Major Problems

- "Cincturing" is the removal of a strip of bark from around the drunk of a tree.
- Tree is susceptible to attach by fungug (*Cephalosporium diospys*).
- Removal of astringency in persimmon fruits: Ethephone @ 500 ppm.
- Calyx cavity dehiscence is a serious problem.
- Line cracking Skin russeting is due to high RH.
- Calyx end cracking.
- Major problem in persimmon: Alternate bearing.

63

Pistachio Nut

- B. N.: *Pistachio vera* L.
- Family: Anacardiaceae
- Chromosome No.: 2n = 30
- Origin: Asia Minor
- Inflorescence: Panicle
- Fruit type: Drupe
- Edible part: Epicarp
- Chilling requirement: 500 hours below 7.2°C

Introduction

In India, pistachio cultivation is mainly grown for its edible nuts and it is a deciduous tree in the family Anacardiaceae. On the tree, pistachios grow in grape-like clusters, and the shell has the pistachio inside. The leaves of the pistachio tree are pinnate and composed of five individual oval leaflets. The pistachio tree flowers are small, brownish-green in colour, and lack petals. Pistachio trees can reach up to 20 feet in the wild but are generally smaller under cultivation. It is a famous dry fruit, widely loved by people in India.

Area and Uses

Pistachios with the highest production in India are grown in Jammu and Kashmir. Pistachio in mainly originated from central Asia and the Middle East country. Pistachio nuts have been used for ice decoration or other foods. In India, the pistachio production areas are Tamil Nadu, West Bengal, Punjab and Kashmir.

Climate and Soil

Weather conditions are very important for cultivating the pistachio crop. These nuts prefer a temperature level of the day that is above 36°C, and in the winter season, it is preferred to maintain a temperature level of 7°C. Pistachio trees do not grow well in high altitudes due to cold temperatures. Jammu and Kashmir is the best location for the cultivation of pistachio nuts. The well-drained deep soil, along with sandy loam, is the best soil for pistachio cultivation. These types of nut trees are quite drought tolerant, whereas some of the areas have high humidity. The high quality of the pistachio tree was produced when the soil pH level was kept between 7.0 and 7.8. High-alkalinity soil is good for pistachio cultivation.

Rootstocks

P. atlantica and *P. terebinthu*

Varieties

Kermen, Peter, Chiko, Red Allepo, Joley.

"Blanks" are fruits without kernels (Parthenocarpic fruit)

Propagation

Pistachio is propagated by budding on to a suitable pistachio rootstock. Rootstock varieties change widely between growing areas. Budding is carried out in the fall, with the budded tree being planted the same year on the following year depending on the seedling size.

Planting Time and Spacing

You can plant pistachio trees twice a year, in the months of June - July, just before the arrival of the monsoon. In the irrigated orchard, pistachio trees can be spaced at 6 x 6 meters. In arid regions where supplemental irrigation is not used, trees may be spaced 8–10 metres apart.

Harvesting and Yield

Pistachio trees take longer time to produce nuts these budded trees become ready to produce fruits for 5 years and will not reach maximum production until the 12th year of planting. The easy identification of maturity is when the hull separates easily from the shell. The skin changes from translucent to opaque, and the mesocarp becomes loosened from the shell. Yield of the pistachio nuts mainly depends on climate, cultivar and crop management practices. On an average a yield of 8 to 10 kg per fully matured tree.

64

Plums

- B. N.: *Prunus domestica* (European plum), *P. saliciana* (Japanese), *Prunus americana* (American plum), *Prunus simony* (North America)
- Family: Rosaceae
- Sub-family: Prunoideae
- Chromosome No.: 2n = 32
- European Plum: 2n = 16
- Japanese Plum: 2n = 48
- European plum: Caspian Sea
- Japanese plum: North America
- Inflorescence: Fasicle
- Fruit type: Drupe
- Edible part: Epicarp and mesocarp
- **Cilling requirement:** European plum- 1000-1200 hr below 7.2°C. Japanese plum- 700-1000 hr below 7.2°C

Introduction

Plum is an important temperate fruit, which is used as a fresh fruit as well as in preserved form. It is next to the peaches in economic importance amongst the stone fruits. It requires a certain period of chilling during winters to break dormancy, so it is cultivated in areas where winters are cool. Fruits are a rich source of minerals, vitamins, sugars and organic acids, in addition to protein, fat and carbohydrates. The types of plums with high sugar content above 18% are known as prunes which can be dried with pit. In India, plum was introduced by Alexander Coutts in 1870 in Himachal Pradesh. After evaluation, only Japanese plum has been recommended for commercial cultivation in the mid hills of the temperate region of the north-western Himalayas. Some low-chilling varieties of plum were also introduced at Punjab Agriculture University, Ludhiana.

Area and Production

Plums are grown in all the temperate zones of all the countries in the world. The main plum and prune producing countries in the world are China, Romania, Serbia, Chile, Iran, Africa, Algeria, Morocco. South Africa, North and Central America. In India, these fruits are grown on a commercial scale in the mid-hills of Uttarakhand, Himachal Pradesh, Jammu and Kashmir, and but on a limited scale in the north-eastern states. In Himachal Pradesh, plums are mainly grown in Solan, Sirmour, Shimla. Mandi districts. Low chill varieties are also grown in Punjab, Haryana and Eastern U.P.

Composition and Uses

Rich in minerals, vitamins, sugars, organic acids, protein, fat and Carbohydrates, jams, jellies, and marmalades. An excellent source of vitamin A, calcium, magnesium, iron, Pottasium Fruit TSS at time of maturity -12.5°Birx.

Morphological Characterization

Tree medium to large, upright growth and deciduous. Leaves alternate, serrate, sharp pointed, medium sized and glabrous. Flowers are produced three in a bud on a one-year shoot or on a spur. Flowers are perfect, solitary or raceme, with sepals 5, petals 5, usually white, stamens numerous, perigynous, pistil 1 with elongated styles, 2 ovules, and a fruit drupe that is usually single seeded. The plum tree grows 5 to 7 m and has greenish white flowers.

Climate and Soil

Plum requires varying types of climate and is grown from the subtropical plains to the temperate high hills. European plums require a temperate climate and are grown in high hills at elevations ranging from 1300-2000 MSL. Plums can be grown in areas where winters are cold and summers are hot. Cold, wet and windy weather during bloom is detrimental to good fruit set as spring frost injury causes damage to bloom. A northern slope is preferred, particularly for Japanese plums, which tend to delay the bloom period and thus avoid early frost injury. Plum requires 90–110 cm of well-distributed rainfall throughout the year. Prolonged drought during fruit growth and development and excessive rains during fruit maturity hamper fruit quality. Although plum can grow on a wide range of soils, deep, fertile and well-drained loam soils with a pH of 5.5-6.5 are the most suitable. Hard pan, water logging, and excessive salts should be avoided in the soil. Very heavy or light soils are not suitable. The Japanese plums do well on average soils that have a high pH. Japanese plum - frost sensitive (spring flowering); good if there is no spring frost.

Species

Cultivated plums belong to two species.

1 ***Prunus domestica* (European plum):** Most common plum, it is a hybrid of diploid myrobalan plum (*Prunus cerasifera*) and tetraploid black thorn (*Prunus spinosa*). Fruits are larger in size than Japanese plums. The fruit is oval or round, with a yellow or green ground colour and a red or blue skin. The cultivated varieties of European plum are classified into three main groups:

   a) **Prunes: -** Fruit is oval in shape with a bulging ventral side and is compressed bilaterally. They are blue or purple in colour and high in sugar content, which makes them suitable for drying without removal of the pit. Varieties include Italian prunes, Giant prunes, and President.

   b) **Reineclaude and green gauge plum: -** This is a hybrid of *Prunus domestica* with *Prunus insititia*. The fruit is greenish yellow in colour and round in shape having yellow skin and flesh. Important varieties are Golden Drop, Green gage, Golden transparent,

   c) **Lombard plum: -** The colour of fruit is purplish red. The varietiess are Lombard and Victoria.

2 ***Prunus salicina* (Japanese plum): -** Originated in China but introduced in Japan, from where it is disseminated around the world. The American plum is more vigorous, productive, precocious and resistant to diseases than the European plum. The fruits are large and heart-shaped with a pronounced apex. A few cultivars are oblate or round. (more popular in India).

3. ***Prunus insititia*:** Origin-Western Asia. This is a small-fruited European plum, hexaploid and grown wild in Europe and Western Asia. The plums of the species are known as damson and mirabelles. The fruits are small and purple (damson) or yellow (mirabelles). Plants are small and compact and form excellent hedge rows. European plum cultivars include California Blue, Washington, French Prunes, Early Italian, Stanley, Grand Duke, Victoria, Damson. Japanese plum Cultivar Beauty, Methley, Santa Rosa, Kelsey, Mariposa, Satsuma, Burbank Red Beaut, Fronteir

Prune: a plum with high sugar content, dried through fermentation. Dried plums are used for chutney preparation Wine and brandy from varieties of high sugar and sorbitol. Seed oil has a 40–50% cosmetic and medicinal value. dried whole without fermentation (high sugar). Production,

Varieties

1. *Prunus domestica* is originated through the amphi-diploidisation (diploidisation) of the hybrid between *P. cerasifera x P. spinosa.*
2. European plum cultivar: California Blue, Yellow Egg, Washington, French Prunes, Early Italian, Stonley, Grand Duke, Damson, Diamond, Tragedy, Starking Delicious and Imperial brooks.
3. In India Japanese plum are more popular compared to than European plum.
4. Japanese plum bears heavy crops and needs pruning fruit thinning.
5. Plums are available in the market from second week of May (Titron)to third week of July (Jamuri).
6. Plums can be dried successfully with or without stone are referred as prunes.
7. Santa Rosa is commercial cultivar which is Japanese plum hybrid. It is the result of cross between *P. salicina* as female parent with *P. simoni* and *P. americana* as pollen parent.
8. Burmosa, Frontier, Nubiana and Frier are imporoved cultivars of plum for hills.
9. Mariposa, Burbank, Aldorado, Kelsey, Formosa. Eldorado, Wickson, Larado and Burmosa are self-unfruitful cultivars of plum.
10. Titron, Alucha Black, Alu Bokhara, Kala Amritsari, Peshawari Kala, Alucha Yellow are low chilling cultivars of plum.
11. Japanese plum Cultivar: Methley, Santa Rosa, Kelsey, Satsuma, Burbank, Red Beauty, Frontier, Sweet Early, Titron, Beauty Aloocha, Elephant Heart and Mariposa are cultivars of Japanese plum.
12. Alu Bokhara- the fruit is large having yellow skin tinted with red colour Pulp is juicy and sweet.
13. Early-Methley, Kelsey. Sant Rosa, Beauty, Settler, Cloth of Gold Early Subza.
14. Mid-Season-Mariposa, Red Ace, Late Yellow, Grand Duke, Silver wilkson
15. Low hill- Alucha purple, Titron, Satluj purple. Alubukhara.

Interspecific Hybrids

1. Plumcot- Plum x Apricot, Plouts- Plum x Apricot x Plum,
2. Apriums- Apricot x Plum x Apricot, Santa Rosa- *P. salicina* x *P. simoni or P. Americana*

3. Introduction Western Himalayas: Santa Rosa, Sutlij Purple: Suitable for midhills of North
4. Subtropical plum varieties: Sultlej Purple, Kala, Mariposa
5. Beauty, Kelsey, Santarosa and Mariposa are self-fruitful cultivars and used as pollinisers.

Rootstocks

1. The plum grown mostly on plum rootstocks though seedling of peach, apricot and almond can be used.
2. In India plum is propagated mainly on seedlings of wild apricot (Chuli)
3. Clonal Rootstocks for plum: **Myrobalan B**, Myrobalan 29 C, Myrobalan – 2261, Mariana GF 8/1, Mariana 2624, Mariana 8-6. Mariana 9-25, **Pixy, St. Julien A,** GF-667 and GF 1246
4. Famous plum breeder: Burbank, California, USA

Propagation

Plums are propagated vegetatively by budding and grafting on rootstocks. These can also be propagated by hard wood cutting and by leafy soft wood cuttings under intermittent misting. Cuttings taken from hard and semi-hard wood are treated with IBA (2000-5000 ppm) for better rooting. For raising the seedlings, the seeds of wild apricot are stratified under alternate layers of moist sand for 45 to 50 days at a temperature of 3–5°C to break the rest. The stratified seeds are sown in nursery beds 6–10 cm deep in rows 25–30 cm apart with a distance of 10-15 cm from seed to seed. The seedling attains graftable size in a year. In Punjab, own rooted plants of Kala Amritsari are generally used for planting. Clonal rootstocks of plum are multiplied commercially by layering. It has been observed that most clonal rootstocks of plum are difficult to root. Application of 2500 ppm IBA to stool shoot helps to improve rooting. Clonal rootstocks are also propagated by hardwood cuttings. The hardwood cutting should be taken during the dormant season and dipped in a 2500 ppm IBA solution for 30 seconds. Then these cuttings are planted in mist propagation chambers for rooting.

Propagation of Scion

Seedlings as well as clonal rootstocks which are 0.8 to 1.2 cm in thickness are grafted in February with the tongue and cleft method of grafting The seedlings which do not attain graftable size in February should be budded with T-and chip methods of budding in June-july. For grafting, the scion wood should be collected from healthy, disease-free, true-to-variety trees during January. The scion wood is collected from one-year-old shoots. They are packed in moist sphagnum moss, which is properly labelled for variety. These bundles of scion wood are stored in cold storage or buried deep in the soil in a shady place till grafting is done. The best times for grafting these fruits are February at lower elevations and March at higher elevations. Chip budding can also be done in March, July and September.

Planting

Planting of plums is done in December-January when the plant is in dormant conditions. Before planting, the site of an orchard should be properly cleared of the bushes and weeds. It is advisable to plough the plantation area. In hilly areas, terraces should be kept inwards to facilitate soil conservation.The orchard area should be properly laid out about two months before planting. On sloppy land, the layout of an orchard should be done with a contour or terrace system, while on flat land, a square system is followed. The planting distance varies according to the varieties, rootstocks and fertility of the soil. Plums are generally planted at a distance of 6 m x 6 m. In high density plantations, the plum plants raised on semi-dwarfing clonal rootstocks should be planted at a 4 m x 4 m distance. Pits of 1 m x 1 m x 1 m size are dug in October – November. Pits should be filled with fertile top soil mixed with 40 kg of well rotten FYM and 1 kg of single superphosphate. To avoid any damage from insects, the pits are drenched with 10 litres of chlorpyriphos (4 ml/L) solution. The graft union should be kept 10-15 cm above the ground level at the time of planting to avoid any scion rooting. Young plants should be watered regularly and the basin area mulched with 15 cm of dry grass.

Pollination and Fruit Set

Most native American plums are self-fruitful and need pollinizers be interplanted in the orchard. The partial self-fruitful cultivars such as Santa Rosa and Beauty are also benefited in fruit set with the pollinizers. The European plum in view of the fruit set are categorized in to self-fruitful varieties in which 30% of the flowers set fruit and self-unfruitful varieties in which 1.5 to 2 % of the flowers set fruits which need pollinizer. For proper pollination and to obtain a good harvest, at least every third tree in every third row should be planted with at least two strong beehives per acre. A pollinizer branch should be grafted onto every tree if the cultivars produce little or no pollen. Pollinizer should be planted in every 3rd row as the 3rd tree. One bee colony per acre improves fruit set. Flower initiation Flower initiation in one season and flowering seen in the next year, flower development greatly affected by age of wood, position of bud on tree, temperature, water, CH_2O, N and others, PGR. Flowering takes place on one-year-old shoots and two-year-old spurs. In plum, self-incompatibility is mediated by multiple alleles.

Training and Pruning

Plums are generally trained on the open centre system or on a modified central leader system, depending upon the varietal growth habits. The varieties with spreading habits of growth, as in Japanese plum, should be trained to have an open centre system. In HP, an open centre system is followed irrespective of variety. However, in the plains, where plenty of sunlight is available, trees should be trained in the form of a modified central leader system with 4-5 scaffold branches. Training systems: Branching at 1 m height is practiced. The training system is determined by the type of growth habit; spreading type-open center; upright type-central modified leader; open center–more common in Europe in India, modified leaders are more common. Other systems Hedged rows of pyramids for mechanical farming Palmette Pruning Intensity differs with varieties of Japanese plum over bearing heavy pruning. Japanese plum varieties are mostly self-fruitful and require pollination. Plum is trained to a central leader system in the case of the spreading type and a modified central system for upright-growing cultivars (Santarosa, Stanley and Wickson). The open centre system works well with Japanese plums. The modified central system is most commonly followed in India. Plum's oldest training system is the open center.

Manures and Fertilizers

Plums require an adequate amount of nutrients for better growth and quality fruits. The application of manures and fertilizers is determined by soil fertility, soil type, topography, tree age, cultural practices, and crop load. The requirement for fertilizers varies from region to region. The farmyard manure, along with a full dose of P and K, should be applied during December and January. Half the dose of N is applied in the spring before flowering and the remaining half a month later.

Irrigation

Plums are mostly grown under rainfed conditions. However, in order to produce fruits of good fruit size and better quality, irrigation is essential. Various methods of irrigation are adopted to irrigate plum orchards, but in the hills, basin and drip irrigation methods are more popular and are widely used and recommended. After fruit setting, the plum trees are irrigated at weekly intervals, and 6 to 8 irrigations are recommended for higher production of quality fruits in the Santa Rosa plum. The peak water requirement period in plum is May to June. The most common floor management practice is clean basin + permanent sod in orchard.

Orchard and Weed Management

In plum orchards, sod culture and mulching of the tree basin area with hay mulch or black alkathene mulch is the most common method of orchard floor management. During the pre-bearing stage, intercropping with leguminous and vegetables is also done in orchards planted on flat and less sloppy land. Weed removal is done manually and is one of the practices employed in the orchards besides inter cultivation and cover crop growing. In the rainy seasons, the weeds in plum orchards are controlled with post-emergence sprays of glyphosate at the rate of 800 ml/hectare.

Crop Regulation and Quality Improvement

Generally plum tends to bear heavy crops and bear under sized fruits of low-quality, thinning therefore, is necessary to increase the fruit size and uniformity in colour of fruit and to stimulate flower initiation for the regulation of next year's crop. Various methods viz; hand, mechanically and use of chemicals are used but chemical thinning has superiority with respect to thinning cost, fruit size and quality. NAA at 20 to 40 ppm sprayed after petal fall resulted in good fruit thinning.

Maturity

Among various indices of maturity, the most commonly used are colour development, flesh firmness, 5.9 ± 0.45 kg, days from full bloom (94 ± 3), TSS 13.5-14.5 and TSS acidity ratio 1.2: 1.5. It is Climacteric fruit.

Harvesting and Yield

It has been observed that plums usually ripen unevenly over the tree. Fruits, therefore, are harvested in two or three pickings and it is very important to find out the exact stage of picking when they are mature. The fruits are harvested with the stalk intact, avoiding any skin injury. Fruits are very delicate and perishable, so picking baskets should be lined with soft material on the inner surface. Immediately after plucking, the fruits should be kept in the shade of the tree to remove field heat. Post-harvest management for distance markets, harvest in the early morning or in the afternoon and leave it for over-night cooling.

Grading and Packing

To obtain a high price in the market the grading of the fruits is done to have a uniform size and better quality. The packing and grading standards of the plum. The plum is graded into 3 grades.

Major Problems

- Major problem in tropical cultivars: Sun burn
- In plum heavy bearing is a problem.
- Fast track plum breeding started to reduce the juvenile period in temperate fruits.
- Transgenic early continuous flowering (ECF) gene induce early flowering.
- Deficiency of boron in plum result in misshapen fruits.

65

Strawberry

- Strawberry: *Fragaria x ananasa*
- Family: Rosaceae.
- Chromosome No. 2n = 56
- Origin: Man - made hybrid
- Fruit type: Eterio of achenes
- Edible part: Fleshy thalamus
- Cultivated varieties of strawberries are allo-octoploid

Introduction

The cultivated strawberry is one of the world's most important soft fruits. Fruit is technically known as an accessory fruit in that the fleshy part is derived not from the ovaries (achenes) but from the receptacle. Strawberries are a very nutritive fruit. It is a rich source of vitamins and minerals. Strawberry fruit contains about 0.5% pectin. The flavour of the fruit is due to the presence of volatile esters. Most of the fruits are consumed fresh and some are processed into jams, chutneys, squash, and other foodstuffs. It is used in large-scale ice cream making. The fruits are canned and shipped in frozen conditions to western countries. It is a fruit plant that is monoecious, has a short day, and grows quickly. Strawberry industries in America were started at the beginning of the 18th century, and today nearly 50% of the world's production is in Europe.

Area and Production

The cultivated strawberry (*Fragaria x ananassa*) was first tried in Europe in the early 18th century and represents the accidental cross of F. virginiana from eastern North America, which was noted for its fine flavor and *F. chiloensis* from Chile, which noted for its large size. The strawberry is commercially grown on a large scale in Greece, Uzbekistan, chile Among the European countries, France, Italy, Poland and Spain. In India, strawberries are commercially grown in the plains, mainly for fruit, and in the hills for raising planting material production. The major strawberry-growing states in India are Haryana,

Maharashtra (Mahabaleshwarm), Jammu and Kashmir, Punjab, Uttar Pradesh, Himachal Pradesh, Uttarakhand. The United States is the largest producer. Low perennial herbs, fruit crops for the kitchen garden. unique place among cultivated berry fruits. The fruit of the strawberry is a complete fruit with 98% of edible portions.

Morphological Description

The strawberry is a perennial herb, the petiole mostly long and channled above, the stipules adnate at the base of the petiole. Leaves are 3-foliate, with leaflets sharply dentate but entire at the more or less wedge-shaped base. The flowers are polygamo-dioecious, rarely hermaphrodite, the male flowers are larger, all 5-parted, and the central flower opens first. Calyx lobe form flat hypanthium, stamens about 20 or less, filament mostly shorter than receptacle, anthers oblong, receptacles roundish or conic bearing numerous pistils with lateral styles. The flowers are high in vitamin C and iron, and they are white in color. The edible portion of a strawberry is the succulent thalamus of the flower. Flavour of strawberry fruits due to ethyl esters *i.e.,* Ethyl butanoate and ethyl hexanoate. Strawberry is 1st fruit crop micropropagation studied. Strawberry genome sequenced in 2011 (*Fragaria vesca,* 2n = 14). Size of the genome 250Mb. Flowers are borne in small clusters and are white in colour. type of inflorescence: dichotomic raceme. Pollination method: both self and cross-pollination, pollinator honey bees

Species

- Artificial man made hybrid (*Fragaria ananasa)*: *F. chilonensis* x *F. virginiana.*
- The cultivated strawberry (*Fragaria* **x** *ananassa*) was first tried in Europe in the early 18th century and represents the accidental cross of *F. virginiana* **from eastern North-America, which was noted for its fine flavor and** *F. chiloensis* from Chile and noted for its large size.
- Progenitor of strawberry: *Fragaria chilonensis and fragaria virginiana (*American octoploid)
- Hermaphroditism is much more common in *Fragaria chilonensis*
- Related species *F. moschata, F. vesca* (wild Indian strawberry) and *F. viridis.*

Climate and Soil

Strawberries can be grown in all types of climates but thrive best in temperate climates. It is grown in temperate to sub-tropical areas. Some cultivars (Gorella, Chandler, Senga Sengana) are adapted to septentrional culture, *i.e.,* short days in autumn and a hard winter, or to meridional conditions, *i.e.,* long days in autumn and a moderate winter (cv. Tioga). The day-neutral cultivars (Selva and Fern) can be grown all year round. Strawberry requires a certain minimum amount of dark period (8 hours) for flower bud initiation, but better flowering and fruiting can be achieved if plants are chilled. Under normal conditions, flowers are initiated in autumn. A temperature of about 22–30°C is considered optimum for plant growth. In cold climates, frost damage and winter injury are major constraints for strawberry cultivation. Blossoms of most June bearing cvs are damaged at -2°C. The optimum temperature for the period between the onset of flowering and the onset of ripening should be 14 to 16°C. Strawberries thrive best in light soil as heavy soil inhibits root growth and development. Sandy loam soils rich in organic matter (4-5%) are best for strawberry cultivation. Most of the roots are found in the top 15-20 cm of soil. Therefore, it is essential that the top layer of soil be porous and rich in humus. Strawberry is not very sensitive to soil reaction, yet it prefers light acidic soils (pH 4.6 to 6.4).

Varieties

1. The varieties recommended for Indian conditions are: Chandler, Camarosa, Sweet Charlie, Ofran, Oso Grande, Gorella, Selva., Sanga Sangana, Belrubi, Shasta, Fern, Tioga, Pajaro, Premier, Red Coat, Local Jeolikot, Dilpasand, Bangalore, Florida-90, Katrain Sweet, Blackmore, Pusa Early Dwarf, Phenomenal, Majestic, Sujatha, Labell, taliman and torry.
2. Pajaro: most successful under summer system and tolerant to viruses
3. Chandler most popular acceptable variety in North Indian plains and resistant to viruses, resistant to physical damages caused by rain and suitable for fresh market and processing.
4. Hybrid: Arking- Cardinal X Ark-5431
5. Ideal varieties for processing- Midway, Midland, Hood, Red chief and Beauty
6. Suitable varieties for ice cream making- Olympus, Hood and Shuksan (High flavour and bright red colour)
7. The day neutral cultivars (Selva and Fern, Tribute, Brighton, Hecker, Muir,)

Propagation

Strawberry is commercially propagated by runners. Nowadays, tissue culture techniques have been adopted for mass multiplication. Propagation by seed is not suitable as the seedlings do not come true to type. Runners produced from stolons of established plants are commonly used for the propagation of strawberries. Although propagation by runners perpetuates the character of the parent, viral diseases are also transmitted through runners from infected mother plants. Therefore, virus-tested clonal material should be selected and planted at a 1.2 x 1.2 to 1.8 x 1.8 m distance on land where strawberries have not been grown during the past few years. For higher runner production, the inflorescence must be removed by hand in order to prevent fruiting. It is a short-day plant and is commercially propagated through one-year-old runners. Induction of runners' treatment by: IBA @ 100 ppm.

Planting

Land to be used for strawberry cultivation should be prepared by repeated ploughing and removal of weeds and stubble. The roots of strawberries are confined to the top 40 cm of the soil. Therefore, the 40 cm of soil surface should be fertile and friable. A good practice in preparing the soil for strawberry planting is to grow green manuring crops like clover, soyabean and plough them into the soil, which helps to increase organic matter. Manure and fertilizers should be mixed during land preparation. High level of phosphorus and potash are desirable for best fruit production. Strawberries can be planted on flat beds, in the form of hill rows or matted rows, or can be planted on raised beds of 4x4 meters. Strawberry is planted at a spacing of 25-30 cm plant to plant and 45-50 cm apart in rows from September to mid of October. Plants are to be set with their crowns at the soil surface. If the plant is set too shallow, the roots tend to dry out and the plant may die. Irrigation with sprinklers is given after planting. Runners produced from stolons of established are commonly used for propagation of strawberries. Application of GA3 at 50-100 ppm and at 60-120 ppm increases runner production. Strawberries can be planted on flat beds, in the form of hill rows or matted rows, or can be planted on raised beds of 4x4 meters. Strawberry is planted at a spacing of 25-30 cm plant to plant and 45-50 cm apart in rows from September to mid of October. Ideal time of planting runners or crown in hilly regions: September to October.

Manures and fertilizers

Strawberries require both organic and inorganic fertilizers for growth and fruit production. Apply 50-60 tones/ha of well rotten FYM at the time of preparation (pre-planting) along with 40 kg of P_2O_5 and 30 kg of K_2O/ha. Apply 80 kg of N/ha in two split doses. Half N should be applied after a month of planting and the remaining half before blossoming. Foliar application of urea (0.5-1.0 %) and micronutrient formulation (0.3%) should be applied 3–4 times before flowering, fruit initiation and fruit enlargement.

Irrigation

It is a shallow-rooted plant and grown in sandy loam soil, so it requires more frequent light irrigation. Excessive irrigation is, however, detrimental, which encourages growth of leaves and stolons at the expense of flowers and fruit and also increases the incidence of Botrytis rot. Irrigation is applied in furrows below the rows. The use of sprinklers and the drip method of irrigation are more convenient for irrigating strawberries. The results of various studies reveal that irrigation should be given at 100% ET with the drip method.

Mulching

After one month of planting, the plants are mulched with black polythene or grass mulch, which will protect the berries from direct contact with the soil. Mulching of strawberries in the fall has been found useful as it minimizes winter freezing injury to plants and suppresses early spring growth, thereby reducing fruit bud frost injury. Mulching also suppresses weed growth and reduces soil erosion. Poly tunnels can also be used to protect the strawberry fruit from frost in winter. Wheat straw and paddy straw can also be used as mulch materials. Black polythene mulch is very beneficial and is placed after planting by creating the holes at the point of the plant. Alkahene mulch can also be used over the beds to maintain a more favourable soil environment and to protect the plants from cold injury. Mulching is the most important cultural operation in strawberry production.

Weed Control

Weed control is one of the most serious and expensive management practices in strawberries. The weeds pose a great problem and sometimes become unmanageable, especially during the rainy season. Both chemical and hand weeding are used to control the weeds. Application of venzar (lenacil) at 1.5-2.0 kg/ha as pre-planting gives good control of weeds. Use of mulch also helps in controlling the weeds in strawberry plantations. Hand weeding should be done at regular intervals to keep the field free from weeds. There are 4 training systems, among which the matted row system is commercially followed in India.

Flowering

Flowering is controlled by photoperiod. Increasing the light intensity in greenhouse- forcing cvs, reducing the time of flowering, increasing the number of inflorescence and fruit set. The everbearing cvs (Ozark Beauty, Geneva, and Superfaction) initiate and produce flowers throughout the growing season. Day neutral strawberries (Tribute, Brighton, Hecker) have gained popularity in different geographical regions. Strawberry production may be limited by low temperature during autumn, winter, or spring in many temperate regions. Low temperature during bloom and in the fall and winter can destroy the flowers and buds. Flowering in short-day cvs. can be induced on long days so long as the temperature is sufficiently low. Some cultivars have specific temperature requirements for flower bud initiation. Purner et al. (1984) studied the effects of photoperiods and temperature on day neutral, June bearer (Red Chief and Guardian) and everbearer (Qurown and Ozark Beauty) strawberries. They were grown at a constant temperature of 21 °C 16-hour long days (LD), 9-hour short days (SD) or 9 hours with the dark period interrupted (NI) by 3 hours of low-level incandescent radiation. Flowering in day neutral was unaffected by photoperiod. In June, bearing flowering was inhibited with NI and LD compared with SD. Flowering of ever bearer was promoted by LD compared with SD. Pollen is carried.

Harvesting

Fruits are usually harvested when half to three-quarters of the skin develops colour. Harvesting stage: ½ to ¾ th of skin develops colour. A delay in picking increases the percentage of overripe and rotten berries. It is conventional practice to harvest the strawberries every third day. The ever bearing cvs (Ozark Beauty, Geneva, and Super faction) initiate and produce flowers throughout the growing season. Strawberry production may be limited by low temperature during autumn, winter, or spring in many temperate regions. Low temperature during bloom and in the fall and winter can destroy the flowers and buds.

Pests and diseases

i) **Red spider mite (*Tetranychus urticae*):** They are red in colour, present on the underside of the leaves. They are controlled with the spray of Chlorpyriphos (1ml/L) and acracide like dienochlor and amitraz.

ii) **Leaf roller:** *Eiphyas postvitlana* and *Ctenopsenstis obliquana* are common leaf roller, affecting strawberries.

1. The cater pillars fold the leaves across the midrib.
2. Two sprays of endosulfan (1ml/L) in August September controls the pest.
3. Light brown soft spot appear on fruits, the berries dry out and covered by a dusty fungus growth.
4. Physiologicalo disorder: Albinsm- Lack of fruit colour.
5. Serious disease of cultivated strawberry: Grey mould or Fruit rot *(Botrytis cinerea)*

66

Walnut

- Walnut/King of nut/Nut of god
- B. N.: *Juglans regia*
- Family: Juglandaceae
- Chromosome No.: 2n = 32
- Origin: Central Asia (Iran)
- Inflorescence: Catkin (male) and raceme (female)
- Fruit type: Nut
- Edible part: Lobed cotyledons

Introduction

The wild seedling of walnut has been found growing in west regions right from the Carpathian Mountains in Eastern Europe across Turkey, Iraq, Iran, Southern USSR and Afghanistan to the North-Western Himalayas. It is believed to have originated in Iran and the areas surrounding it. The Returning army of Alexander brought it to Europe from Iran. Almost all parts of the walnut tree are utilized. The fruit has excellent flavour and is mainly consumed as a dried fruit, eaten for table purposes. The kernels contain about 60–75% fat and oil, 16% carbohydrate, and 15% protein. Jammu and Kashmir is the largest walnut producing state in India and the USA is a leading producer of walnuts.

Area and Production

It is grown extensively in the USA, China, France, Italy, Turkey, Poland, Yugoslavia, Romania, and Iran. In India, it is grown in the states of Jammu and Kashmir Uttarakhand and Himachal Pradesh. The most valuable nut crop in terms of exchange value.

Climate and Soil

Walnut is one of the world's most important nut fruits, grown primarily in semi-cold climates. Walnut is grown successfully in all parts of the Himalayan region between the elevations of 1200 to 2150 MSL. The plants can tolerate as low as –11°C during deep dormancy without serious damage, but as soon as growth commences after dormancy, the temperature even 2 or 3°C below freezing point (0°C) kills leaves, shoots and flowers, thus resulting in crop failure. Like wise the plants which continue to grow till late during fall are subjected to serious foliage damage by frost. High temperature more than 38°C cause sunburn of hulls and shrivelling of kernels, resulting in blank nuts. The damage is further aggravated if the humidity is low and temperatures exceeds above 40°C. Persistently lower summer temperatures are not favourable for proper filling of the nuts. The cool growing season delayed the harvest. Like other temperate fruits, walnuts require certain amount of chilling period to break bud dormancy. In the absence of adequate chilling, the bud opening and blooming are irregular and delayed, resulting in a poor crop. The chilling requirement varies with the cultivars and ranges from 700 to 1500 hrs below 7°C. Walnut does well in a mild climate with moderate rainfall. An annual rainfall of about 80–90 cm, well distributed throughout the season, is considered sufficient for the cultivation of walnut. Walnut grows well on fertile, well-drained sandy loam or silt loam soil with a depth of 5–6 feet. The soil pH should be 5.5 to 6.5.

Varieties

The walnut plantations in India are of seedling origin, but some cultivars introduced from the USA and France are under test. The survey of walnut germplasm in the states of Jammu and Kashmir, Himachal Pradesh and Uttarakhand has resulted in identifying potential seedlings which have been released as cultivars.

Exotic varieties- Hartley, Payne, Franquette, Serr, Sunland, Chico, Vina, Howard, Chandler, Tehana, Tulari, Lara.

Local selections- Gobind, Pratap Roopa and Karan selection, Solding selection, Kotkhai Selection, Kashmir budded, Wilson Wonder, Chakrata Selection, Sulaman and Hamdam, Lake English, Eureka, placentia, Wilson.

Rootstock

Walnut seedlings of *Juglans regia* are generally used as a rootstock for walnut in all over the world, however, in USA seeds of *J. hindsi*, Paradox (*J. hindsii x J. regia*) are used for raising seedling rootstocks. Traditionally, walnut in India is propagated by seed. Although sexual method is the most efficient and economical but not satisfactory, because the seedlings are highly variable, having long juvenile period, producing the nuts with inferior quality and with low shelling percentage. In India, Persian walnut known as **akhrot** is most common source of seed. The seeds are stratified for 90-110 days at low temperature 4-5°C to break the seed dormancy. Various pre-sowing treatments with growth regulators, scarification and water soaking etc. not only reduce the stratification requirement and improve seed germination but also enhance seedling growth. Soaking of seed in 750- 1000 ppm GA_3 or 1000 ppm etherel solution for 24 hours after stratification is very effective in stimulating seed germination. The stratified seeds are sown in line 30 cm apart at a distance of 15-20 cm and at a depth of 10-15 cm in Feburary – March. The nursery beds are covered with mulching and regular irrigation and weeding is done so that seedling become graftable or buddable in one year.

Propagation

Walnut is commercially propagated by patch budding. The seedlings are grafted with the tongue method during February in mid-hill and in March in high-hill conditions. Side veneer grafting gives best results when performed during July- August. Walnut is very difficult to propagate and gives low success with grafting. Budding with the Chip method in mid-May to the first week of June, and annular budding and patch budding performed at the end of June to mid-July gave 80-85% bud-take success.

Planting

Planting distance and density varies with the soil, rootstock, topography and nature of bearing habit of varieties. On flat land, square or rectangular system and on sloppy land contour and terrace systems of layout are adopted. The planting can be done any time from December to March but the establishment of plants is better in early planting. In general, the planting is done at a spacing of 10–12 metres for grafted plants and for seedling plants, a spacing of 15–15 metres is recommended. However, lateral-bearing varieties raised on clonal rootstocks like Paradox are planted at a spacing of 8 x 8 metres.

Training and Pruning

Walnut trees are mainly trained according to a modified leader system. After planting, the plants are headed back to 60–70 cm above the graft union. The most vigorous upright limbs are selected after one year of growth, removing the other shoots but leaving one or two small lower limbs on the south-west side to protect the tree from sunburn. The first main frame work branch should be 4-5 feet from the ground. The framework of the tree is made in such a way that branches are spaced widely, both vertically and horizontally. To reduce fruit and increase vigorous shoot growth throughout the tree periphery, a large number of new shoots should be headed back to the periphery in the cultivar that fruits laterally. In contrast, on the terminal-bearing cultivars, pruning consists of heading back selected framework branches and thinning out competing limbs. Shading becomes a problem in a mature tree, so thinning out of limbs at the top and sides of the tree should be done. The best method of pruning a tree of any age below 16–35 years old is to cut back to 4-5-year-old shoots.

Manures and Fertilizers

The walnut is grown as scattered trees and not in the form of a well laid out regular plantation. As such, they are not fertilized. There is virtually no information available on the fertilizer requirements. However, the walnut trees should be fertilized with ½ kg of 15:15:15 NPK fertilizer mixture/year/tree up to 16 years' age. The fertilizers dose is stabilized after 16 years and a full-bearing tree should be fertilized with 8 kg of NPK mixture along with 100 kg of FYM. The full dose of manure and fertilizer is applied during December – January.

Orchard Soil Management

Sod plus clean basin management is the best orchard floor management in walnut. The basin area can also be mulched with dry grass to conserve moisture and control weeds. Walnut plants exhibit allopathic effects and do not allow weeds or other vegetation beneath them.

Irrigation

Walnut trees require adequate moisture, especially for the 5–6 weeks immediately following bloom. A water deficit prior to shell hardening results in small nuts and in mid-summer results in a stick-tight hull. The critical period for water requirements is from full bloom to shell cracking. Therefore, walnut trees should be irrigated at weekly intervals during the summer months for higher fruit production.

Pollination

Winds play a significant role in pollination and fruit set. Pollen of the *Juglans regia* variety is capable of fertilizing the eggs of the same variety and any other variety as well. In walnut, dichogamy is the main problem, which is further aggravated by the short period of both pollen viability and stigma receptivity. However, pollen germination on stigma occurs 1-3 days after pollination. Pollen viability last for 100 hours at 14-15°C and 55 hours at 22-25°C, however above 25°C and humidity below 33% during pollination results in pollen sterility. Adverse weather conditions like snowfall, hail, frost and rain during flowering result in crop failure due to inadequate pollination. Improvement in pollination and fruit set can be made by (1) hanging of catkins of male flowers and (2) planting of 3–4 varieties in the orchard. Catkins (inflorescence carrying male flowers) are born terminally and female inflorescence laterally.

Maturity

Harvesting at the proper maturity and subsequent handling of the produce are of prime importance. Any delay in picking after the maturity of kernels deteriorates the quality and increases the incidence of mould and pests. The maturity indices are commonly used for determining the harvesting dates of walnut are cracking of hulls from nut and change in colour of packing tissues between the When about 80% of the hulls have cracked from the nuts, it is a time of walnut picking. Secondly, the proper maturity is also assessed by observing the packing tissue between and around the kernel halves, which turn brown on maturity.

Harvesting

Depending upon the climate, the harvesting is usually done after two weeks of browning. The kernel becomes dark if the further harvesting is delayed. kernels. The spray of ethephon 200-500 on kernel maturity advances harvesting by 10 days, makes harvest of an entire crop at a time and promotes hull splitting. Walnuts are normally harvested from August to October. Maturity is earlier in lower altitudes but late at higher altitudes. Walnuts drop naturally over about a one-month period after the splitting of the hull in a natural way, while others are forced to drop down by splinting with long poles. The nuts should be gathered, hulled and dried immediately. The fallen nut is piled high in the orchards and covered with wet leaves and grass to ferment, which takes about 10-15 days. Dehulling, washing and cleaning of nuts are done manually in India and mechanically in other advanced countries. Application of nuts in ethephon at 1500 ppm when the hull is still on will loosen the hull in 3–4 days. Harvesting: PTB stage (packing tissue turns brown). Reduction of walnut hull dehiscence: Ethephone @ 2000ppm.

Post-Harvest Management

After cleaning, the nuts are dried with the help of a drier at 32-30°C temperature till the moisture content in the nuts is about 8%. The nuts can be dried in the sun for many days by spreading them on a canvas sheet or hand floor. After drying, the nuts are started in the gunny bags in small ventilated rooms. Blank nuts are due to the hot summer with low humidity.

Section 4
Plantation Crops

67

Plantation Crops

Plantation crops: Plantation crop are a group of commercial crops that are perennial in nature, cultivated extensively in tropical and subtropical situations that need employment of labour throughout the year and whose products are usually consumed after processing. ex. Tea, Coffee, Coconut, Rubber, Cocoa, Areca Nut, and Oil Palm.

Important Research Stations on Plantation and Beverage Crops

1. **Coconut :** Central Plantation Crop Reserch Institute , Kasargode, Kerala
2. **Arecanut:** CPCRI, Regional Research Station, Vittal, Karnataka
3. **Cocoa:** CPCRI, Regional Research Station, Vittal, Karnataka
4. **Rubber**: Rubber Research Institute of India, Kottayam, Kerala
5. **Cashew:** Directorate of Cashew Research, Puttur, Karnataka
6. **Oil Palm:** National Research Centre, Elur, Pedavegi, Andhra Pradesh
7. **Palmyrah:** Srivaliputtur, Tamil Nadu
8. **Tea:** Tea Research Institute, United Planters Association of South India (UPASI), Valparai, Tamil Nadu
9. **Coffee:** Central Coffee Research Institute, Balehonnur, Karnataka

68

Arecanut

- Arecanut/Betelnut/Supari
- B. N.: *Areca catechu*
- Family: Palmae / Arecaceae
- Chromosome No: 2n = 32
- Origin: Malayan /Indonesia
- Inflorescence: Spadix
- Fruit type: Drupe
- Edible part: Endosperm

Introduction

The Arecanut palm is cultivated primarily for its kernel obtained from the fruit, which is chewed when it is tender, ripe or processed. Despite being native to the Malayan Archipelago, the Philippines, and other East Indian Islands.Only in India is it under scientific cultivation. India is the world's largest producer of arecanut, contributing nearly 74 % of world production. India is the world's largest consumer of arecanut (accounting to 89 % of world production). Other uses for arecanut other than chewing are negligible. Its export prospects are also very limited. Therefore, the present policy is not to expand the area under arecanut but to adopt intensive cultivation and take up replanting of the aged and unproductive gardens. Inter and mixed cropping in arecanut gardens is advocated to augment the income from the existing arecanut groves. Alkaloids present in arecanut, **Arecoline** is the main and physiologically most active alkaloid present to an extent of 0.07 to 0.1 per cent.

Climate and Soil

Arecanut cultivation is predominant in gravelly laterite soils of the red clay type of southern Kerala and coastal Karnataka. Arecanut needs deep and well-drained soil, preferably not less than 2 meters, for the development of its root system. Laterite, red loam and alluvial soils are the most suitable. It is cultivated in fertile clay loam soils in the plain region or Maidan in Karnataka. In areas where tank irrigation is common practice, the soils may have a mixture of tank silt. Rich palm growth was supported by deep black fertile clay loam soils. The temperature should be a minimum of 4°C (at Mohitnagar in West Bengal) and a maximum of 40°C (at Vittal in Karnataka and Kannara in Kerala). However, the palm flourishes well within a temperature range of 14°C to 36°C. Arecanut requires abundant and well-distributed rainfall. It grows well in tracts of very high rainfall, where annual showers may go up to or even more than 4500 mm. But it also survives in low rainfall areas having 750 mm of annual precipitation. During prolonged dry spells, palms should be irrigated.

Varieties

- Mangla- Introduction from China (VTL-3), Semi-tall variety
- Sumangla- Selection from Indonesia (VTL-11), Tolerant to burrowing nematode (Rad)
- Sreemangla- Selection from Singapore (VTL-17), Tolerant to burrowing nematode
- Sreevardhani- Indigenous to Maharashtra, commercially grown in coastal Maharashtra
- Sarwamangala- Mohitnagar x HD hybrid, High yielding dwarf variety ***Areca catechu x A triandra***
- Mohitnagar- Indigenous to West Bengal, Popular in West Bengal
- SAS-1- (Sirsi) Arecanut Selection-1, Suitable for processing, Popular in hill regions of Karnataka
- CAL-7(Calicut-7) Popular in Andaman and Nicobar Islands
- VTLAH-1(Hirehali dwarf x Mohitnagar) Arecanut hybrid dwarf variety
- Shreewardhanee- Selection from Shriwardhan Local
- Kahikuchi- Assam
- Samrudi-
- Thrithahalli and South Kanara- Malnad area Karnataka
- **Kahikuchi Tall:** A high yielding variety, Kahikuchi (VTL-64) for Assam and NEH region
- **South Kanaralocal/Kasargod Local:** It is largely grown in Dakshina Kannada district of Karnataka and Kasaragod district of Kerala.
- **Calicut-17**: The striking features of this cultivar is its consistent and high yielding potential
- **Shriwardhan:** Raigad and Rathnagiri districts of Maharashtra and Coastal Karnataka.
- **Sagar:** It is mainly grown in Uttara Kannada and Shimoga districts of Karnataka.

- **Thirthahalli:** It is grown extensively in Malnad areas of Chikmagalur and Shimoga
- **Vittal Areca Hybrid-1:** It is released for coastal Karnataka and Kerala.
- **Vittal Areca Hybrid-2:** It is released for coastal Karnataka and Kerala. It was released by CPCRI.

Propagation

Being a perennial, seed-propagated crop, much care has to be bestowed on the production and selection of planting materials. The different stages in the production of quality planting materials include the selection of mother palm, seed nuts, seedlings, and the adoption of proper nursery techniques.

Selection of Seed Plantation

The plantation should be disease free and have a higher proportion of regularly high-yielding palms.

Selection of Mother Palms

One of the established methods of selection is to select palms possessing characters of high heritability. Earlyness in bearing and percentage of fruit set of the several characters of mother palm studied, age at first bearing and percentage of fruit set were found to be correlated with yield and are having high heritability. The heritability of age at first bearing is 72%, and the genotypic correlation with yield (number of nuts) is -0.55. Similarly, the percentage of nuts set has a 0.88 genotypic correlation with yield and 33 per cent heritability. Hence, it is suggested to select palms that are early bearers and have a high percentage of fruit set.

Selection of Seed Nuts

Heavier nuts (> 35 g fresh nut weight) that are fully ripe are selected. Undersized and malformed nuts should be rejected. Heavier nuts in a bunch not only give a higher percent of germination but also produce vigorous seedlings of better quality. More than 25 per cent of nuts in a bunch are light in weight.

Selection of Seedling

Twelve to eighteen-month old seedling are transplanted in the main field. Nuts germinate in 53 to 94 days (2-3 months) after sowing.

Planting Time

May–September, in clay soils subject to water logging in heavy rainfall areas.

Spacing

Different methods of planting, viz., square, rectangular and quincunx are in vogue. a spacing of 2.7 m x 2.7 m. Arecanut needs adequate protection from the Southwestern sun (October to January) to minimize sun scorching of the stem. Growing of tall and quick-growing evergreen perennial shade trees on the western and southern borders. E.g. *Acacia auriculiformis*, Jack, Kokum, neem, Casurina, Pongamia etc.

Manures and Fertilizers

FYM/Compost/Green leaf/Cattle manure 20 kg per palm per year. Recommended dose 100 g N: 40 g P_2O_5 140 g K_2O per palm per year applied in two equal splits *i.e.,* Pre monsoon May- June, Post monsoon September – October.

Irrigation and Drainage

The Arecanut palm is very sensitive to drought. The palm should be irrigated during hot and dry weather at 4–7 day intervals at 175 litres water/palm depending on the soil type. When there is a shortage of water, use drip irrigation. Application of organic mulch around the base of the palm helps to conserve soil moisture. Proper drainage should be ensured by constructing a drainage channel 25–30 cm deep for every 2 rows of palms. They are to be cleared at the beginning of the monsoon every year.

Harvesting

In arecanut the stage of maturity of nuts at harvest varies depending on the type of produce that is to be prepared. Based on the method of processing, there are two types of arecanut; Kottapak is primarily consumed in North India in the Gutkha industry. Kali pak consumed mainly in South India, Ripe nut harvesting for chali/ Kottapak preparation. From blooming to maturity, it takes 9 to 11 months and these nuts are harvested (usually from November to March), sundried, and dehusked to prepare chali. Ripe nuts, *i.e.,* when green nuts turn orange-yellow and the husk becomes soft.

Yield

Depending on the cultivar, about 2 kg of chali can be obtained per palm. Kottapak is consumed mainly in North India in the Gutkha industry, while Kali pak is consumed mainly in South India.

Processing

Processing of ripe nuts in arecanut (Kottapak) and chali (dried ripe nut): It is a dried ripe whole nut. In chali preparation, only ripe nuts are harvested. The outturn of Patora or Koka (lower grade produce) will be more if unripe nuts are harvested, which will fetch a low price in the market. Fully ripe nuts are harvested (more than 9 months' stage of maturity depending on the agro-climatic conditions) from November to February and are sundried for about 40 to 45 days. Dried arecanuts are dehusked. Proper drying of the nuts is important to prevent fungal infection of the nuts in the drying yard. Chali preparation is done mainly in Kerala, Karnataka, Assam and Maharastra.

Characters of Good Kottapak

Absence of immature nuts,

Absence of surface cracking

Absence of husk sticking

Free from fungus and insect attack: Inadequate/improper drying leads to fungal infection and poor quality produce.

Good cutting feel, inside structure and taste

Processing

Nuts are harvested at 6 to 7 months' stage of maturity. At this stage, the outer skin of the husk is dark green, the nuts are soft and finger nails can be pressed into them.

Dehusking

Separation of husk from kernels cutting soft nuts into pieces.

Boiling

Cut pieces of nuts are boiled in water in a container till the water becomes a thick syrup. We can also use a diluted extract from the previous batch of boiling.

Kali or Chogaru Coating

Kali is the extract obtained during the boiling of tender nuts. Usually the same water is used for boiling 3 to 4 times. At this stage, the water becomes concentrated, called kali. After boiling, the arecanuts are coated with kali (to improve color). The Kali coating is repeated to get a glossy appearance. Kali contains many polyphenols.

Drying: It can be dried in the sun or in an oven after draining the adhering liquid.

Colouring

Faulty drying, exposure to rain, or delay in boiling results in bleatched appearance to nuts due to lack of proper colouring. A thick syrup of kali or chogaur is used to colour these dried nuts.

A. Pests

1. Mites (Red and white)
2. Spindle bug (*Carvalhoia arecae*): This is a serious pest multiplying rapidly with close of monsoon.
3. Root grub (*Leucopholis lepidophora*): The grubs attack and feed on roots of both young and old palms.
4. Tender nut drop
5. Thrips

B. Diseases

1. **Koleroga, mahali or fruit rot**: Caused by *Phytophthora palmivora*
2. **Anabe roga/Foot rot:** Caused by **Ganoderma**
3. **Yellow Leaf Disease:** Caused by Phytoplasma and spread through plant hopper
4. **Leaf blight:**
5. **Bud rot:** Caused by *Phytophthora palmivora*
6. **Inflorescence die back and button shedding :**

C. Physiological Disorders

1. Sunscorch/stem breaking
2. **Nut splitting: Boron deficiency**

69

Betelvine

- B. N.: *Piper betle*
- Family – Arecaceae
- Chromosome No.: 2n = 26, 32
- Origin: Central and Eastern Malaysia
- It is a dioecious crop, Vitamin A -9600
- It is trained on Bareja Boroja

Introduction

The betelvine is a perennial dioecious climber and is believed to have originated in Central and Eastern Malaysia. It is mainly cultivated in South East Asian countries, viz., India, Bangladesh, Sri Lanka, Malaysia, Thailand, Philippines, etc. It is one of the important commercial crops grown mostly by small and marginal farmers in the states of Assam, West Bengal, Bihar, Andhra Pradesh, Karnataka, Tamil Nadu, Maharashtra, Madhya Pradesh, Kerala, Uttar Pradesh and Orissa with an estimated area of 53,539 ha (Ray, 2008). Annual production is worth Rs. 9000 million and estimated that about 20 million people earn their livelihood directly or indirectly partly or fully from production, processing, handling, transportation and marketing of betel leaves in India. Betel leaves worth 30-40 million rupees are exported to countries like Bahrain, Canada, Great Britain, Nepal, Pakistan, Saudi Arabia, and many other European countries.

Climate and Soil

It is successfully grown in a temperature range of 18°C to 40°C and a rainfall of 1250–4750 mm. It comes up well at an elevation of up to 1500 MSL. It requires partial shade for good growth. Clay loam soil rich in organic matter and having adequate drainage is well suited for its cultivation.

Propagation

Terminal cuttings having 3-5 nodes are most suitable for planting. It can also be raised in polybags and later shifted to the main field. If there is a scarcity of planting material, single-node cuttings with a mother leaf can also be used.

Training and Pruning

One month after the planting, young shoots start appearing and these young shoots are trained along the support and tied with them using banana fibre or jute fibre once every 15 to 20 days.

Manures and Fertilizers

Oilseed cakes like castor cake, linseed cake, sesamum cake or neem cake are applied as manure @ 15 Q/ha. The cake is first soaked in water in a big earthen pot for a week. Then this slurry is applied at frequent intervals. Oil cakes in powder form are also applied in the rainy season. Nitrogen @ 200kg/ha/year as farmyard manure or oil cakes should be applied. A dose of 100 kg each of P_2O_5 and K_2O/ha/year is also recommended. The fertilizers and insecticides should be applied in 4–6 split doses at 2–3 month intervals.

Irrigation

It grows well under moderate temperature and humid conditions. Hence, frequent irrigation is very essential to maintain high moisture in the soil. The frequency of irrigation depends on rainfall in the rainy season. However, regular irrigation is given at a 5-7-day interval during the summer months. Farmers are using drip irrigation to save irrigation water nowadays. Crops are very sensitive to water logging and hence, regular drainage channels are to be provided to drain out the excess water.

Harvest

Leaves are harvested with a portion of petiole by hand or with an iron nail specially made for harvesting betel leaves. Leaves are plucked from side shoots (laterals). Harvesting is done at 30-45 day intervals, and 8 to 9 harvests are taken in a year.

Yield

Generally, leaves are harvested 12–18 months after planting. About 20-30 lakh leaves/ha can be obtained per year.

Pests and Diseases

Important pests and diseases noticed in Betelvine garden are as follows.

1. **Leaf and Foot rot** (*Phytophthora capsici*): It is a serious disease during monsoon period. Drenching the soil with 1% Bordeaux mixture and foliar application at 0.5% is followed for control of foot rot of Betelvine.
2. **Anthracnose** (*Colletotrichum capsici* Syd): Foliar application of 0.5% Bordeaux mixture or 0.2 % copper oxychloride to control this disease.
3. **Powdery mildew** (*Oidium piperis*): Wettable sulphur at 0.2 % is sprayed to control this.
4. **Red Spider mite** (*Tetranychus* Sp.): It is a serious pest in betel garden during summer which is controlled with application of 0.2 % of wettable sulphur or 5 % NSKE.

70

Cashewnut

- Plough crop/Gold mine of waste land/Dollar earning crop/Wonder nut crop
- B. N.: *Anacardium occidentale*
- Family: Anacardiaceae
- Chromosome No.: 2n = 42
- Origin: Brazil
- Inflorescence: Panicle (polygamomonoecious)
- Fruit type: Drupe
- Edible part: Cotyledon
- Evergreen tree, hardy tropical plant

Introduction

Cashew is native to South Eastern Brazil, from where it was introduced to the Malabar coast of India in the sixteenth century to cover bare hills and for soil conservation. It gained commercial importance in 1920. African countries produce large quantities of cashew nuts, but they are not processed into consumable products because of difficulties in organizing the native labour. India imports raw cashew nuts which are processed and converted into cashew kernels and cashew shell liquid and re-exported to countries like the USA, Canada, the United Kingdom, Australia and Russia. The cashew kernel of India is of the highest quality in the world. India is the largest producer and exporter of cashews in the world. India stands first in area and production in the world. Indian cashew export: 63 per cent of global cashew trade. The national average is 835 kg per ha (*i.e.,* around 5.50 kg per tree) and this is far less than the achievable level of 2000 kg per ha (*i.e.,* around 12.50 kg per tree). Maharashtra state is having the highest productivity of about 1337 kg per ha (mainly due to adoption of drip irrigation and mulching techniques), while productivity in West Bengal has come down (1140 kg/ha), mainly because of old plantations. Due to its nutritive value and commerce, it is designated as "zero cholesterol nut or white gold". The cashew nut is a unique combination of fat (47 %), **protein (21 %),** carbohydrates (22 %) and vitamins (10 %). India contributes about 65 % of the world's exports. The USA is the largest importer of cashew kernels.

Climate and Soil

It is a hardy tropical plant and does not require a very specific climate. It can come up in places situated within 350 latitudes on either side of the equator and also in the hill ranges up to 700 m MSL. It can grow well in places receiving rainfall of 50 cm to 250 cm and tolerate a temperature range of 25 to 49°C. It requires bright weather and does not tolerate excessive shade. Cashew is cultivated on a wide variety of soils in India, like laterite, red and coastal sandy soil. To a limited extent, it is also grown on black soils. It can also be grown on hill slopes in virgin organic matter-rich soils. They do not prefer waterlogged or saline soils.

Varieties

- **Cashewnut Research Station (CRS), TNAU, Vridhachalam, Tamil Nadu.**
- Hybrids are **Vengurla 3** (Ansur 1 x Vetore 56), **Vengurla 4** (Midnapur Red x Vetore 56) and **Vengurla 5** (Ansur Early x Katekar).
- BPP-1,2,3,4,5,6- **Released from** Bapatala, Andhra Pradesh.
- Vengurla-1,2.3,4,5,6,7- **Released from,** Vengurla, Maharashtra.
- VRI-2: Failure variety is due to chaffy and light in weight and small nuts.
- VRI-1,2,3,4- **Released from,** Vridhachalam, CRS, TNAU.
- Ullal-1,2,3,4, Chintamani-1, UN 50- **Released from** Ullal, Karnataka.
- Anakkayam-1, BLA-39-4, K-22-1, Madakkathara-1,2 **-Released from,** Kerala.
- Bhaskara, NRCC Selection-1, NRCC Selection-2- **Released from** NRCC, Puttur, Karnataka.
- Kanaka, Priyanka, Amrutha, Akshaya.
- Damodar: Anakkayam -1x H-313.
- Priyanga- Export variety.
- Jhargam-1, Commercially grown in West Bengal.
- Chintamani Cashew -1: Recommended for Southern Dry region of Karnataka.
- Among vengurla varieties V-4 is popular in Khanapur area of Karnataka.

- **Varieties of Kerala** Anakkayam -1, Madakkathara -1, Dhana (H-1608), Kanak (H-1598), Amrita, Priyanka (H-1591), Dharashree (H 3-17), Anagha (H8-17), Sulabha, Hybrid Dhara, KGN -1 (K.G. Nair (ex CEPC Secretary) -1 a dwarf cashew collected from Brazil which can be planted at 3 m x 3m spacing), etc.
- Priyanka and Anagha are from Annakayam research station.
- Cashew varieties with bigger and bold nuts are UN-50 ,Vengurla -7, Anagha, Priyanka, Akshaya.
- **Varieties suited:** Dwarf varieties with compact branches are well suited. Ex. Anakkayam -1 and Dhana.

Propagation

Among the vegetative propagation methods, **soft wood grafting** is more successful in most of the cashew growing areas.

Planting

Pits of 45 x 45 x 45 cm are dug and filled with a mixture of top soil, 10 kg of farm yard manure and one kg of neem cake at a distance of 7 m x 7 m either way during June – July and planted. In the case of seedlings, 45-day old seedlings are transplanted.

Training and Pruning

All the side shoots must be removed up to a height of at least 2 m from the ground to cause the branches to form and spread out from the upper section of the trunk. Periodical pruning of dead wood and criss-cross branches during the month of July is recommended to minimize the losses through disease such as dieback and to increase the yield.

Manures and Fertilizers

The fertilizer recommendations of the National Research Centre for cashewnut Puttur in Karnataka are 500 g N, P_2O_5 and K_2O 125 g per tree annually to be applied in 2 split doses before and after the south west monsoon.

Flowering and Fruiting

It starts flowering after 3–5 years of planting. It bears terminal flowers consisting of 200 – 600/panicle. The ratio of a perfect flower is 0.45 to 24.9. Under south India conditions, 3 growth flushes take place and flowering takes place from the middle of September to the end of February. The mid-flowering period is in October–November. The maximum number of hermaphrodite flowers is recorded in the month of January. Pollination takes place through insects and wind. Fruit setting varies from 4–8 % under different conditions, with spraying of 100 ppm. NAA provides a 100 % fruit setting. It requires about 2000–2400 hours of sunshine per year (> 6 hours per day). It requires about 1285 hours of sunshine in the flowering/fruit set period (Nov-March). The predominant acid in cashew apple: Malic acid. cross-pollinated crops, pollinatiors by insects (flies, bees) and wind. *Anacardium pumilum* is a cashew dwarf rootstock. Intensive branching or high yielding tree > 60% and for low yielding tree (<20%).

Harvesting

The maximum size of nuts takes place within 30 days of flowering and hardens after 10 days' size retained. The total days required for ripening is 60 days. The peak time of harvesting is March – April. Harvested nuts should be collected and stored for drying and processing. Well dried raw nuts (moisture content of 8 to 9%) could be stored for up to one year without any quality deterioration. Cashew harvesting, yield and processing in Cashew. Cashew pre-flowering age is 3 to 4 years. Full reproductive age of 10 to 12 years and an economic life of more than 50 years (possibly up to 60 years).

Yield

2.1 kg - 5.5 kg kernels per tree per year and the total seed is 386–455 kg/acre/ year.

Pests in Cashew

Tea mosquito bug ***(Helopeltis antonii)***, thrips, leaf minor and leaf blossom webber are important pestsof cashew. The tea mosquito and stem borer are the most damaging to cashew. Stem and root borer: *Plocaederus ferrugineus* L.

Diseases

Dieback or pink disease (*Pellicularia salmon color*), Leaf spots Powdery mildew, Root and seedling rot Fruit Rots Cashew declines. Inflorescence blight, Leaf and nut blight diseases Anthracnose, Cashew wilt

Grading in Cashew

Based on counts number of kernels per pound *i.e.,* 453.5 g CEPC (Cashew Export Promotion confirms) grade specification and recognizes 24 different grades of kernels. Grades of Whole kernels in cashew (Wholes are the kernels which have no split). These are again separated in to 6 grades as.

Processing in Cashewnut

Harvested nuts will be dried for 2 to 3 days. Nuts, after drying, can be stored in gunny bags well protected from rodents and stacked on a platform above the ground level, leaving space on all the sides of the room. Well dried raw nuts (moisture content 8 to 9%) could be stored for up to one year without any quality deterioration Processing in cashew refers to the recovery of kernel from mature nuts by manual or mechanical means. Mechanical processing involves moisture. conditioning (moisture content in the nut is increased to 15 to 25 %) Roasting - Shelling - Peeling (Removal of seed coat or testa) -drying - Grading - Packing (Vacuum packing extends shelf life in cashew by an additional one year) - Marketing Wholes (No split on the kernel) grade contain 6 grades

Roasting (Makes shell brittle Methods of roasting)

Drum roasting: Results in high percentage of wholes. Dis advantage here is loss of CNSL.

Oil bath roasting: Shell gets heated and shell wall gets separated releasing oil into bath. i.e. CNSL is recovered. Steam boiling Shell contains 35 % CNSL and CNSL contains anacardiac acid to an extent of 90 %.

Cashewnut Shell Liquid (CNSL)

This raw material is used for a number of polymer-based industries like paints and varnishes, resins, industrial and decorative laminates, break linings, and rubber compounding resins. The nut has a shell of about 1/8-inch thickness inside which is a soft honeycomb structure containing a dark reddish brown viscous liquid. It is known as cashew nut shell liquid, and it is present in the pericarp of the cashew nut. In India, annual production of CNSL will be around 15,000 tones, whereas the potential available is around 45,000 tonnes and the world production of CNSL will be around 1.25 lakh tones. For more details, contact the Regional Research Laboratory, Trivandrum, Kerala.

71

Cocoa

- Cocoa/Food of god
- B. N.: *Theobroma cocoa*
- Family: Sterculiaceae
- Chromosome No.: 2n = 20
- Origin: Amazon valley of South America
- Fruit type: Berry

Introduction

The Greek names Theos and Broma mean "Gods" and "Food," respectively. Cocoa consumption is in temperate countries. Cocoa is a beverage crop introduced for cultivation in India in the early 1965. The cocoa plant is a perennial dicotyledonous which has been under cultivation since pre-historic times. shade-loving and evergreen trees. It is native to the Amazon valley of South America and is now cultivated largely in Cote d'Ivoire, Ghana, Nigeria, Indonesia, Peru, Sierra Leon Cameroon, Equator, West Indies and Malaysia. In India, cocoa is mainly grown in Andhra Pradesh, Tamil Nadu, Kerala, Karnataka and. It is a rich source of **fat (37%)** and protein (7%). It also produces theobromin. Cocoa is used in manufacturing chocolate, cocoa powder, cocoa butter and other nutritious foods.

Climate and Soil

Cocoa is a crop in the humid tropics requiring well distributed rainfall. A minimum of 90 to 100 mm of rainfall per month with an annual precipitation of 1500–2000 mm is ideal. However, it can also be grown in other regions by supplementing rainfall with irrigation during dry periods. For successful cultivation, the dry months should not exceed 3 to 4 months. Cocoa tolerates a minimum temperature of 15°C and a maximum of 40°C, but a temperature around 25°C is considered the optimum. Cocoa grows on a wide range of soils, but loose soils that allow root penetration and movement of air and moisture are ideal. It should retain moisture in the soil during the dry season, as cocoa requires a regular supply of moisture for proper growth. Though cocoa can be grown in soils ranging in pH from 4.5 to 8.0.

Varieties

There are Three Main Types of Cocoa *viz.*

Criollo Group

Pods are red before ripening, varying in shape and turning yellow on ripening. Its fruits have furrows on their fruit surface and have a rough, warty fruit surface. It is generally a poor yielder, with a slow growing habit and small leaves. It is susceptible to various diseases.

Forestero Group

It is the commercially cultivated type of cocoa in major cocoa-growing countries. Here, pods are green before ripening and turn yellow on ripening. Pods have a smooth surface and have shallow furrows on them. Seeds have dark purple cotyledons and the seeds are more or less flat. When compared to Criollo, it is more vigorous and high-yielding. It has a good plain chocolate flavour and is good for milk chocolate.

Trinitarios Group of Cocoa

It is originally selected from Trinidad. It is a heterozenous type, probably resulting from Foretero and Criollo cross. It has botanical features of an intermediate nature between the Criollo and Forestero types. The products of this group are also of intermediate quality.

Amelonado: A forestero type bean with a melon-shape pod

Amazon: Vigorous growing and high yielding type

Kerala Agricultural University has released 7 improved clones of forestero type namely CCRP-1, 2, 3, 4, 5, 6 and CCRP – 7 and three hybrids CCRP -8, 9, 10.

Central Plantation Crops Research Institute, Vittal has released one clone named as **VTLCC-1** and 4 hybrids namely VTLCH-1 ,2, 3 and VTLCH-4.

Propagation

Cocoa can be planted Seedling and vegetative propagated material.

Seed Propagation

At present, commercial propagation of cocoa is through seedlings. Use ripe pods and seeds will lose viability soon after they are taken out of pods. Hence, fresh beans are used for sowing. Seed viability can however be enhanced for several weeks when kept mixed with charcoal. Pods as such stored in polythene bags and dry charcoal powder retained viability up to a period of 30 days.

Vegetative Propagation

Cocoa is a cross-pollinated plant whose seedlings show considerable variability. Vegetative propagation is through (i) cuttings, (ii) budding, and (iii) soft wood grafting.

Soft Wood Grafting

Vegetative propagation is important since true-to-type trees are produced. Soft wood grafting is a more preferred vegetative method for the production of planting material. The material consists of cleft grafting of scions to 2–3-month old seedling root stocks. The length of scion sticks is 12–14 cm. The scion sticks should be 12–14 cm long and secured to the root stock cleft by 1.5 cm wide polythene tape. Graft union takes place within one month. The grafts will be ready in 3 months for planting.

Planting

In arecanut gardens where the spacing is 2.7 x 2.7 m, cocoa is inter-planted in alternate rows at a spacing of 5.4 x 2.7 m. In coconut gardens, they can be planted 2.7 m apart in a single row. Under the double hedge system, cocoa is planted in two rows, adopting a spacing of 2.7 m within the row and 2.5 m between rows of coconut planted at a normal spacing of 7.5 x 7.5 and above.

Manure and Fertilizers

It is recommended to apply 100 g N, 40 g P_2 O_5, and 140 g K_2 O per tree per year in two spittle doses. During the first year of planting, the plants may be given one third of the above dose, while in the second and third year and above, the full dose of fertilizers are applied. The fertilizers are applied in two splits, the first dose in March – April and the second dose in September - October. Fertilizers may be applied uniformly around the base of the tree up to a radius of 75 cm and forked and incorporated into the soil.

Irrigation

Under conditions of well-distributed rainfall, cocoa grows well as a rainfed crop and irrigation is not necessary. In areas where the rain-free period extends to 4-6 months, there is a need for supplemental irrigation. Irrigation may be given once in 5 days. It helps in the better growth of plants and the precocity of bearing.

Pruning

The cocoa trees should be pruned regularly to develop a good shape. The chupon, or vertical growth of the seedlings, terminates at the jorquette, where four or five fan branches develop. Further chupon develops just below the jorquette and continues its vertical growth till another jorquette forms and so on. Cocoa grows in a series of storeyes. When the first jorquette develops at a height of about 1.5 m, the canopy will form at a convenient height for harvesting and other operations. Hence, all the fan branches arising from the main step are nipped off up to a height of about 1.0–1.5 m or cut in the initial years of their growth. The branches badly affected by pests and diseases also need to be removed. Bearing habit of cocoa is cauliflorus, shade loving.

Jorquette

It is the point where chupon (vertical/orthotropic growth of seedlimgs) terminates to produce fan branches.

Chupon

It is the main stem coming from the seedling i.e., vertical/ orthotropic shoots. It is determinate in growth. Achupon or vertical growth of the plant, terminates at the jorquette, where 4-5 fan branches develop. Furtherchupon develops just below the jorquette and continues vertical growth till another jorquette is formed.

Harvesting

Cocoa flowers in the second year of planting and the pods take about 140–160 days to mature and ripen. Each pod will have 25–45 beans embedded in white pulp (mucilage). Generally cocoa gives two main crops in a year, *i.e.*, September–January and April–June. Off-season crops may be seen almost all through the year, especially under irrigated conditions. Only ripe pods have to be harvested without damaging the flower cushion at regular intervals of 10 to 15 days. The pods are harvested by cutting the stalk with the help of a knife. The harvested pods should be kept for a minimum period of two to three days before opening for fermentation. A pod will have about 30–45 seeds covered with pulp or mucilage.

Yield

The average yield of dry beans is about 600 kg per hectare when grown as a pure crop. When grown under mixed cropping conditions with arecanut and coconut, the plants commence flowering one year after planting and give a high yield of about one ton of dry beans per hectare.

Processing of Cocoa

Harvesting of cocoa consists of picking and breaking ripe pods, removing the beans and transporting them for fermentation and drying. Only ripe pods are harvested. kept for a minimum period of 2 days before opening the pods for fermentation (should not be kept beyond 4 days). Break the pods and collect the beans for fermentation.

Purpose of Fermentation in Cocoa Includes

1. To remove the adhering mucilaginous pulp.
2. To develop chocolate flavor and aroma precursors and to reduce bitterness.

Chocolate flavor is developed by the process of fermentation and drying at producer level and roasting them by manufacturer. The correct fermentation of cocoa beans is very important since, no subsequent process will correct the bad effect created at this stage.

Process of Fermentation in Cocoa

The process of fermentation lasts for about 6 days (Forestero type), during which period the pulp or mucilage adhering to the beans disappears and the cotyledons change colour. For example, in Forestero, the colour changes from purple or violet to dark chocolate brown. Fermentation of cocoa beans is essential to remove the mucilaginous pulp, to develop flavour and aroma precursors, reduce bitterness, kill the germ of the seed and loosen the testa. Among the various methods adopted for fermentation in different cocoa producing countries, the heap, box tray, and basket methods are considered the standard methods.

Heap Method

This method involves keeping a mass of not less than 50 kg of wet beans over a layer of banana leaves. The banana leaves are spread over a few sticks to keep them a little raised over the ground level to facilitate sweating. The leaves are folded and kept over a heap of beans with a few wooden pieces kept over it to keep the leaves in position. The heaps are dismantled and the beans mixed on the third and fifth days. It needs about six days for the completion of fermentation and the beans can be taken out for drying on the seventh day. Even though the minimum quantity of beans required for effective fermentation is 50 kg, a further increase in the quantity of beans in a heap will be beneficial. However, heaps of more than about 500 kg may be difficult to handle.

Tray Method

Wooden trays of the size of 90 cm x 60 cm x 13 cm, with battens or reapers fixed at the bottom with gaps in between, are filled with beans. Each tray can contain about 45 kg of wet beans. Six such trays are stacked one over the other and an empty tray is kept at the bottom to allow for drainage of sweat. After stacking, the beans in the topmost tray are kept covered with banana leaves. After 24 hours of setting, the stack of trays is kept covered with gunny sacking to conserve the heat that develops. There is no need to mix the beans and fermentation will be completed in four days. On the fifth day, the beans can be taken out for drying. The minimum number of trays required to be stacked is about six, but as many as 12 trays can be used simultaneously.

Box Method

Wooden boxes of 1.2 x 0.95 m x 0.75 m with holes at the bottom and sides of the box are filled with wet beans. These boxes can hold one MT. of wet beans. The beans are to be mixed on alternate days. As the quantity of beans is high, this is best done by changing the beans from one box to another at the time of mixing. This would necessitate having a minimum of three boxes. Wet beans taken for fermentation should be sufficiently ripe so as to separate the beans from the polacuta and husk easily. The minimum quantity of wet beans for a normal fermentation is about 100 kg. The duration of fermentation is commonly 3-5 days, i.e., 72-120 hrs. Fermentation over 120 hours will cause a loss of chocolate flavour and the development of an off flavour.

Drying: The fermented beans can be dried either in the sun or by artificial means.

Pests and Diseases

- Black pod disease *(Phytophthora palmivora)*-Pods turn chocolate brown to black
- Serious pest: Cocoa mealy bug *(lilacinus)*
- Cherelle wilt: Caused by many Factors-Insects, fungi.
- Cocoa canker is caused by fungus *Phytophthora.*
- Stem borer *(Zeuzara aurantii)* is a serious pest of cocoa.

72

Coconut

- King of Species/Tree of Heaven/Kalpavriksha/Tree of life
- *B. N.: Cocos nucifera*
- Family: Araceae/Palmae
- Chromosome No.: 2n = 32
- Origin: South East Asia
- Inflorescence: Spadix
- Fruit type: Single seeded drupe
- Edible part: Endosperm
- World Coconut Day: **2nd September**

Introduction

It is considered as Kalpavriksha as it provides the basic necessities of life. Coco- Spanish word meaning monkey face (probably refers to the three scars on the base of the shell resembling two eyes and a nose on a monkey's face). World Vietnam, India, Guyana, Jamaica, Indonesia, Papua New Guinea, Philippines, Sri Lanka, Thailand, Malaysia, Fiji, etc. Vietnam has the highest productivity. In India, coconut is mainly grown in Kerala, which stands first in India in both area and production of coconut, while productivity is highest in Tamil Nadu.

- World Coconut Germplasm (WCG) is located at Sipoghat, Andaman Nicobar Islands
- Coconut Development Board (CDB) is located at Cochin, Kerala
- CDB comes under Ministry of Agriculture established at Kochi in 1981, Kerala

Composition and Uses

Every part of the coconut palm is useful to mankind in one way or other. It supplies food, fuel, drink, oil and shelter to us. The following are the uses of the various parts of the coconut palm

- **Nut:** Coconut water, Kernel, Shell, Husk
- **Coconut water:** Diuretic, Increases blood circulation in kidneys, Eliminates mineral poisoning, Good substitute for saline glucose under conditions of gastroenteritis.
- Kernel/ meat:
- Culinary preparations, Composition of fresh coconut, Water -45 per cent, Protein - 04 per cent, Fat - 37 per cent, Minerals - 4 per cent, Carbohydrate - 10 per cent.
- Copra: Dried kernel Kerala's richest source of vegetable fat, containing 60 to 67% oil 60 to 65 per cent of the total coconut produced is converted into milling copra.

There are Two Types of Copra

- **Ball copra:** Whole - It is prepared by storing fully mature nuts (unhusked) for 8 to 12 months.
- **Cup copra:** Split into two halves - Cup copra is used for household purposes in North India since fresh nuts are not available for edible purposes. Coconuts are the chief source of vegetable fats. Vegetable fats are solid or semi-solid at ordinary temperature. Oil- 60-67 percent.

Husk

Husk forms the basic raw material for the coir industry. It employs over half a million people and earns foreign exchange. Stage of nut maturity for best quality coir 10th month stage when ripe, the husk in coconut accounts for 30 to 45 percent of the nut weight. In the husk, about 30 per cent is fibre and 70 per cent is coir dust. World coir production is 0.30 million tons, of which 50 per cent is from India alone.

Rubberized Coir Industry

Mattresses (Gadi), Pillows, Cushions, Folding car seats etc

Toddy: Sugar containing juice, called toddy is collected on tapping un opened spadix *i.e.,* before the flowers fully develop (Sap of the coconut palm). Sweet toddy is the unfermented fresh juice obtained by tapping. Toddy on fermentation becomes an alcoholic drink.

Snow Ball Tender Nut (SBTN)

Snow ball tender nut is a tender coconut without husk, shell and testa which is ball shaped and white in colour. A coconut of eight months of age is more suitable for making SBTN in which there is no decrease in the quantity of tender nut water and the kernel is sufficiently soft.

Climate and Soil

The coconut palm is a tropical palm that prefers a humid tropical climate. Altitude is generally up to 600 m. can be grown at a higher elevation (1000 m.). Temperature 27 to 32°C, mean annual temperature of 27°C with variations of 6 to 7°C is ideal. (Hence, in North India, we can't grow coconut). Rainfall is 1,800 to 2,500 mm per year. It requires 7.1 hours of sunlight per day and at least 2000 hours per year. Humidity, coconut likes tropical humid climate and 80 to 90 per cent relative humidity (RH) is ideal. The soils of coconut growing areas range from littoral sands to the heaviest clays. Ideal soil for better growth should be well drained. Deep, at least 1 to 1.5 m deep, $_{p}$H 5.20 to 8. It is rich in organic matter and has good water holding capacity.

Cultivars and Hybrids in Coconut

The major classification of coconut based on stature or height is as follows:

1. Tall palms (mostly cross pollination) and **2.** Dwarf palms (self-pollination)

Varieties (Tall): (Pre – bearing age: 6 – 10 years)

1. West Coast Tall (WCT)
2. East Coast Tall (ECT)
3. Lakshadweep Ordinary (Chandra Kalpa)
4. Lakshadweep Micro (LM)
5. Andaman Ordinary (VPM- 3)
6. Spicata
7. Benaulim Tall (Pratap)
8. Phillippines Ordinary (Kerachandra

Varieties (Dwarf): (Pre – bearing age: 3 – 4 years)

1. Choughat Dwarf Orange
2. Chowghat Dwarf green,

3. Malayan Dwarf yellow,
4. Malayan Dwarf Green,
5. Malayan Dwarf Orange and
6. Gangabondam (grown popularly in AP).
7. Coconino (grown in phillippines)

Hybrid

The first hybrid between Tall and Dwarf coconut was released in the year 1932. The inter-varietal hybrids produced for commercial plantings are T X D and D X T, with different parental combinations.

Hybrids	Parents
Kalpa Sankar	CGD X WCT
Kalpa Samrudhi	MYD X WCT
Kalpa Sreshta	MYD X TPT
Kera Sankara	WCT X COD
Chandra Shankar	COD X WCT
Chandra Laksha	LCT X COD
Laksha Ganga	LO X GB
VHC - 1	ECT X MGD
VHC -2	ECT X MYD
VHC -3	ECT X OD
Kera Ganga	WCT X GB
Kera Sree	WCT X MYD
Kera Sowbhagya	WCT X SSAT
Ananth Ganga	ECT X GB
Godavari Ganga	ECT X GBGD
Konkan Bhatye	GBGD X ECT

CPCRI Varieties

Tall cultivars: Kalpa Mitra, Kalpa Pratibha, Kalpa Thenu, Kalpa Tharu, Kalpa Haritha, Chandra Kalpa.

Dwarf cultvars: Kalpa Raksha, Kalpa Sree (Oil purpose), Kalpa jyoti (yellow), Kalpa Surya (Orange).

Kalpathru: Suitable for ball copra production.

Kalpa Sankar: Hybrid suitable for root (wilt) diseae affected tracts.

Kalpa Raksha and Kalpa Sree dwarf cultivars: highly resistance to root wilt disease

Kalpa Haritha: Tolerant to eriophyde mite.

Planting Material

Coconut is a cross pollinated palm and selection at various stages is very important to eliminate poor quality seed nuts and seedlings. Coconut being perennial, the performance of palm can be judged only after 15 years of planting. Coconut ripen in 12-13 months from the opening of the inflorescence.

Selection of Seed Plantation

- Having high proportion of heavy bearing palms
- Free from major pest and diseases
- Avoid small plantation maintained under very favorable conditions as we cannot assess the inherent yield potential.
- Avoid palms located near the cattle shed and compost pits.

Vertical Method

Vertical method of sowing seed nuts: It is planting with stalk end up and practiced only in India.

Advantages

Easy bundling /packing and transportation of seedlings. *i.e.,* handling becomes easier.

Disadvantages

Nut water does not remain in close contact with germinating embryo.

Horizontal Method

with the stalk end of the nut placed horizontally. It is the common and recognised method of sowing in other coconut-growing countries.

Advantages

high percentage of early and total germination, vigorous seedling with thicker girth at the collar, possibly because of nut water being in close contact with the embryo.

Disadvantage

There is a possibility of damage to the seedlings during transportation.

Raising a Nursery

Sow seed nuts in beds of 1 m width and convenient length at a spacing of 40 cm x 30 cm either vertically or horizontally in 20 to 25 cm deep trenches and polybag nursery for coconut sowing.

Advantages of Poly Bag Nursery

- Reduced transplanting shock due to the absence of root damage.
- Ease of irrigation and fertilizer application in the bag.
- Early germination.

Planting

- Talls - 8m x 8 m or 8.5 m x 8.5 m.
- Dwarfs and hybrids -7 m x 7 m or 7.5 m x7.5 m.
- Planting along the boundary of plantation - 7 m x 7 m.

Manures and Fertilizers

Organic manure: 50 kg per palm per year. The addition of neem cake @ 1 to 5 kg per palm is also beneficial/recommended. RDF for coconut (4th year onwards): 500: 330: 1200 g NPK per palm per year.

Irrigation

Irrigation requirements for coconut depends on the method of irrigation, whether/season and climatic conditions apart from soil characters. Apply 100% of the recommended irrigation during the summer, 60% during the winter, and 40% during the rainy season's rainless period.

Cover Cropping

Leguminous cover crops recommended in coconut plantations are

- *Mimosa invisa*
- *Stylosanthes gracillis.*
- *Calopogonium muconoides*
- *Macuna bracteata*
- Valvet bean: *Macuna* spp.

Multiple Cropping

The monoculture of coconut as a monocrop does not fully utilize the basic resources such as soil and sunlight available in the plantation. Based on the growth habit of the palm and the amount of light transmitted through its canopy, the life span of a coconut palm could be divided into three distinct phases from the point of view of intercropping.

Intercropping

- Tropical Tuber crops like tapioca, elephant foot yam, sweet potato, colocasia, greater yam, lesser yam
- **Rhizome spice crops**: Ginger and turmeric
- **Cereals:** Upland rice varieties, Maize, pearl millet, finger millet.
- **Vegetable:** in maidan areas of Karnataka- Chillies, potato, French beans, Kasaragod (Coastal tract) conditions- Coccinia, snake gourd, bottle gourd, amaranthus, brinjal, bitter gourd.
- **Pulses:** Cowpea, black gram, green gram, red gram, Bengal gram, soya bean.
- **Oil seeds:** ground nut.
- **Fruit crops:** Banana and pine apple.and Papaya can also be intercropped with coconut.
- **Floriculture:** Orchids, anthuriums, and other cut flowers and ornamentals can be successfully grown in coconut plantation.
- **Medicinal and aromatic plants:** Lemon grass, Kacholam, dioscorea.

Mixed Cropping

Growing of perennial crops in association with coconut palm is referred to as coconut based mixed cropping. Eg. Cocoa, clove, nutmeg, cinnamon, pepper, betel vine, jack, bread fruit, cardamom, coffee, vanilla and mango.

Harvesting

Usually, nuts are harvested 6 to 10 times in a year. Harvesting season: year-round harvesting inflorescence is produced in every leaf axil (12–14 leaves per year), leading to year-round harvesting. However, the main harvesting season is summer.

Yield

80 to 100 nuts per palm per year. (The national average yield is 44 nuts per palm per year). Hybrids: 100 to 130 nuts per palm per year,

Pests

1. Rhinocerous beetle or Perisierola *(Orycetes rhinoceros)*: Characteristic geometric cut (inverted 'V' shape)
2. Leaf eating caterpillar in coconut is *Opisina arenosella.*
3. Red palm weevil *(Rhynchophorus ferrugineus)*: Oozing out of a viscous brown fluid and longitudinal splitting of leaf bases
4. Root grubs
5. Mites, **Eriophid mite**
6. Mealy bugs
7. Scales
8. Coreid bugs
9. Rats etc

Diseases

1. Bud rot stem blending *(Phytophthora palmivora*): Yellowing of two young leaves and emitting foul odour
2. Anabe roga.
3. Pestalotia leaf spot /Grey leaf spot.
4. Thanjavar wilt / Ganoderma wilt of coconut and bud rot of arecanut is caused by *Gonodema lucidrm.*
5. Root (wilt) disease.
6. Mahali or fruit rot.
7. Button shedding.
8. Production of barren nuts.
9. Tatipaka in AP.
10. Leaf yellowing.

73

Coffee

- King of beverage crop/Coffee
- B. N.: *Coffea* spp
- Family: Rubiaceae
- Origin: Ethiopia
- Fruit type: Drupe

Introduction

Coffee is the second important beverage, ranking second among traded commodities. Its dried beans are roasted, ground, and brewed to make a stimulating and refreshing beverage. Arabica coffee originated in Ethiopia, from a place called Caffa. The word "coffee" is derived from a place called "Caffa." Here it occurs naturally in the forest between 1,400 to 1,800 feet of elevation. Nearly 80 percent of the world's coffee is produced by *Coffea Arabica*. Robusta coffee is thought to have originated in Central Africa (Congo and Zaire). Arabica coffee was introduced in 1670 by Muslim pilgrim Baba Budan. He brought seven seeds from Yemen and cultivated them in Chikmagalur, Karnataka. The original seeds he brought were probably Mokka seeds. **Robusta coffee** was introduced from the Indochina region at the close of the 19th century for planting in the states of lower elevations. World: Brazil, Vietnam, Colombia, Indonesia, India, Ethiopia, Peru, Mexico, Guatemala and Honduras. It is also growing in India in Karnataka, Kerala, Tamil Nadu, Andhra Pradesh and Orissa.

Important Differences Between Arabica, Robusta and Liberica Coffee

Arabica	**Robusta**	**Liberica**
Coffea Arabica	*Coffea robusta*	*Coffea liberica*
Tetraploid, 2n = 44	Diploid, 2n = 22	Diploid, 2n = 22
Self-pollinated	Cross pollinated	
Ethiopia	Central Africa	

Small tree if not pruned, a bush under training, high elevation.	Bigger bush than Arabica, lower elevation.	Tree coffe
Small and dark green leaves	Large and pale green	Dark green, 2 -3 times bigger than Arabica
Flower buds in clusters in leaf axils (10 – 20)	Flower buds in large clusters than in Arabica	Flowers are large
Flowering Sep. - March	Nov. - Feb	
Flowers open in 9 – 10 days after blossom shower	After 7 – 8 days	
Self-fertile	Self-sterile	
Take 8 – 9 months to maturity	10 – 11 months	One year
Berries bigger in size	Smaller than Arabica	Smaller
Dark red berries	Light red	Yellow – reddish brown

Climate and Soil

Coffee is usually cultivated in tropical and subtropical latitudes. It prefers rich soils and high rainfall, with a seasonally dry period. The conditions prevailing in southern India; summer temperatures combined with poor subsoil moisture can be a limiting factor, whereas at northern latitudes, cold winter temperatures can be equally limiting. Soil should be deep, friable, open-textured, rich in plant nutrients, humus-rich, and slightly acidic (pH 4.5 to 6.5).

Varieties

- **Blue Mountain-** famous variety.
- **San Roman- Mutant variety.**
- Sln. 795- Resistant to leaf race 1, 11 and most popular variety of Arabica coffee.
- Sln. 8- Spontaneous hybrid of robusta-arabica origin, highest vertical resistance to leaf rust.
- Sln. 9 and Sln. 10- Drought hardy, suitable for different coffee zones.
- Cauvery (Caturra x Hybrido-de-Timor)- Resistant to all races of coffee rust variety.
- **Cauvery hybrid** is a F4 Cross between Caturra (A semi dwarf variety of arabica) X Hybrid de Timor (A semi-dwarf hybrid).
- **Features of Cauvery Coffee** Suitable for High, Profuse branching, Precocious, Reaction to leaf rust Quality: Medium sized beans with 64 % A grade and 15 % B grade beans.

- **Chandragiri coffee:** It is a newly released coffee in 2007-08 by Coffee Board with the original source from Portugal. It was introduced in the year 1975 to CCRI Balehonnur from Portugal. Farm trials and intensive research trials were take up at CCRI Balehonnur.
- Features Bushy growth with slightly bigger leaves than Cauvery coffee.
- **Bigger sized berries:** It produces 25 per cent bigger sized berries compared to other varieties.
- **Resistant to leaf rust:** Lower (5 to 7 %) leaf rust incidence in this variety is reported compared to other varieties (20 to 40 %). Tolerant to drought.
- **Improvement of robusta coffee (*Coffea canephora*).**
- Robusta selections from CCRI Balehonnur are.
- Selection –1 R (S-270 and S-274) ,2) Selection –2 R (BR series 9,10 and 11).
- Selection – 3 R (C X R coffee).
- Unlike arabica coffee robusta coffee has long productive life of 70 to 80 years and hence, farmers' donot replace the old varieties frequently.
- **Selection – 1R (S-270 and S-274):** It is a seeding progenies of two individual mother palm identified in robusta gene pool in India. It has 35 to 50 fruits per cluster.
- **Selection 2 R (BR series 9,10 and 11)** Performance of clones BR (*Balehonnur Robusta*).
- It is important to note that only seed mixture of these three clones BR –9, BR-10 and BR-11 should be used for planting. Individual clones should not be planted separately which may result in no fruit set due to incompatibility problems.
- **Selection – 3 R, (C x R):** It is an inter specific hybrid between *Coffea congensis* x *C. canephora* (Robust coffee, Selection- 274), F1 (C x R) F1 is back crossed with both the parents (*Robusta* and *congensis*) *Coffea congensis* is a species closely related to both arabica and robusta with compact bush.

Propagation

Coffee is propagated **by seeds**

Preparation of Seeds

Healthy and well developed, fully ripe berries are harvested from specially identified plants for use as seed bearers. After discarding the floats, the sound fruits are de-pulped and mixed with sieved wood ash and dried in the shade.

The beans are then cleaned and graded to remove any cut or defective beans. Prior to planting, the seeds are treated with Agrosan or any organomercurial compound to prevent fungal infection.

Nursery Practices

Select light loamy soil of good drainage with high organic matter content and provide water and shade facilities. Form raised beds of 15 cm in height, 1 m in width and at a convenient length. Incorporate 30 - 40 kg of well decomposed compost/FYM, 2 kg of finely sieved agricultural lime and 400 g of rock phosphate to a bed of 1 x 6 m size. In heavy soils, it is necessary to add coarse sand for drainage and aeration.

Sowing

Pre-sowing seed treatment with fungicide. Seeds are sown in December - January in the bed at 1.5–2.5 cm apart. Then they are covered with a thin layer of fine soil and a layer of paddy straw. Water the beds daily and protect them from direct sunlight with an overhead pandal. Seeds germinate in about 45 days; after which they are transplanted to a secondary nursery bed for raising balls or bag nursery.

Bag Nursery

Polythene bags with an adequate number of holes in the bottom half are taken and filled with a prepared mixture containing jungle soil, FYM and sand in the proportion of 6:2:1. An area of 12 x 8 m can accommodate 5000 seedlings. Seedlings are planted in polythene bags.

Beneficial Effects of Shade in Coffee

1. **Quality improvement:** Shade trees form natural canopy and it improves aromatic properties.
2. **Improved foliar health:** Shade helps in getting glassy elegant leaves and controls pre mature yellowing of leaves, defoliation and dieback.
3. **Mulching effect:** Leaves shed from the shade trees acts as soil mulch which in turn held in conserving the soil moisture and prevents soil erosion and improves soil organic matter status.
4. **Temperature regulation.**
5. **Improved organic matter status of soil**
6. **Disease incidence:** some of the diseases viz., **Cercospora leaf spot** and ***Colletotrichum*** etc are minimized under shaded situations

7. **Yield regulations**: Shaded situations prevents over bearing in any of the particular year and results in less variations / fluctuations in annual yields.
8. **Wind breaks:** Shade trees serve as wind belts and protects from hailstone damage and damage due to cold wave

Types of Shade Tree

It can be provided by providing a lower canopy (temporary) or an upper canopy (permanent shade trees). Temporary shade trees /Lower canopy shade trees: Temporary shade tree spacing: 6 to 8 m (one shade for every two coffee plants). Temporary shade trees are lopped periodically to maintain the required light.

Examples of Temporary Shade Trees

1. *Erythrina lithosperma*: Dadap or halwana
2. *Grevillea robusta*: Silver oak

Tree spices: Nut meg, clove and cinnamon: Spaced at 5 m apart.

Planting Shade Trees

A dadap is commonly used as a lower canopy shade. Two-metre-long stakes are planted for every two plants of coffee. Silver Oak and Dadaps are planted during June when the rains of the South-West monsoon commence. During summer, the stems of young dadaps are painted with diluted lime or wrapped in agave leaves or polythene sheets in order to prevent them from sun scorching. Regulate shade by cutting criss-cross branches during the monsoon season. Silver oak trees are planted for permanent shade.

Training and Pruning

restriction: Main purpose of training in coffee is to restrict the plant's growth at a desirable height for better management.

Frame work: To attain the proper shape, a strong framework in coffee.

Bearing wood: To promote the production of bearing wood. In coffee, by pruning, old unproductive wood is removed by encouraging the growth of new branches. These new branches/growth would become next year's cropping wood.

Light pruning: In coffee trained on a single stem, regular light pruning is desirable. Time of pruning: starts after the harvest of coffee (December to February) and continues till the onset of the monsoon. However, it is better to prune after a few **Periodicity of medium to severe pruning:** It is done once in four years, usually to replace the lateral.

Preparation of Field

Terracing should be done in deep slope areas. After the summer showers, pits of 45 cm x 45 cm x 45 cm are dug at 1.25–2.5 m apart. The pits are left open for weathering and then filled and heaped for planting. At the time of filling, apply 500 g of rock phosphate per pit along with top soil. Planting is done along the contour in the slope area.

Spacing

Arabica Coffee 1.5 m to 2.0 m either way. Dwarf varieties, Sanraman, 1 m x 1 m. Robusta coffee, 2.5 m either way.

Irrigation

It is generally grown as a rainfed crop. But irrigation with sprinklers during March-April increases blossoming and results in higher yields.

Harvesting

Harvest starts in November and harvesting extends up to February. Coffee fruits should be harvested as and when they become ripe. Coffee is just ripe when, on gently squeezing the fruit, the beans inside come out easily. Unripe fruits should be scrupulously sorted out before using the fruits for pulping. They may be tried separately as cherry.

Fly picking: Small scale picking of ripe berries during October to February.

Main picking: After fly picking- 4-6 pickings @ 10-15 days' interval. Well-formed and ripened berries are harvested during December. Bulks of the yields are obtained from this picking.

Stripping: Picking of all the berries left irrespective of ripening.

Cleanings: This is collection of fruits that have been dropped during harvesting. Unripe fruits should be scrupulously sorted out before using the fruits for pulping. They may be tried separately as cherry.

Yield

750 - 1000 kg dry parchment /ha.

Pests and Diseases

- Coffee leaf rust *(Hamelia vestatrix)* introduced from Sri Lanka.
- White stem borer *(Xylotrechus quadripes)*: Wilting and yellowing of leaves, major pest of India.
- Coffee berry borer *(Hypothenemus hampei)*: Most serious pest in the world.

- Coffee berry borer is controlled by *Beauveria bassiana* (Fungal bio-control agent).
- Shot hole borer *(Xylosandrus campactus)*: Extensive tunneling.
- Coffee nematodes (*Pratylenches coffeae*): Dieback.

Diseases in Coffee

Leaf rust, Blackrot, Pinkdisease, Anthracnose, Root diseases etc.

Non parasitic disorder: Kondli (Copper toxicity) Bean disorder in coffee.

Rust in coffee (*Hemelea vastatrix*) It is an oriental leaf disease. It is considered as one of the seven classic diseases of the world since, It ruined the economy of Sri Lanka rendering it bankrupt and it changed the social habit of people from coffee drinking to tea drinking.

Preparation of Parchment Coffee

Pulping ⇨ Demucilaging (by Natural fermentation, reatment with alkali, Friction ⇨ Drying.

Wet method – to prepare plantation or parchment coffee.

Dry method – to obtain cherry coffee.

74

Oil Palm

- Small Holders Irrigated Crop/Oil Palm/Crop for future
- B. N.: *Elaeis guineensis*
- Family: Araceae
- Chromosome No.: 2n = 32
- Origin: West Africa
- Inflorescence: Spadix
- Fruit type: Drupe
- Humid tropical palm, Introduced in India: 1960

Introduction

Oil palm is recognized as the highest and cheapest edible oil-yielding crop. Palm oil is extracted from the fruit's fleshy mesocarp, which contains 45-55 percent oil. The crop is unique in producing 2 distinct types of oil palm oil and palm kernel oil, which are used for culinary and industrial purposes. Being rich in beta carotene, palm oil contributes substantially to the nutritional and energy requirements of the masses. It is also grown in India Andhra Pradesh, Karnataka, Tamil Nadu, Mizorum, Orissa.

Climate and Soil

Oil palm grows best in areas with a mean maximum temperature of 30–32°C on an average of at least 5 hours of sunlight. Seedling growth is arrested below 15°C. It can be grown in areas which receive an annual well-distributed rainfall of 200 cm or more. However, it tolerates 2–4 months of a dry period. Oil palms grow on a wide range of tropical soils. Moist deep, loamy soils, rich in humus and with good water permeability, are ideal for crop growth. Adequate soil moisture is more important than nutrient supply. The adult palm can withstand occasional waterlogging, but it should avoid frequently waterlogged, extremely sandy, and highly lateritic soils. Oil palms can be grown in a variety of soils with pH levels ranging from 4-6. The elevation for growing oil palm is generally accepted as 300 m above mean sea level. Higher elevations are not favourable due to limitations in temperature, moisture retention and capacity.

Varieties

1. Parton dolly – dwarf, having large fruit
2. Gangavathi
3. **Dura:** Has a low to medium mesocarp (35 to 55 %) and it contains a thick shell around the kernel. It is not preferred for commercial cultivation. Shell thickness Mesocarp 35–55% (54 % by weight), Shell 30% by weight, Kernel 16 % by weight, Shell thickness 2 to 8 mm.
4. **Tenera:** It is a hybrid, *i.e.,* Dura (female) x Pisifera (male) Tenera (F1 hybrid). It is commercially/ widely cultivated all over the world due to its high proportion of mesocarp (60 to 95 %). It is characterized by the presence of a thin shell. Tenera is characterized by the presence of a distinct ring of fibres embedded in the mesocarp near to and encircling the seed. Tenera fruits have a lot of pulp. Thin shell and a big kernel. Mesocarp 60 to 95% (74% by weght), Shell 10 % by weight. 16% kernel.
5. **Pisifera:** It is a shell-less fruit with a pea-like kernel inside. Embryo abortion is common in this variety and often the kernel is also absent. The presence or absence of a shell is genetically controlled. Dura has a genetic constitution of Sh+ Sh+ while Pisifera is Sh-, Sh- and hybrid is Sh+Sh-. On selfing or intercrossing, the hybrid fruit forms segregate into 25% Dura, 50% Tenera and 25% Pisifera. Mesocarp accounts for 90% of the total weight. Kernel = 10%.

Hybrid

1. DD x AVROS,
2. DD X Dumpyavros, 109 D X 291 P

Cover Crops

- Pueraria phaseoloides
- *Calopogonium mucunoides*
- *Centrosema pubescens*
- *Mimosa invisa*
- *Mucuna* sp.

Propagation

Oil palm is propagated through seeds. Tenera hybrids, which are a cross between Dura and Pisifera, are used for commercial planting. Elite dura and pisifera palms are selected based on progeny trials. **Seed rate of 2 to 3 kg per ha.**

Nursery

A nursery is raised by planting germinated sprouts initially in a pre-nursery bed or in polybags in a primary nursery and transplanting them at five leaf stage to a secondary nursery of large sized polybags. Raising seedlings in large polybags without a pre-nursery stage is also being done. The potting mixture is made by mixing top soil, sand and well-decomposed cattle manure in equal proportions. Smaller polybags of 250 guage and 23 x 13 cm size, preferably black are used for raising primary nurseries. These bags are filled with the potting mixture, leaving one cm at the top of the bag. A healthy germinated sprout is placed at the centre at a 2.5 cm depth. While placing the sprout, care must be taken to keep the plumule of the sprout facing upwards and the radicle downwards in the soil. It is better to plant sprouts soon after the differentiation of the radicle and plumule. The seedlings are to be watered daily. Application of a fertilizer mixture containing one part of ammonium sulphate, one part of super phosphate, one part of muriate of potash and two parts of magnesium sulphate is recommended at 15 g at one-month stage, 45 g at three months' stage and 60 g at six months' stage per seedling. This has to be applied 6–8 cm away from seedlings during the first application, 10–12 cm away during the second application, and 15-20 cm away during the third application in the primary nursery. Surface soil is slightly scratched at the time of fertilizer application.

Planting

Oil palm is planted in the main field in a triangular system at a spacing of 9 m, accommodating 140 palms/ha in pits of 60 cm x 60 cm x 60 cm size. Planting is done preferably at the onset of the monsoon during May - June.

Intercropping

The most suitable crops are vegetables, bananas, flowers, tobacco, chillies, turmeric, ginger, pineapples, etc. While growing intercrops in mature oil palm gardens of 8–12 years' age or where palms have attained a height of 3 meters, intercrops should be able to grow under partially shaded conditions and should not compete with oil palm for water, sunlight and nutrients (e.g. cocoa, pepper, heliconia and ginger lilly).

Irrigation

Oil palm requires sufficient irrigation as it is a fast-growing crop with high productivity and biomass production. Do not grow oil palm if assured and adequate irrigation facility is not available. For grown-up yielding palms of 3 years' age and above, a minimum of 150 to 200 litres of water per day is required. However, in older plantations during the hot summer, this quantity may be increased up to 300 litres.

Manuring

Fertilizers are applied in 2 equal split doses in May and September within 2 m. diameter around the palm and forked in. Supply of compost at 50 kg/palm before and after planting is advantageous, especially where the soil is poor in organic matter content. Magnesium application is necessary at 175 g/palm for palms up to 7 years and 350 g/palm for palms above 7 years.

Flowering

Oil palm comes into flower 14–18 months after planting. It produces both male and female flowers separately on the same palm. Male and female phases do occur naturally in consequent cycles in a palm. Type of pollination, cross **pollination** through weevil (*Elaeidobius kamerunicus*).

Ablation

Ablation is the removal of male and female flowers produced in the early stages of plantation. **Purposes of ablation:** uneconomic processing by collection of a few bunches in the initial years of bearing. **Period of ablation: 14** to 26 months after planting. Ablation can be commenced after about 14 months of fieldplanting and continued till 26 months when about 70 per cent of palms are producing inflorescences (at an interval of 4 to 5 months).

Harvesting

It takes about 180 days (6 months) from pollination to maturity. Fully ripe fruit bunches are harvested. when the colour of the fruit changes from black to orange, red, or yellowish orange. While harvesting, a stalk length of 5 cm alone should be left. Harvesting should be done at a 10–12 day interval. During the rainy season, harvesting should be done at a closer interval of 6-7 days as ripening is hastened after a hot summer. In young plantations, we get more bunches with less bunch weight and in adult plantations the bunch weight is more but the bunch number is less.

Yield

The average weight of harvested fruit bunches will be 30 kg. The average number of bunches per palm is 10 to 12. average bunch weight of 30 to 40 kg/bunch, average FFB yield per year per ha, 12 tonnes of oil recovery out of FFB **(Fresh Fruit Bunch)** of 18 to 21%, *i.e.,* extraction ration from oil to bunch of 20 %.

Palm Oil

- **Sterilization of FFB:** After harvesting bunches are sterilized at 130ºC for one hour under pressure of 2 kg per cm2. Inactivation of lipase and glycolytic enzyme activity *i.e.,* fat splitting enzyme which are responsible for increase in FFA.
- **Stripping:** Fruits are separated from the bunches.
- **Fruit digestion:** During digestion process there will be release of oil from the pulp *i.e.* conversion in to oily slurry (mesocarp).
- **Pressing:** Separation of liquid component from the solid.
- **Clarification:** Oil is cleaned of water, cell debris and particles of fibre and shell.
- Crude palm oil - Refining – Palmolein- Further purified Palmolein has thick consistency. red colour, Production of Edible oil < 2 per cent free fatty acid content.

75

Palmyrah

- State Tree of Tamil Nadu/Kalpaka virucham/Kalpaga thara/Tree of Life
- B. N. *Borassus flabelifer*
- Family: Arecaceae
- Chromosome No.: 2n = 36
- Origin: Tropical Africa
- Inflorescence: Spadix
- Fruit type: Drupe

Introduction

The palmyrah is believed to be a native of tropical Africa. It grows extensively in drier parts of India, Sri Lanka, Myanmar, Thailand, Vietnam, Malaysia and Indonesia. It is recognised as the state tree of Tamil Nadu in 1978, recognising its importance. It has been estimated that there are about 8.6 crores of Palmyra palms existing in India, of which 5.02 crores of palms are alone spread over Tamil Nadu and the remaining 3.58 crores of palms are in Andra Pradesh.

Climate and Soil

It can come up in dry areas with moderate to low rainfall. It grows from sea level up to an altitude of 750 m. Palmyrah can be grown in soils of all descriptions but is found to grow well in deep, sandy and loamy soil.

Species

Borassus aethiopium	- African Palmyra Palm (tropical Africa)
Borassus flabellifer	- Asian Palmyra Palm (southern Asia)
Borassus heineanus	- New Guinea Palmyra Palm (New Guinea).
Borassus madagascariensis	- Madagascar Palmyra Palm (Madagascar).
Borassus sambiranensis	- Sambirano Palmyra Palm (Madagascar)

Varieties

SVPR-1 semi-dwarf variety, Palmyrah Research Station, Srivilliputhur, TNAU.

Planting Material

The mother palm selected for seed nut collection should be free from pests and diseases, high yielders of padaneer and fruits, dwarf in stature, early and regular bearers, and well matured fruit bunches indicated by the yellow tinge in the stylar. Some of the fruit may be harvested for seed collection. The selected seeds are stored in the shade for 3 weeks. Shrunken, weight-less, and bored seeds may be rejected. The seeds may be directly sown ***in situ*** or sown in a nursery to raise the seedlings. For direct sowing, 3 to 4 whole fruits are planted in pits and then covered with dried leaves (mulch) to the same depth. Sowing may be taken up during the rainy months.

Nursery Transplanting

If transplanting is to be undertaken, seeds can be sown in a mound formed by heaping sand to form a bed of 1 m broad 30 cm height and of convenient length or in a masonry nursery bed built with bricks of 2 m broad, 60 cm height and of convenient length. In both cases, seeds are sown 10 cm from the space and covered with 5 cm of sand. About one-year-old seedlings are lifted from the nursery and containerized in the polythene bags. After roots are formed, the seedlings can be successfully transplanted into the main field. Shading for the seedlings should be provided using plaited coconut or palmyrah fronds. Spacing: **3 m x 3 m (450 plants/ha).**

Manuring

Farmers generally adopt sheep penning for their palmyrah plants since the soils are usually sub-fertile. Application of 10 kg of FYM/pit before planting the nuts is advantageous. The dosage may be increased biannually till reaching 60 kg FYM/pit/year.

Mixed Crops

With semi-arid zone fruit crops like Anona, Ber, Amla, Pomegranate, Moringa, Carissa, West Indian Cherry, Guava, etc, it appears to be good.

Pruning

Pruning of 30 % leaves at a rate of 10 % in a phased manner increases the yield by about 28%. **The flowering season** starts in January and continues till August, with a peak during May.

Harvesting

All the parts of the palmyrah palm are useful. Besides, edible and non-edible products are made from its different parts.

Yield

The trees continue to yield for 30 to 40 years, depending on economic life.

Plant part	Value added products
Edible value added products	
Inflorescence sap	Toddy, Jaggery, Sugar, Honey, Wine
Fruit	Toffee, Spread, Jam, Pickle, Sweets (Burfi), Beverages (RTS, Squash, Nectar)
Kernel	Canned products
Non-edible value added products	
Leaf	Mats, Baskets, Fans, Hats, Umbrellas, Buckets, Writing Material, Fence, Fibre extracted is used to make brushes and handicrafts
Fruits	Fibre extracted is used to make toys and fancy items
Stem	As poles for sheds construction and as timber source

Pests

As in coconut rhinoceros' beetles, red palm weevils and black-headed caterpillar Eriophid mites often cause damage. Bud rot, stem bleeding, grey blight and tuber rot.

Bud rot: The palm is liable to infection *Phytophthora palmivora* Buttler, causing a serious fungal disease.

Tapping

The extraction of sap from the inflorescence is called "tapping". It varies with the sex of the palm and the age of inflorescence. Age at first tapping 10 to 12 years after planting if properly managed it is about 15 years.

Method of Extraction of Tapping in Male Palm

Male Palm

The sheath covering the young inflorescence (two weeks old) is removed and allowed to dry for three days. The end is cut every time and pot is tied to the inflorescence. Thus in case male palms the flowering shoots are tapped.

A. Male palms: Tapping period short period (Dec- Feb)

Aripanai Method

In male palm sheath of young inflorescence (1-1½ months old inflorescence) is removed and allowed to dry for three days. The end is cut every time and pot is tied to the inflorescence. It is practiced for 1 to 1 ½ months.

Vellupanai Method

(Using a one-month-old comparatively old inflorescence) Here also, male palms are selected but inflorescences are one month old. Each male spike (bearing sessile flowers) is pretreated by pressing and stroking (pettu/hodeta). Three to six such spikes are brought together, wrapped with aplmyrah leaves and fitted to a pot.

Female Palm

The tissues of the young female inflorescence are softened by hitting the main axis of the inflorescence with an iron rod and the fork is used to press the regions from which the fruits develop. Thus, in the case of female palms, the fruiting branches are tapped when the drupes are still very small. Tapped for a longer period, i.e., April-December. It gives 30 to 50 per cent more sap yield. Normally, female palms are tapped for a longer period (April–Dec.) compared to male palms (Dec–Feb).

Thattipalai

Young female inflorescence is tapped in this method. The tappers soften the tissues by hitting the inflorescence main axis with the iron rod and fork is used to press the region from which fruit develop.

Kaivetty

When female inflorescence is about 2 to 3 months old. Here the inflorescences are matured and will be having fruits. In this method fruits are sliced as the tapping progress.

Neera or padaneer contains sugar (12-16%), Vitamin-C and B-complex

Palm sugar: Boiling of stained and clarified sap to 108-110°C @ 85°Brix.

76

Rubber

- Para rubber/Natural Rubber
- B. N.: *Hevea brasiliensi*
- Euphorbiaceae
- Chromosome No.: 2n = 36
- Origin: Brazil

Introduction

The rubber tree is the most important commercial source of natural rubber, an industrial raw material of strategic importance, recovered from its latex. Hevea was first used as a source of rubber by the French. The name rubber is derived from the quality of material in rubbing pencial marks. It was introduced to tropical Asia in 1876 through Kew Garden in the UK with seeds brought from Brazil by Sir Henry Wickham. It is now distributed in the tropical regions of Asia, Africa and America. The major rubber-growing countries are Thailand, Indonesia, Vietnam, Cote d'Ivoire, China and India Thailand has the maximum area under rubber, but China has taken the credit for maximum rubber production. India is the second largest producer and consumer of natural rubber in the world. In India, rubber is grown over an area of 7.57 lakh hectares with a production of 12.57 lakh tonnes, 9% of the total world natural rubber output. In India, Kerala is the predominant rubber-growing state. International Rubber Research and Development Board (IRRDB), Kuala Lumpur, Malaysia.

- Association of Natural Rubber Producing Countries (ANRPC), Kuala Lumpur, Malaysia
- Rubber Research Institute of India (RRII-1955) is located at Kottayam, Kerala

Climate and Soil

A tropical climate with an annual rainfall of 2000–4000 mm is suited for cultivation. Minimum and maximum temperatures should be between 21 and 35°C, with an ideal relative humidity of 80 percent. Regions prone to heavy winds should be avoided. It requires deep and lateritic fertile soil with an acidic pH of 4.5 to 6.0.

Classification of Rubber Clones

- **Primary clones:** When mother trees are selected from existing seedling populations of unknown parentage and are multiplied vegetatively. Eg: **Tjir –1** (Tjirandji –1 of Indonesia), G.T-1 (GodngTapen of Indonesi), G.I –1(Glenshiel-1 of Malaysia) and PB-86.
- **Secondary clones:** When the mother trees are evolved by cross pollination between two primary clones and are then multiplied vegetatively, they are known as secondary clones. Eg: **1)** RRIM - 600 = TJIR –1 x PB – 86 and **2)** PRIM - 628 = TJIR –1 x PRIM – 527
- **Tertiary clones :** Are produced by controlled pollination of two existing clones, but they differ from secondary clones in that at least one of parents or both the parents are of secondary clones. Eg: RRIM – 703 = RRIM – 600 x RRIM –500
- **RRII 105** is very high yielding clone and its parents are **Tjir-1 and GL-1.** Occupying 80 per cent of the area under rubber.
- RFU-112, 600, 605, 703, 513, 625.
- PB-86,206,213,235.
- RRII-105,118, 203, 287.

Mostly Used Cover Crops in South India

- Pueraria phaseoloides
- Centrosema pubescens
- Mimosa invisa
- Calapogonium muconoides

Propagation

In India, Hevea seeds normally ripen during July-September, when the seeds are collected and seedlings are raised. All earlier plantations were raised from unselected seeds. The production of those plantations was poor. Selection work on Hevea with a view to improving the planting materials and the introduction of vegetative propagation by bud grafting. It is commercially propagated by forket **budding.**

Season

June - July is optimum for cultivation

Method of propagation

Propagated by Bud grafting is done on the seedlings when they are 2–8 months old, using green or brown dormant bud patch collected from selected scion clones.

Brown Budding: Collected from bud wood of about 1-year age and grafted on to seedlings at about 10 months.

Green Budding: Seedling of 2–8 months old and budwood that is 6–8 weeks old.

Nursery

Bed size: 60-120 cm width and of convenient length. Spacing: Seedling stumps – 23 x 23 cm, 30 x 30 cm and 34 x 20 cm. **Budded stumps** – 30 x 30 cm. **Stump budding** – 60 x 60 cm. **Bud wood nursery** – 60 x 90 cm or 60 x 120 cm.

Planting

Pit size of 1m^3 are dug and filled up with top soil and compost

Planting material	Spacing (m)	Population / ha
Budded plants		
Hilly areas	6.7 X 3.4	445
Plains	4.9 X 4.9	420
Seedlings		
Hilly areas	6.1 X 3.0	539
Plains	4.6 X 4.6	479

Manuring

For immature rubber trees at the pre-tapping stage. Apply 12 kg of compost or FYM and 120 g of rock phosphate in each pit before planting. Apply 10:10:4:1.5 NPK and Mg as per the schedule given below. Apply 400 kg of mixture/ha in 2 doses, one in April/May and another in September/October from the 5th year till the tree is ready for tapping.

Months after planting	Period of application	Quantity per plant	
		10:10:4	**12:12:6**
3	September/October	225 g	190 kg
9	April/May	445 g	380 kg
15	September/October	450 g	380 kg
21	April/May	450 g	480 kg
27	September/October	550 g	480 kg
33	April/May	550 g	380 kg
39	September/October	450 g	380 kg

Matured Rubber Trees under Tapping

For matured rubber trees under tapping, apply NPK 10:10:10 grade mixtures at the rate of 900 g/tree (300 kg/ha) every year in two split doses. If there is a magnesium deficiency, add 10 kg of commercial magnesium sulphate for every 100 kg of the above mixture if there is magnesium deficiency.

Tapping

S/2 d/2	Half spiral, alternate day for 6 months and rested for 3 months	100% Intensity
S /2 d/2	Hhalf spiral, alternate day for 6 months and rested for 3 months	67 intensity
S /2d/3	Half spiral, third day	67 intensity
S /1 d/4	Full spiral, fourth day	100% intensity

Ethrel Treatment

Ethrel is recommended to increase the latex yield of trees. It is applied at 5% a.i. concentration. The first application may be done after a drought period, preferably after a few pre-monsoon showers and subsequent applications may be done in September and November. However, continuous application of Ethrel is not recommended for periods of more than 3 years at a stretch.

Tapping Panel Dryness (Brown bast)

Syndrome characterized by prolonged dripping of latex with a gradual decline in volume yield, precoagulation of latex and partial or complete drying of the tapping area (10–25 per cent).

Control

Reduce tapping intensity or give tapping a rest for 3 to 12 months.

Yield

Rubber yield steeply increases year by year, reaching a peak after 14 years of planting. In South India, the annual yield of rubber is 375 kg/ha from seedling trees, whereas budded plants yield 800–1000 kg/ha.

Pests and Diseases

- Abnormal leaf falls (*Phytophthora palmivora)*: Infected leaves fall in large number prematurely.
- Pink disease is caused by *Coriticium salmonicolor.*
- Powdery mildew (*Oidiumhaveae*): Ashy coating noticed on tender leaves.
- Scale insects (*Saissetianigra*): Severely affected portion dry up and die due sucking of sap.
- Mealy bug (*Perrisianavirgata*): Severely affected portion dry up and die.

1. **Clonal seeds (Highly heterogenous):** Seeds of clone are termed clonal seeds
2. **Commercial propagation:** Forket buddin, Spacing: 4.9m x 4.9m
3. **Tappable stage:** 7th years onwards (75% of trees attains tappable girth). The tapping technique in Rubber is devised by **Ridley.**
4. **Ideal for tapping:** Seedling: 55 cm girth and 50 cm height (From the ground).
5. Budded trees: 50 cm girth and 125-150 cm height (From the bud union).
6. **Tapping depth:** 1 mm close to cambium.
7. **Tapping slope:** 25°- Seedlings tree and 30°- Budded plants.
8. **Slaughter tapping:** Intensive tapping prior to felling of the old trees.
9. **Swelings:** Callus formation at tapping place.

77

Tea

- Tea/Golden leaf/Queen of Beverage Crop
- *B. N.: Camellia sinensis*
- Family: Theaceae
- Chromosome No.: 2n = 30
- Origin: China
- Inflorescence: Cymose
- Fruit type: Capsule

Introduction

Tea is the most popular and cheapest beverage consumed by two-thirds of the world population. The stimulative effect of tea is due to the presence of caffeine (1.25–4.5 %). Panthothenic acid, caffeine and theophylline present in tea stimulate the nerves and are good for the heart. Tea is grown in 50 countries, predominantly in Asia, Africa and Europe. Of the major tea producers, India, China, Sri Lanka, Kenya, Russia and Indonesia contribute the maximum share to global production. Though the share of India in global tea production has declined over the years, it still retains its status as the largest tea producer with 30 % of world output (2,780 million kg). India can also boast of producing many varieties of tea. Well-known varieties like Darjeeling, Assam and Nilgiris are well-known for their distinctive characters. Assam (24.86 %), West Bengal (15.20 %), Tamil Nadu (6.28 %) and Kerala (2.28 %) are major tea producing states in the country.

Climate and Soil

The performance of tea is excellent at elevations ranging from 1000–2500 m. The temperature range is 16-32°C. Annual rainfall: 125-150 cm. Humidity is always around 80 %. Tea is a calcifuge crop, meaning it has a low amount of calcium but a high amount of potassium and silicon. Tea requires well-drained soil with a high amount of organic matter and a pH of 4.5 to 5.5.

Nursery

The nursery soil should be well drained and deep loam in nature with a pH of 4.5 to 4.8. The soil and sand used in the preparation of rooting medium should be tested for pH and nematode infestation.

Pre-treatment of Rooting Medium

Treating with aluminum sulphate can reduce soil pH. For this purpose, the nursery soil is formed into beds of one metre in width, about 8 cm in height and of a convenient length. Then the beds are drenched with a 2 % solution of aluminum sulphate applied at 10 litres per 2.5 sq. m of area. Over this another layer of soil of 8 cm in height is spread and again drenched with an equal quantity of water twice. Then the soil is allowed to dry and the pH is checked before being used in the nursery.

Preparation of Sleeves

Polythene sleeves of 150 or 200 gauge, 10 cm in width, and 30–45 cm in length may be used. Drainage holes may be provided at the bottom. The lower 3/4 of the sleeves should be filled with 1:3 sand and soil mixture and the top 1/4 with 1:1 sand and soil mixture and staked in rows. Overhead shade is provided.

Selection of Mother Bush and its Treatment

Healthy and vigorously growing high yielding bushes should be selected. Apply to each selected bush with 40 g of young tea mixture + 60:90 NK mixture up to 5 years. The following mixture has to applied before taking the cuttings.0.5 % $AlSO_4$ + 1 % $MgSO_4$ (before 3 weeks) 2 % Zn SO_4 (before 2 weeks) 1 % Urea (before 1 week)

Preparation of Cuttings

Cuttings are taken on April - May and August - September. Semi hard-wood cuttings are prepared with one leaf and an inter node with a slanting cut at the bottom.

Planting of Cuttings

The sleeves are watered thoroughly and holes are made in the soil. The cuttings are inserted in the hole and the soil around is pressed firmly to avoid airspace followed by watering. Small polythene tents may be provided which maintain high humidity and regulate the temperature inside. Cuttings may take 10 - 12 weeks for rooting. After 90 days *i.e.* when all the cuttings have rooted, the polythene tent may be removed gradually over a period of 10 - 15 days.

Manuring of Nursery

After the tent is removed the cuttings are sorted and staked. 30 g of Nursery soluble mixture of the following composition dissolved in 10 litres of water may be applied over an area of 4 sq.m. This should be done fortnightly.

Composition of the Fertilizer

Ammonium phosphate (20:20)	35 parts by Wt
Potassium sulphate	15 parts by Wt
MOP	12 parts by Wt
Magnesium sulphate	15 parts by Wt
Zinc sulphate	3 parts by Wt
Total	**80 parts by Wt**

Hardening of the Cuttings

Hardening of 4–6 month old young cuttings should be done by removing shade gradually in stages over a period of 4–6 weeks, starting with a few hours' exposure to the sun every day initially and extending the time of exposure gradually.

Botany

- *Camellia assamica* (Assam Jats)
- *C sinensis* (China Jats)
- *C assamica* subspecies lasiocalyx (Indo china)

Character of Assam and China Type Tea

Assam	China
It is a tree	It is shrub
Few robust branches	Branches abundant and whippy (weak)
Leaves are large and glossy	Leaves are small and leathery
Colour of leaves light to medium green	Colour of leaves dark green
High yielding and medium quality	Low yield but good quality
Susceptible to drought and frost	Hardy and resistance agent's drought and frost.
Scattered flowering	Profuse flowering

Varieties

- UPASI stands for United Planter's Association for South India. started in 1893
- Tocklai Tea Experiment Station was established in 1911, Jorhat, Assam
- UPASI, Tea research station, Valparai, Tamil Nadu
- UPASI is an apex body of tea, coffee, pepper and cardamom in the Southern States Viz. Tamil Nadu, Karnataka and Kerala
- UPASI-9 (**Athrey**): Most suitable for higher pH
- UASI-1, 2, 8: Biclonal varieties
- UPASI-10 (**Pandiyan**): Resistant to drought and winds, suitable for higher elevations
- Sundaram is a very high yielding clone.
- Jayaram, Golconda and Brooklands
- UPASI -3, and UPASI -17 as scions and drought hardy clones UPASI-6, UPASI-24, ATK-1, SA-6 and TRI 2025.
- Singara is suitable for high elevation.

Methods of Planting

Single Hedge System

In this method, the spacing adopted is 1.20 x 0.75 m accommodating 10,800 plants/ha.

Double Hedge System

In this method, the spacing adopted is 1.35 x 0.75 x 0.75 m accommodating 13,200 plants/ha.

Season and Planting

May - June or September – October, polythene bag should be opened lengthwise without injuring the roots and planted in the pit and the soil is gently press.

Irrigation: Subsoil irrigation may be given for young tea seedlings during summer months.

Manuring

Manuring should be done 2 months after planting. Phosphorous should be applied at 80 - 100 kg/ha as Rock phosphate once in a year by placement at 15 - 25 cm depth up to the first pruning and thereafter once in two years. N: K ratio 2: 3 should be adapted for the first 3 years.

Fertilizers

Year of application	Total weight kg/ha/annum		Qty/plant (g)
	N	K	Ammonium Sulphate
I year	180	270	13
II year	240	360	23
III year	300	450	29
IV year onwards	300	300	33

Weed Control

Perennial grasses (Forbicot weeds) can be controlled by spraying. Glyphosate 1.75 litres + Kaoline 2 litres + 2 kg of wetting agent in 450 litres of water.

Training Young Tea

Centering

To induce more laterals, centering should be done 3–5 months after planting. The main leader stem should be cut, leaving 8–10 matured leaves.

Pruning

Pruning is done to maintain a convenient height of bush and to remove dead and diseased branches. Pruning should be done in April-May or August-September.

Types of Pruning

1. **Rejuvenation pruning :** The whole bush should be cut near the ground level less than 30 cm, with the view of rejuvenating the bushes.
2. **Hard pruning :** Hard/formation pruning of young tea is done at 30 to 45 cm (12” to 18”) for proper spread of bushes.
3. **Medium pruning :** To check the bush from growing to an inconvenient height, this type of pruning is done in order to stimulate new wood and to maintain the foliage at lower levels, less than 60 cm.

4. **Light pruning :** Pruning depends on the previous history of the bush, raising the height of medium pruning by an inch or less to manageable heights for plucking (less than 65 cm).
5. **Skiffing :** This is the lightest of all pruning methods. A removal of only the top 5–8 cm of new growth is done so as to obtain a uniform level of pruning surface (more than 65 cm).

Tapping

When secondary's reach more than 60cm, they are tipped at 50-55 cm in height by removing 3 to 4 leaves and bud to induce tertiaries.

Shade Regulation

Pollarding of shade trees should be done prior to heavy rains at a height of 8–10 m from the ground level.

Crop Duration and Harvest

Plucking commences when the tea bush is 3 years old. The plucking of the extreme tip of the growing branch consists of an unopened bud together with two leaves and is popularly known as "two leaves and a bud", while fine plucking is anything less than this. In South India, plucking continues throughout the year at weekly intervals during March-May and at intervals of 10-14 days during the other months.

Rush Period

During the rush period, harvesting is done at a 7 to 10 day interval.

Lean Period

During the lean period, harvesting is done at 10–15 day intervals.

Yield

The yield of green leaves is 10 t/ha.

Pests and Diseases

1. **Root diseases:** Blister blight disease of tea and its control
2. **Blister blight: (*Exobasidium vexans*) a fungal disease: It** is commonly found fungi in South India, Darjeeling and Assam but moderate in Himachal Pradesh. This disease is endemic to the growing areas of Asia.

Types of Shoots in Tea

1. **Aperiodic shoot / (Primary shoot):** From pruned sticks. Are the ones arising from the buds on the pruned frame/sticks? It produces large leaves and the production of banji buds is only occasional. When these aperiodic shoots or primaries grow above a predetermined height, they are cut or tipped. The branches of primaries are the first order laterals, which when plucked give rise to the second order laterals. These again produce the third order laterals.
2. **Periodic shoots / Flush shoots/Crop shoots:** Arising from the axils of leaves It constitutes the crop shoots. Periodic shoots exhibit a cyclic pattern, producing 7–8 leaves (including scale leaves- fig -24 of Wealth of India book) in a cycle before terminating in a "Banji" whose growth and expansion ceases temporarily.

Processing of Tea

CTC (Crush, Tear, Curl): Withering⇨Rolling⇨Fermentation⇨Drying⇨Grading

Orthodox method: Withering⇨Rolling⇨Fermentation⇨Drying

Fermented tea 'dhool' contains 55-60% moisture

Made tea contains 2.5-3% moisture

Economic age of tea bushes: 40-60 years

Important

- **Compound responsible for colour of tea** –Theaflavins and Thearufigens.
- **Calcifuge crop**, Tea 1st introduced by Mr. Robert Kyd (980)
- Study of tea is known as **"Tsiology"**
- **Cataphylls or Janams:** First 1-2 scals leaves
- **Fish leaf:** Above scale leaves
- Major phenolics in green tea: Catechins
- Health properties in tea is due to catechins (*i.e.* epigallocatechin-3-gallate-EGCG)
- Stimulative effect of tea is due to caffeine (1.25-4.5%)
- Tea astringent taste is due to tannis (Polyphenols)
- Soil amendment for maintaining soil PH: Agricultural lime ($CaCO_3$), dolomite lime ($CaMgCO_3$)
- Tea is processed by CTC mthod (cut, tear and curl) or orthodox method

- **Phyllocron or leaf expansion time:** Number of days taken from unfolding of successive leaves (1Leaf + bud in 2 leaves bud)
- **TES** - Tea Experiment Station, Tocklai, Jorhat, Assam.
- **Tea Board:** Calcutta.
- **Commercial propagation: Single node cuttings**
- **Collar pruning:** severe most pruning
- Banjii shoot means dormant shoots
- **Green tea:** Produced without fermentation. Green tea is un-withered, unfermented tea prepared by drying tea leaves either by steaming or passing hot air over leaves.
- **Black tea:** Fully fermented. About 73 per cent of global output is black tea
- **Oolong tea:** Partially fermented.
- **Orthodax method:** Withering Rolling Fermentation Drying.
- **CTC (Cutting, Tearing and Curling) method:** Withering- Rolling -Fermentation - Drying - Grading.
- **Laurie Tea Process (LTP):** LTP is essentially a pulverizing machine. It carries many sets of beaters.

78

Fruit and Plantation Crops in the World

Top Five Countries Production and Share of Fruit in the World (Year-2020)

Country Name	Production mm tons	Share (%)
China	242.79	27.37
India	105.97	11.95
Brazil	39.75	4.48
Turkey	24.15	2.72
Mexico	23.83	2.69
World	**887.02**	-

Fruit Crop Area, Share and Productivity in the World (Year 2020)

Country Name	Area ha mha	Area Share (%)	Productivity Tonne/ha
China	**16.59**	**25.43**	16.40
India	**6.98**	**10.70**	13.03
Brazil	**2.33**	**3.57**	**17.02(IIIrd)**
USA	1.19	1.82	**22.73(2nd)**
Mexico	1.42	2.18	15.13
Spain	1.63	2.50	11.66
Indonesia	0.77	1.18	**24.17 (Ist)**
Philippines	1.68	2.58	9.72
Italy	1.14	1.75	15.79
Turkey	1.38	2.12	15.79
Others	28.60	43.84	11.21
World	**65.24**	-	**13.59**

Fruit Production in the World (Year 2020)

Fruit name	Production in million metric tons	Share (%)
Bananas	119.83	13.51
Apples	86.44	9.74
Grapes	78.03	8.80
Oranges	75.46	8.51
Mangoes, Mangosteens & Guavas	54.83	6.18
Plantain & other fruits	43.12	4.86
Tangerines, Mandarins, Clementines & Sutsumas	38.60	4.44
Pineapple	27.82	3.14
Peach & Nectarines	24.57	2.77
Pears	23.11	2.61
Lemon & Limes	21.35	2.41
Papaya	13.89	1.57
Plum	12.23	1.38
Dates	9.45	1.07
Grape fruits & Pumelon	9.34	1.05
Strawberry	8.86	1.00
Kiwifruits	4.41	0.50
Persimmons	4.24	0.48
Apricots	3.72	0.42
Cherries & sour cherries	4.09	0.46
Cashew apples	1.35	0.15
Figs	1.26	0.14

Top Five Countries Production and Share of Fruit Crops in the World (Year-2020)

Sr No.	Country	Productionin (000) MT	Share(%)
1. Banana			
1	India	31504.00	26.29
2	China	11513.00	9.61
3	Indonesia	8182.76	6.83
4	Brazil	6637.31	5.54
5	Ecuador	6023.39	5.03
2. Apple			
1	China	40500.00	46.85
2	USA	4650.68	5.38
3	Turkey	4300.49	4.98
4	Poland	3554.30	4.11
5	India	2734.00	3.16
3. Grapes			
1	China	14,769.09	18.94
2	Italy	8,222.36	10.54
3	Spain	6,817.77	8.74
4	France	5,884.23	7.54
5	USA	5,388.68	6.91
4. Citrus			
1	China	5,867.16	42.72
2	Nigeria	3,981.70	28.99
3	Iran	622.11	4.53
4	Colombia	571.67	4.16
5	Angola	425.36	3.10
5. Mango, mangosteens & guava			
1	India	24,748.00	45.14
2	Indonesia	3,617.27	6.60
3	Mexico	2,373.11	4.33
4	China	2,368.18	4.32
5	Pakistan	2,344.65	4.28
6. Plantain			
1	Uganda	7,401.58	17.17
2	Congo	4,891.99	11.35
3	Ghana	4,668.00	10.83

4	Cameroon	4,526.07	10.50
5	Philippines	3,100.84	7.19
7. Mandarins & Sutsumas			
1	China	23,120.00	59.90
2	Spain	2,172.18	5.63
3	Turkey	1,585.63	4.11
4	Brazil	1,026.64	2.66
5	Egypt	968.05	2.51
8. Pine apple			
1	Philippines	2,702.55	9.72
2	Costa Rica	2,624.12	9.43
3	Brazil	2,455.69	8.83
4	Indonesia	2,447.24	8.80
5	China	2,220.26	7.98
9. Peach & nectarines			
1	China	15,000.00	61.06
2	Spain	1,306.02	5.32
3	Italy	1,015.35	4.13
4	Turkey	892.05	3.63
5	Greece	890.58	3.63
10. Pear			
1	China	16,000.00	69.24
2	Italy	619.47	2.68
3	USA	609.63	2.64
4	Argentina	600.00	2.60
5	Turkey	545.57	2.36
11. Lime & Lemon			
1	India	3,717.00	17.42
2	Mexico	2,879.02	13.49
3	China	2,687.80	12.60
4	Argentina	1,823.23	8.54
5	Brazil	1,585.21	7.43
12. Papaya			
1	India	6,011.00	43.26
2	Dominic	1,271.30	9.15
3	Brazil	1,235.00	8.89
4	Mexico	1,117.44	8.04
5	Indonesia	1,016.39	7.31

13. Plum			
1	China	6,465.22	52.89
2	Romania	757.88	6.20
3	Serbia	582.55	4.77
4	Chile	416.21	3.41
5	Iran	375.87	3.08
14. Graprfruit			
1	China	4,950.00	52.98
2	Vietnam	833.40	8.92
3	SA	518.00	5.54
4	Mexico	490.83	5.25
5	S. Africa	416.12	4.45
15. Strawberry			
1	China	3,326.82	37.56
2	USA	1,055.96	11.92
3	Egypt	597.03	6.74
4	Mexico	557.51	6.29
5	Turkey	546.52	6.17
16. Kiwifruit			
1	China	2,230.07	50.60
2	New Zealand	624.94	14.18
3	Italy	521.53	11.83
4	Greece	307.44	6.98
5	Iran	289.61	6.57
17. Perssimon			
1	China	3,272.35	77.15
2	Korea	198.82	4.69
3	Japan	193.20	4.56
4	Azerbaijan	185.25	4.37
5	Brazil	158.69	3.74
18. Apricot			
1	Turkey	833.40	22.41
2	Uzbekistan	529.11	14.23
3	Iran	334.41	8.99
4	Algeria	187.27	5.04
5	Italy	173.38	4.66

19. Cherry			
1	Turkey	724.94	27.79
2	USA	294.90	11.31
3	Chile	255.47	9.79
4	Uzbekistan	185.07	7.10
5	Iran	164.08	6.29
20. Sour cherry			
1	Russia	254.80	17.23
2	Turkey	189.18	12.79
3	Ukraine	174.63	11.81
4	Serbia	165.74	11.21
5	Poland	153.10	10.35
21. Fig			
1	Turkey	320.00	25.36
2	Egypt	201.21	15.95
3	Morocco	144.25	11.43
4	Algeria	116.14	9.21
5	Iran	107.79	8.54
22. Avocado			
1	Mexico	2,393.85	29.71
2	Colombia	876.75	10.88
3	Dominic Rep	676.37	8.39
4	Peru	660.00	8.19
5	Indonesia	609.05	7.56
23. Date Palm			
1	Egypt	1690.96	17.90
2	Saudi Arab	1541.77	16.32
3	Iran	1,283.50	13.59
4	Algeria	1,151.91	12.19
5	Iraq	735.35	7.78
24. Guava			
1	India	2180.00	39.35
2	China	480.00	8.66
3	Thailand	380.00	6.85
4	Indonesia	310.00	5.59
5	Pakistan	230.00	4.15

25. Quince			
1	Turkey	189.25	27.16
2	China	111.98	16.07
3	Uzbekistan	96.24	13.81
4	Iran	87.80	12.60
5	Morocco	57.70	8.28
26. Raspberries			
1	Russia	436.90	63.55
2	Poland	146.50	21.31
3	Ukraine	25.79	3.75
4	UK	16.38	2.38
5	Germany	11.97	1.74
27. Currants			
1	Russia	436.90	63.55
2	Poland	146.50	21.31
3	Ukraine	25.79	3.75
4	UK	16.38	2.38
5	Germany	11.97	1.74
28. Olive			
1	Spain	596.50	31
2	Italy	219.41	11
3	Morocco	191.22	10
4	Turkey	152.50	8
5	Greece	122.81	6

Top Three Countries Production, Area and Productivity of Fruits in the World (Year-2021)

Crop	Major Producers	Area	Productivity (Yield)
Fruits	China > India > Brazil	China > India > Brazil	Indonesia > USA >Brazil
Banana	India >China >Indonesia	India >Brazil >Tanzania	Costa Rica>Indonesia > Guatemala
Grapes	China >Italy >Spain	Spain > China > France	India > China > USA
Mango	India >Indonesia > Mexico	India > China > Thailand	Brazil > Ezypt>Mexico
Citrus	China > Nigeria > Iran	China > India > Nigeria	Indonesia> South Africa > South Korea
Lemon & Lime	India > Mexico >China	India > Mexico > China	USA > Argentina >Brazil
Papaya	India >Dominic> Brazil	USA >Fiji>Jamica	Costa Rica>Indonesia >Mexico
Pineapple	Philippines >Costa Rica >Brazil	Côte d'Ivoire>Swaziland> Fiji	Indonesia > Costa Rica >Ghana
Grape fruits	China> Vietnam> USA	Zimbabwe> Venezuela> Australia	Cyprus> Israel> Paraguay
Fig	Turkey> Egypt> Morocco	Lebanon> Afghanistan> South Africa	Australia> Portugal>Cyprus
Avocado	Mexico> Colombia> Dominic	Lebanon>El Salvador>Rwanda	Samoa>Brazil>Kenya
Olive	Spain> Italy > Morocco	Spain> Tunisia > Italy	Egypt>USA>Peru
Dates Palm	Egypt> Iran > Saudi Arab	Egypt> Iran > Saudi Arab	Egypt> Iran > Saudi Arab
Litchi	China>India> Vietnam	-	-
Guava	India>China>Thailand	-	-
Mandarins	China> Spain > Turkey	-	-
Apple	China > USA > Turkey	China > India > Poland	Chile > USA > Italy
Peach	China> Spain> Italy	NewZealand> China>Bhutan	Venezuela> France> Montenegro
Pear	China> Italy> USA	China>Croatia>Lebanon	Bhutan> Montenegro> Switzerland
Plum	China> Romania> Serbia	China> Egypt > Tanzania	Slovakia> Lithuania> Estonia
Kiwifruit	China> New Zealand> Italy	South Korea>China>Israel	New Zealand>Iran>Cyprus

Perssimon	China> Korea> Japan	China> Korea> Japan	China> Israel > Brazil
Strawberry	China> USA> Egypt	Greece>Uzbekistan>Chile	USA>Spain>Egypt
Apricot	Turkey> Uzbekistan> Iran	Jordan>Chile>Switzerland	Australia>USA>Albania
Cherry	Turkey> USA> Chile	Croatia>Netherlands>Kyrgyzstan	Latvia>Estonia> Netherlands

79

Scenario of Horticulture Crops in India

(The per capita Availability and Requirement for Fruits are 207.9 and 230 gm/person/day, Respectively)

Total Area, Production Productivity and Growth of Horticultural Commodities (2020-21)							
Horticultural Commodities		**Percent Share to Horticulture production**	**Area (mha)**	**Produc-tion (million MT)**	**Produc-tivity (MT/ha)**	**Growth rate (%) 2020-21 over compared year 2019-20**	
						Area	**Production**
Fruits		30.63	6.93	102.48	14.79	2.30	0.39
Vegetables		59.90	10.85	200.44	18.47	5.32	6.50
Flowers	*Cut*	0.89	0.32	0.83	-------	--	--------
	Loose			2.15	-------	--	---------
	Total	-----	0.32	2.98	9.31	0.31	- 0.67
Honey		0.04		0.125	-------	------	4.17
Plantation Crops		4.97	4.25	16.63	3.91	2.70	3.18
Spices		3.32	4.46	11.12	2.49	3.87	9.67
Medicinal & Aromatic Crops		0.25	0.65	0.82	1.26	1.87	12.39
Total Horticulture Crops		100	27.48	334.60	12.18	3.75	4.40

Area, Production, Productivity and Share of Different Fruit Crops (2020-2021)

Fruits Crops	Area in '000 Ha	Area in %	Production in '000 MT	Production in %	Productivity in MT
Almond	10	0.14	11	0.01	1.10
Aonla	100	1.44	1197	1.17	11.97
Apple	313	**4.52 (4 th)**	2276	2.22	7.27
Banana	924	**13.33(3rd)**	33062	**32.26 (1st)**	35.78
Ber	53	0.76	580	0.57	10.94
Citrus					
(i) Lime/Lemon	327	4.71	3548	3.46	10.85
(ii) Mandarin	477	6.88	6219	6.07	13.30
(iii) Sweet Orange	217	3.13	3988	3.89	18.37
(iv) Other Citrus	77	1.11	489	0.48	6.35
Citrus Total (i -iv)	1097	**15.83(2nd)**	14245	**13.90(3rd)**	12.98
Custardapple	47	0.68	407	0.40	8.65
Grapes	155	2.23	3358	**3.28(6th)**	21.66
Guava	308	**4.44(5Th)**	4582	**4.47(5th)**	14.87
Jackfruit	188	2.71	1893	1.85	10.07
Kiwi	5	0.07	16	0.02	1.20
Litchi	98	1.41	720	0.70	7.34
Mango	2317	**33.43(1st)**	20386	**19.89(2nd)**	8.79
Papaya	146	2.11	5540	**5.41(4th)**	37.94
Passion Fruit	11	0.16	70	0.07	6.36
Peach	18	0.26	111	0.11	6.16
Pear	42	0.61	276	0.27	6.57
Pecan nut	1	0.01	----	--	----
Pineapple	106	1.53	1799	1.76	16.97
Plum	24	0.35	84	0.08	3.50
Pomegranate	288	**4.15 (6Th)**	3271	3.19	11.36
Sapota	79	1.14	822	0.80	91.33
Strawberry	3	0.04	20	0.02	6.67
Walnut	109	1.57	281	0.27	2.58
Other Fruits	448	6.46	7473	7.29	16.68
Total Fruits	**6930**		**102481**		**14.79**

Area and Production Ranks Different Fruits (2020-2021)

Fruits	Area	Mango (33.43%) >Citrus (15.83%)>Banana (13.33%) > Apple (4.52%)>Guava (4.44%) >Pomegranate (4.15 %)
	Production	Banana(32.26%)>Manago (19.89%) >Citrus(13.90%)>Papaya (5.41%)>Guava(4.47%)Grapes(3.28%)

Area, Production, Productivity and Share Area and Production in Percentage of Different Plantation Crops (2020-2021)

Plantation Crops	Area in '000 Ha	Area in %	Production in '000 MT	Production in %	Productivity in MT
Arecanut	794	18.66	1563	9.39	1.96
Cashewnut	1159	27.23	738	4.43	0.64
Cocoa	103	2.42	27	0.16	0.26
Coconut	2199	51.68	14301	86.00	6.50
Total Plantation crops	**4255**		**16629**		**3.90**
Area and Production Ranks different plantation crops (2020-2021)					
Plantation crops	Area	Coconut (51.68 %) > Cashewnut (27.23 %) > Arecanut (18.66 %) > Cocoa (2.42 %).			
	Production	Coconut (6.50 %) >Arecanut (1.96 %) >Cashewnu (0.64 %) > Cocoa (0.26 %).			
Plantation crops production		KL (31.71 %) > TN (24.76 %) > KN (24.56 %) > AP (7.05 %) > MH (2.67 %)			
Plantation crops area		KN (26.76%) > KL (24.87) > TN (15.98%) > AP (8.83%) > Odissa (6.60%)			

Area, Prodution and Share Area and Production of Tea (2020-21)

State	Area (000 ha)	Area share in (%)	Production Qty M.Kgs	Production share in (%)
Assam	337690.35	**26.52(1st)**	667.73	**24.86 (1st)**
West Bengal	148121.74	**11.63(2nd)**	408.17	**15.20(2nd)**
Others North India	49816.95	3.91	32.14	1.20
North India	535629.04	42.07	1108.04	41.25
Tamil Nadu	62885.13	4.94 (**3rd**)	168.67	**6.28(3rd)**
Kerala	35871.16	2.82	61.11	2.28
Karnataka	2171.74	0.17	5.24	0.20
South India	100928.03	7.93	235.02	8.75
All India	1273114.14		2686.12	

Tea Production Rank Wise in the World (2020-21)

Country	China	India	Kenya	Sri Lanka	Vietnam
Production Qty. M. kg	3063.15	1343.06	537.83	299.34	180.0

Export of Tea in the world (2020-21)

Country	Kenya	China	Sri Lanka	India	Vietnam
Export Qty. M. kg	518.92	369.36	282.84	196.54	145.00

Area, Prodution, Productivity and Share Area and Production of Coffee in India (2020-21)

States	Production MTs		Total Production (MTs)	Area (000 ha)	Productivity (MTs)	Production share (%)	Area share (%)
	Arabica	Robusta					
Karnataka	18025	173625	241625	79855	3.02	70.65	21.031
Kerala	1900	68000	69900	77861	0.90	20.44	20.506
Tamil Nadu	12715	5255	17970	18010	1.00	5.25	4.743
NonTraditional Areas (Andhra Pradesh, Orissa & Telangana)	12290	40	12330	196966	0.06	3.61	51.875
North Eastern Region	70	80	150	7005	0.02	0.04	1.845
All India	95000	247000	342000	379697	0.90		

Exports of Coffee from India (2020-21)

Country	Italy	Germany	Belgium	Russia	Libya
Export (%)	20.37	11.72	8.77	4.77	3.32

Area, Prodution and Productivity of Rubber in India (2020-21)

States	Production Tonne	Area (ha)	Productivity (Tonne)
Kerala	**455600**	525408	0.87
Tripura	**37,339**	**63,223**	0.59
Karnataka	20000	32415	0.62
Assam	**5000**	12117	0.41
Tamil Nadu	21500	**19355**	1.11

Prodution and Consumption of Rubber in the World (year-2021)

Country wise Production (000 tonne)		Country wise Production (000 tonne)	
Countries	**Production**	**Countries**	**Consumption**
Thailand	4836	China	5665
Indonesia	3045	India	1257
Vietnam	1260	USA	960
Coted Ivoire	965	Thailand	890
China	851	Japan	678
India	757	Indonesia	584

Area, Prodution and Productivity of Coconut in World (2020)

Countries	Area (000 ha)	Countries	Production 000 tonn	Countries	Productivity (Nuts/ha)
Philippines	3651	Indonesia	16824848	Vietnam	10547
Indonesia	3397	India	14695000	India	9346
India	2173	Philippines	14490923	Guyana	9200
Sri Lanka	444	Brazil	2458839	Jamaica	7118
Papua New Guinea	221	Sri Lanka	2233600	Papua New Guinea	6710

Area, Prodution and Productivity of Coconut in India (2020-21)

States	Area (000 ha)	States	Production million nuts	States	Productivity (Nuts/ha)
Kerala	770.79	Kerala	7448.65	Tamil Nadu	14251
Karnataka	513.85	Karnataka	6773.05	Gujarat	13775
Tamil Nadu	461.06	Tamil Nadu	6570.63	Karnataka	13181
Andhra Pradesh	115.21	Andhra Pradesh	1377.53	West Bengal	12,641
Orissa	50.91	West Bengal	341.68	Andhra Pradesh	11,957

Area, Prodution and Productivity of Areca Nut in World (2020)

Countries	Area (000 ha)	Countries	Production 000 tonn	Countries	Productivity (kg/ha)
India	452.00	India	622.00	China	2728
Bangladesh	165.00	China	121.40	Srilanka	2556
Indonesia	137.00	Myanmar	118.90	Myanmar	2148
Myanmar	55.30	Bangladesh	102.00	Thailand	1688
China	44.50	Indonesia	47.00	India	1376

Area, Prodution and Productivity of Areca Nut in India (2020-21)

States	Area (000 ha)	States	Production 000 tonn	States	Productivity (Kg/ha)
Karnataka	500.00	Karnataka	950.00	Nagaland	5854
Kerala	96.78	Kerala	92.59	Tripura	3167
Assam	67.02	Assam	50.04	Andaman & Nicobar	2486
Meghalaya	18.14	Meghalaya	24.09	Goa	2050
Mizoram	12.28	West Bengal	23.34	Tami Nadu	2023

Area, Prodution and Productivity of Oil Palm in World (2020-21)

Countries	Area (000 ha)	Countries	Production (000 M T)	Countries	Productivity (MT/ha)
Indonesia	12500	Indonesia	46500	Indonesia	3.72
Malaysia	5500	Malaysia	19800	Malaysia	3.60
Nigeria	3000	Thailand	3260	Colombia	3.50
Thailand	1000	Colombia	1838	Thailand	3.26
Colombia	525	Nigeria	1400	Nigeria	0.47

Area, Prodution and Productivity of Oil Palm in India (2020-21)

States	Area (000 ha)	States	Production MT	States	Productivity (MT/ha)
Andhra Pradesh	150530	Andhra Pradesh	240016	Andhra Pradesh	1.594
Karnataka	41431	Telangana	39347	Karnataka	0.06
Tami Nadu	29510	Kerala	4281	Tamil Nadu	0.02
Mizoram	25741	Karnataka	2734	-	-
Odisha	18484	Tamil Nadu	698	-	-

Area, Prodution and Productivity of Cashew in World (2020-21)

Countries	Area (000 ha)	Countries	Production (000 M T)	Countries	Productivity (MT/ha)
India	893.00	Cote d' Ivoire	848700	Philippines	40831
Brazil	758.08	India	772779	Vietnam	28110
Côte d'Ivoire	666.00	Vietnam	348504	Nigeria	17598
Vietnam	340.00	Burandi	300910	Mozambique	11307
Nigeria	330.00	Philippines	255915	Tanzania	9887

Area, Prodution and Productivity of Cashew in India (2020-21) 0.06

States	Area (000 ha)	States	Production 000 MT	States	Productivity (kg/ha)
Orisha	193.99	Maharashtra	269.44	Maharashtra	1337
Maharashtra	191.45	Andhra Pradesh	116.92	West Bangal	1140
Andhra Pradesh	186.78	Orissa	98.59	Nagaland	1080
Tami nadu	142.28	Karnataka	89.45	Assam	1028
Karnataka	129.07	Kerala	88.18	Kerala	962

Area, Prodution and Productivity of Cocoa in World (2020-21)

Area		Production		Productivity (MT/ha)	
Countries	**Area (000 ha)**	**Countries**	**Production (MT)**	**Countries**	**Productivity (MT/ha)**
Côte d'Ivoire	2000.00	Cote d' Ivoire	848700	Malaysia	8828
Ghana	1656.00	India	772779	Indonesia	8000
Nigeria	1370.00	Vietnam	348504	Togo	7599
Indonesia	1000.00	Burandi	300910	Côte d'Ivoire	6108
Peru	6886.00	Philippines	255915	Peru	5246

Area, Prodution and Productivity of Cocoa in India (2020-21)

States	Area ha	States	Production (000 MT)	States	Productivity (MT/ha)
Andhra Pradesh	39714	Andhra Pradesh	10930.20	Andhra Pradesh	950
Tamil Nadu	32080	Kerala	9647.40	Kerala	850
Kerala	17366	Karnataka	3719.10	Karnataka	525
Karnataka	14216	Tamil Nadu	2802.45	Tamil Nadu	350

Horticulture Crops Category Wise (2020-21) Area in '000 Ha

Area and Production Ranking (2020-21)	Area	Production	Productivity
Fruits	MH > AP> UP	AP > UP > MH	AP > UP > GUJ

Area, Production and Productivity of Fruit Crops in Various States in 2020-2021

Sl. No.	States/Uts	Fruits		Productivity TM/Ha
		A (000 Ha)	P (000MT)	
1	Andhra Pradesh	716.88	18004.13	25.11
2	Arunachal Pradesh	48.14	125.84	2.61
3	Assam	168.87	2562.30	15.17
4	Bihar	324.28	4252.91	13.11
5	Chhatisgarh	232.51	2561.30	11.02
6	Gujarat	433.79	9218.86	21.25
7	Haryana	68.52	963.41	14.06
8	Himachal Pradesh	232.14	828.23	3.57
9	Jammu & Kashmir	329.34	2511.37	7.63
10	Jharkhand	105.56	1183.00	11.21
11	Karnataka	373.82	6558.42	17.54
12	Kerala	310.38	1730.89	5.58
13	Madhya Pradesh	375.45	7757.18	20.66
14	Maharashtra	760.19	9998.44	13.15
15	Manipur	47.90	527.97	11.02
16	Meghalaya	37.60	393.51	10.47
17	Mizoram	62.93	347.02	5.51
18	Nagaland	33.69	315.62	9.37
19	Odisha	335.43	2348.02	7.00
20	Punjab	98.68	2087.32	21.15
21	Rajasthan	61.58	994.38	16.15
22	Sikkim	19.54	55.45	2.84
23	Tamil Nadu	306.17	5698.01	18.61
24	Telangana	177.18	1951.41	11.01
25	Tripura	54.81	563.97	10.29
26	Uttar Pradesh	486.65	10852.01	22.30
27	Uttarakhand	180.57	664.89	3.68
28	West Bengal	269.51	3930.96	14.59
29	Others	12.10	82.45	6.81
	All India Total	**6664.17**	**99069.26**	**14.87**

Leading States in Area, Production and Productivity of Various Fruit Crops (2020-21)

	Area	Production	Productivity
Almond	J&K > HP>MH	J&K > HP>MH	J&K > HP>MH
Aonla	UP > MP > TL	UP > MP> TN	AP > TN > Assam
Banana	KN> Kerala > TN	AP > MP > GJ	GJ > MP> PB
Lime/Lemon	AP > TL > KN	AP > GJ > MH	KN > AP > TL
Mandarin	MP > PB > MH	MP > Pb> Mh	KN > PB > RJ Assam
Sweet Orange	AP > TL > MH	AP > MH > TL	AP > TN > MP
Grapes	MH > KN > Mizoram	MH > KN > TN	PB > MH >KN
Guava	UP >BH >MP	UP > MP > BH	MP> PB> TL
Mango	AP>UP>OD	UP > AP > KN	UP > PB > RJ
Papaya	GJ >MP >WB	AP > GJ >MH	TN > AP > TL
Pineapple	Assam > Mani> Tripura	WB > Assam > KN	KN > TN > WB
Pomegranate	MH > KN > GJ	MH >GJ > KN	TN > GJ > TL
Sapota	KN > GJ > TN	GJ >>AP > MH	TN >TL > AP
Bael	OD>JH>MP	OD>JH>MP	JH>MP>OD
Litchi	BH>WB>JH	BH>WB>JH	PB>JH>HR
Plum	UK>HP>J&K	UK>HP>J&K	Pb>HR>TN
Ber	MP>GJ>OD	MP>GJ>CH	MP>GJ>CH
Custared apple	MH>Ch > MP	MH>MP > GJ	MH>MP > GJ
Passion Fruit	Nag > Mizo> Mani	Mani>Nag>Mizo	Mani> Kerala> Nag
Dates	GJ>RJ >TN	GJ>RJ	GJ>RJ
Jack Fruit	Kerala>Assam>WB	Tirpura >OD >Assam	KN>AP>TIRPURA
Apple	J&K > HP > UK	J&K > HP > UK	J&K > HP > UK
Strawberry	Mizoram > MEG >HR	HR > MH > J&K	HR>MH>HP
Walnut	J&K > UK > HP	J&K > UK > HP	J&K > UK > HP
Peach	UK>HP>JK	UK>Pb>J&K	Pb>HR>Nagaland
Pear	JK> UK> HP	JK> UK> PB	TN>PB>MP
Kiwi Fruit	Ar P>Mizo >Nagaland	Ar P>Mizo >Nagaland	Nagaland>Sikkim>MZ
Pecan Nut	HP >J&K	HP >J&K	HP >J&K

Export Data of Fruit Crops

Fruitc Crop	Export countries
Apple	China >Italy > UK
Banana	UAE >Oman > Saudi Arabia
Oranges (Fresh/ Dried)	Bangladesh >Tukeyl >USA
Grapes	Netherland > Russia > UK
Guava (Fresh/ Dried)	Bangladesh > Netherland> UAE
Litchi	UAE > Nepal > Thailand
Mango (Fresh)	UAE > UK >Qatar
Papaya (Fresh/ Dried)	Nepal > SUAE> Qatar
Pineapple (Fresh/ Dried)	Netherland > UK> Chana
Sapota (Fresh)	UAE > Bahrain > Oman
Walnut	UAE >France >UK

Annexure-I

Journals Name

S.No.	Name	Publisher	Periodicity
1	Acta Horticulture	International Society for Horticultural Science	-
2	Scientia Horticulture	Elsevier BV	
3	Progressive Horticulture	Indian Society for Horticultural Research & Development (ISHRD)	Bi-annually
4	Indian Journal of Horticulture	Indian Society for Plantation crops	Quarterly
5	Indian Journal of Agricultural Science	Indian Council of Agricultural Research (ICAR) New Delhi	-
6	Journal of Plantation crops	Indian Society for Plantation Crops	Thrice a year
7	Journal of Agricultural Issues	Indira Gandhi Krishi Vishwavidyalaya	Bi-annually
8	Indian Journal of Plant Genetic Resources	Indian Society of Plant Genetic Resources (NBPGR)	Thrice a year
9	The Madras Agricultural Journal	Tamil Nadu Agricultural University Coimbatore	-
10	Postharvest Biology and Technology	Elsevier BV	-

Magazines Name

S. No.	Name	Publisher	Periodicity
1	Indian Horiculture	Indian Council of Agricultural Research (ICAR) New Delhi	Bi-monthly
2	Indian Farming	Indian Council of Agricultural Research (ICAR) New Delhi	Monthly
3	Kheti	Indian Council of Agricultural Research (ICAR) New Delhi	Monthly
4	Phal Phool	Indian Council of Agricultural Research (ICAR) New Delhi	Bi-monthly
5	Kisan Bharti (1st Farmer magazine)	GBPUAT, Pantnagar	-

International Horticultural Research Centres

1. Global Horticulture Initiative (GHI), Rome, Italy
2. Horticulture Research International (HRI), Wellesbourne, United Kingdom: 2004
3. International Network for the Improvement of Banana and Plantain (INIBAP), Montpellier
4. Biodiversity International, Rome, Italy
5. Royal New Zealand Institute of Horticulture (RNZIH), Canterbury, New Zealand

Horticultural Societies

1. Royal Horticultural Society (RHS) was founded in London, England: 1804
2. Agri-Horticultural Society of India (AHSI) 1820, Kolkata, India
3. American Society of Horticultural Science (ASHS): 1903, Duke Street Alexandria, United States of America
4. Horticultural Society of India (HSI): 1942, Pusa New Delhi
5. International Society of Horticultural Science (ISHS), Leuvien, Belgium: 1959
6. Indian Society for Plantation Crops (ISPC), Kasargod, India: 1971

Horticultural Institutes

1. National Bureau of Plant Genetic Resource Centre (NBPGR), New Delhi
2. Central Institute of Horticulture (CIH), Mediziphema, Nagaland
3. Institute of Horticulture Technology (IHT), New Delhi
4. Indian Institute of Horticultural Research (IIHR), Hessarghatta, Bangalore, Karnataka
5. National Horticultural Borad (NHB) established in 1984, Headquarter in Gurgaon, Haryana
6. National Hoticultural Mission (NHM): 2005-2006
7. The International Year of Horticulture (IYH) 2012, announced by the Ministry of Agriculture
8. The International Year of Fruits & Vegetables (IYFV): 2021

Pomological Research Institutes

1. Central Institute of Subtropical Horticulture (CISTH), Lucknow, Uttra Pradesh
2. Central Institute of Temperate Horticulture (CITH), Srinagar, Jammu and Kashmir
3. Central Arid Zone Research Institute (CAZRI), Jodhpur, Rajasthan, 1959
4. Central Institute of Arid Horticulture (CIAH), Bikaner, Rajastahn
5. Central Citrus Research Institute (CCRI), Nagpur, Maharashtra
6. National Research Centre for Banana (NRCB), Trichy, Tamil Nadu
7. National Research Centre of Grapes (NRCG), Pune, Maharashtra
8. National Research Centre for Pomegranate (NRCP), Solapur, Maharashtra
9. National Research Centre for Litchi (NRCL), Muzaffarpur, Bihar
10. National Research Centre for Makhana (NRCM), Darbhanga, Bihar
11. Central Plantation Crops Research Institute (CPCRI), Kasargod, Kerala
12. Central Coffee Research Institute (CCRI), Chikmagalur, Karnataka
13. Central Arecanut and Cocoa Marketing and Processing Co-operative Limited (CAMPCO), Mangalore, Karnataka
14. Directorate of Cashew Research (DCR), Puttur, Karnataka
15. Directorate of Oil Palm Research (DOPR), Pedavegi, Eluru, Andhra Pradesh
16. Directorate of Cashew and Coconut Development (DCCD), Cochin, Kerala
17. Directorate of Arecanut and Spices Development (DASD), Calicult, Kerala
18. Tea Research Instituite (TRI), Nira Dam, Valparai, Tamil Nadu
19. United Planters Association of Southern India (UPASI), Glenview, Conoor, Nilgiris district, Tamil Nadu

Boards of Plantation Crops

1. Coffe Board of India (1942) Bangalore, Karnataka
2. Indian Rubber Board (1947) Kottayan, Kerala
3. Coconut Development Board (1981) Kochi, Kerala
4. Tea Board of India (1954) Kolkata

Name of the Horticultural Universities

1. Dr. YSR Horticultural University, West Godavari, Andhra Pradesh
2. Dr. Yashwant Singh Parmar University of Horticulture & Forestry, (1st December 1985) Solen, Himachal Pradesh
3. University of Horticultural Sciences, Bagalkot, Karnataka
4. Uttrakhand University of Horticulture and Forestry, Pauri Garhwal, Uttrakhand

National Bureaus

1. National Bureau of Plant Genetic Resources, 1976, New Delhi
2. National Bureau of Animal Genetic Resources, Sep. 1984, Karnal (Haryana)
3. National Bureau of Fish Genetic Resources, Dec. 1983, Lucknow (Uttra Pradesh)
4. National Bureau of Agriculturally Important Micro-organisms, 2001, Mau (U.P.)
5. National Bureau of Agriculturally Important Insect, 1993, Bangalore
6. National Bureau of Soil Survey and Land Use Planning, Nagpur, (Maharashtra)

References

Bal, J.S. 2012. Fruit growing. Kalyani Publishers, B-I/1292, Rajinder Nagar, Ludhiana-141008.

Bijendra, S. 2015. Cultivation of fruits and plantation crops. Rama Publishing House Agarwal Colony, Opp. Ramlila Maindan Delhi Roard, Meerut-250002.

Bose, T.K., Mitra, S.K. and Sanyal, D. (2000). Fruits: Tropical and Subtropical Vol. II, Pub. Naya Udyog, 206, Bidhan Sarani, Calcutta, India. PP. 643-56.

Gupta, S.N. 2015. Instant horticulture. Jain Brother, New Delhi.

Handbook of Horticulture. Published by Director of Information and Publication of Agriculture, ICAR, Krishi Anusandhan Bhavan, Pusa, New Delhi-110 012.

Kumar, N. JBM Md. Khader, A., Rangaswami, P. and I. Irulappan, 2010. Introduction to spices, plantation crops, medicinal and aromatic plants. Oxford & IBH Publishing Company Pvt. Ltd. 113-B Shahpur Jat, Asian Village Side New Delhi 110 049, India.

Kumar, V. and Kumar, S. 2018. Ready Reckoner for Pomology. New Vishal Publications, E-153, West Patel Nagar, New Delhi-110 008.

Muthukumar, P. and Selvakumar, R. 2013. Glaustas horticulture. New Vishal Publications, E-153, West Patel Nagar, New Delhi-110008.

Prabhakar, S. and Vikas, R. 2018. Comprehensive fruit science. New Vishal Publications, E-153, West Patel Nagar, New Delhi-110008.

Salaria, A.S. Horticulture at a glance. Jain Brothers, 16/873, East Park Road Karol Bagh, New Delhi-110005.

Singh, J. 2012. Basic horticulture. Kalyani Publishers B-I/1292, Rajinder Nagar, Ludhiana- 141008.

Index